Transport, Land-Use and the Environment

Transportation Research, Economics and Policy

VOLUME 4

The titles published in this series are listed at the end of this volume.

Transport, Land-Use and the Environment

edited by

YOSHITSUGU HAYASHI
Nagoya University

and

JOHN ROY
University of Tokyo / CSIRO

KLUWER ACADEMIC PUBLISHERS
DORDRECHT / BOSTON / LONDON

A C.I.P. Catalogue record for this book is available from the Library of Congress

ISBN 0-7923-3728-X

Published by Kluwer Academic Publishers,
P.O. Box 17, 3300 AA Dordrecht, The Netherlands.

Kluwer Academic Publishers incorporates
the publishing programmes of
D. Reidel, Martinus Nijhoff, Dr W. Junk and MTP Press.

Sold and distributed in the U.S.A. and Canada
by Kluwer Academic Publishers,
101 Philip Drive, Norwell, MA 02061, U.S.A.

In all other countries, sold and distributed
by Kluwer Academic Publishers Group,
P.O. Box 322, 3300 AH Dordrecht, The Netherlands.

Printed on acid-free paper

Printed in the Netherlands

CONTENTS

Preface ix

Contributors xi

Part A. Setting Up the Problems

1. Economic Development and its Influence on the Environment: Urbanisation, Infrastructure and Land Use Planning Systems
   (Yoshitsugu Hayashi) 3
   Japan

2. Congestion and Pollution in a Rapidly Expanding City of South-East Asia: The Case of Bangkok
   (Kunchit Phiu-nual) 27
   Thailand

3. Better Cities - Better for the Environment
   (Barry Telford) 47
   Australia

4. Higher Density Housing and Transport in Australian Cities
   (John Black) 55
   Australia

5. Urban Productivity and Sustainability–Impacts of Technological Change
   (J. F. Brotchie, M. Anderson, P.G. Gipps and C. McNamara) 81
   Australia

Part B. Policy Implications in Modelling

6. Reduction of CO_2 Emissions of Transport by Reorganisation of Urban Activities
(Michael Wegener) 103
Germany

7. SUSTAIN - A Model Investigating Sustainable Urban STructure And Interaction Networks
(John R. Roy, Leorey O. Marquez, Michael A.P. Taylor and Takayuki Ueda) 125
Australia / Japan

8. Educating Planners in Testing of Alternative Strategies for Cities of Different Sizes
(William Young and Kevin Gu) 147
Australia

9. NO_x Reduction Effects of the Policy to Reduce Diesel Automobiles and its Influence on Price Change
(Hisayoshi Morisugi and Eiji Ohno) 171
Japan

10. Short-term Impact Analysis of Pricing Strategies on VMT Reduction
(Tschangho John Kim and Paul F. Hanley) 191
USA

11. Spatial Impact of Road Improvement in the Suburban Recreational Area
(Kenji Doi and Naohisa Okamoto) 213
Japan

Part C. Alternative Modelling Approaches

12. Analysis of Urban Environmental Policies Assisted by Behavioural Modelling
(Francisco J. Martínez) 233
Chile

13. An Analysis System for Integrated Policy Measures Regarding Land-Use, Transport and the Environment in a Metropolis
(Kazuaki Miyamoto and Rungsun Udomsri) 259
Japan / Thailand

14. Optimising Techniques in Activities-Transport Models
(Tomás de la Barra) 281
Venezuela

15. Modelling the Implications of New Public Transport Technology: An Approach Using Artificial Intelligence
(Roger L Mackett) 297
UK

16. GIS Integrated System for Urban Transport and Development Planning
(Takayuki Ueda, Hideo Nakamura and Eihan Shimizu) 317
Japan

17. Incorporating Environmental Planning Decisions in Transport Planning: A Modelling Framework
(Michael A.P. Taylor) 337
Australia

18. Using Combined Network Equilibrium Models for Environmental Assessments of Land Use-Transportation Scenarios
(Lars Lundqvist) 359
Sweden

19. A Sequential Land Use/Transportation Model with Externalities: Linking the Dynamics of Regional Economic Growth and Urban Spatial Structure
(Jong Gook Seo, James E. Moore II and Peter Gordon) 383
Korea / USA

Part D. Summing Up

(John R. Roy and Yoshitsugu Hayashi) 415

Index 421

PREFACE

Coordination of land use and transport is one of the most important issues in urban planning from the viewpoint of transport infrastructure supply and amenity in urban space. There has been, therefore, much research conducted in the fields of empirical analysis and theoretical and mathematical modelling of the mechanisms of land use-transport interaction.

The members of the Transport and Land Use SIG (Special Interest Group) of the WCTRS (World Conference on Transport Research Society) have conducted extensive research in these fields. Leading on from the activities of ISGLUTI (International Study Group on Land Use-Transport Interaction) chaired by Dr. Vernon Webster, its output was published as a book "Land Use-Transport Interaction / Policies and Models".

Concurrently with this ongoing research, energy consumption in the transport sector has been increasing rapidly and become a crucial issue from the viewpoint of global environmental conservation. An emerging research need is to examine and structurally identify the mechanisms of the influence of land use-transport interaction on energy consumption and environmental damage, both locally and globally.

The SIG held a seminar in December 1993 in Blackheath, Australia which was the first meeting where world class land use-transport experts gathered to discuss the above topic, covering fact finding, scenario analysis and modelling. This book contains selected papers from the seminar. The Australian Government, CSIRO (Australia) and the Asahi Glass Foundation (Japan) supported the seminar. The book was edited with an enormous and patient help by Dr. Omar Osman at Nagoya University.

Nagoya/Melbourne, August 1995

Yoshitsugu Hayashi, Chairman, Transport and Land Use SIG, WCTRS
John R. Roy, Co-organiser, Blackheath Seminar of SIG

CONTRIBUTORS

Miles Anderson, CSIRO Division of Building Research, Construction and Engineering, P.O. Box 56, Highett, Victoria 3190, Australia

Tomás de la Barra, Modelistica-Planification Sistemas, Apartado Poastal 47709, Caracas 1041- A, Venezuela

John Black, School of Civil Engineering, University of NSW, P. O. Box 1, Kensington, NSW 2033, Australia

John Brotchie, CSIRO Division of Building Research, Construction and Engineering, P.O. Box 56, Highett, Victoria 3190, Australia

Kenji Doi, Department of Mechanical and Environmental Informatics, Tokyo Institute of Technology, Meguro-ku, Tokyo 152-2, Japan

Peter Gipps, CSIRO Division of Building Research, Construction and Engineering, P.O. Box 56, Highett, Victoria 3190, Australia

Peter Gordon, School of Urban and Regional Planning and Department of Economics, Von Kleinsmid Center, Room 351, University of Southern California, Los Angeles, CA 90089-0042, USA

Kevin Gu, Department of Civil Engineering, Monash University, Wellington Road, Clayton Vic 3168, Australia

Paul Hanley, Department of Urban and Regional Planning, University of Illinois at Urbana-Champaign, 907 1/2 W Nevada, Urbana, IL 61801, USA

Yoshitsugu Hayashi, Department of Geotechnical and Environmental Engineering, Nagoya University, Chikusa-ku, Nagoya 464-01, Japan

Tschangho Kim, Department of Urban and Regional Planning, University of Illinois at Urbana-Champaign, 907 1/2 W Nevada, Urbana, IL 61801, USA and Department of Urban Engineering, Seoul National University, Sinlim-Dong, Kwanak-ku, 151-742, Seoul, Korea

Kunchit Phiu-nual, Department of Civil Engineering, Faculty of Engineering, Chulalongkorn University, Bangkok 10330, Thailand

Lars Lundqvist, Department of Regional Planning, Royal Institute of Technology, S 10044, Stockholm 70, Sweden

Roger Mackett, Transport Studies Group, University College London, Gower Street, London WC1E 6BT, UK

Leorey Marquez, CSIRO Division of Building Research, Construction and Engineering, P.O. Box 56, Highett, Victoria 3190, Australia

Francisco Martínez, Department of Civil Engineering, University of Chile, P.O. Box 228/3, Santiago, Chile

Cheryl McNamara, CSIRO Division of Building Research, Construction and Engineering, P.O. Box 56, Highett, Victoria 3190, Australia

Kazuaki Miyamoto, Department of Civil Engineering, Tohoku University, Aoba aza Aramaki, Aoba-ku, Sendai 980-77, Japan

James Moore II, School of Urban and Regional Planning, Department of Civil Engineering, Von Kleinsmid Center, Room 351, University of Southern California, Los Angeles, CA 90089-0042, USA

Hisayoshi Morisugi, Department of Civil Engineering, Gifu University, 1-1 Yanagido, Gifu 500-11, Japan

Hideo Nakamura, Department of Civil Engineering, University of Tokyo, Hongo 7-3-1, Bunkyo-ku, Tokyo 113, Japan

Eiji Ohno, Institute of Socio-Economic Planning, Tsukuba University, 1-1-1 Tenodai, Tsukuba 305, Japan

Naohisa Okamoto, Department of Civil Engineering, Tokyo Institute of Technology, Meguro-ku, Tokyo 152-2, Japan

John Roy, CSIRO Division of Building Research, Construction and Engineering, P.O. Box 56, Highett, Victoria 3190, Australia

Jong Gook Seo, Social Science Institute, Sung Kyun Kwan University, 53, Myungryun-dong 3 Ga., Chongro-Ku, Seoul, Korea

Eihan Shimizu, Department of Civil Engineering, University of Tokyo, Hongo 7-3-1, Bunkyo-ku, Tokyo 113, Japan

Michael Taylor, School of Civil Engineering, University of South Australia, The Levels, Pooraka SA 5095, Australia

Barry Telford, Department of Housing and Regional Development, GPO Box 9834, Canberra ACT 2601, Australia

Rungsun Udomsri, Department of Civil Engineering, Yokohama National University, 156, Tokiwadai, Hodogaya-ku, Yokohama 240, Japan

Takayuki Ueda, Department of Civil Engineering, Gifu University, 1-1 Yanagido, Gifu 500-11, Japan

Michael Wegener, Institute of Urban & Regional Planning, University of Dortmund, Postfach 50 05 00, D-4600 Dortmund 50, Germany

William Young, Department of Civil Engineering, Monash University, Wellington Road, Clayton, Victoria 3168, Australia

Part A.
Setting Up the Problems

CHAPTER 1

ECONOMIC DEVELOPMENT AND ITS INFLUENCE ON THE ENVIRONMENT: URBANIZATION, INFRASTRUCTURE AND LAND USE PLANNING SYSTEMS

Yoshitsugu Hayashi

1. Introduction

A child waits for a bus before dawn in a suburb of Bangkok, the capital of Thailand, so as to arrive at school in time to start at 8:30 a.m. (Photo 1). Many commuters leave home between 5:00 and 6:00 a.m. Not surprisingly, as a consequence of heavy traffic congestion, there are a large number of city dwellers who spend over 8 hours a day in commuting. What a huge socio-economic loss!

Photo 1: A Child Stands at a City Bus Stop in the Morning Dark on his Way to School

Source: *The Bangkok Post*, 6 Sept. 1993.

Examining overall energy consumption in Figure 1, transport energy consumption has continued to increase in every country, though industrial use has reached the saturation point or is declining in certain developed countries.

Y. Hayashi and J. Roy (eds.), Transport, Land-Use and the Environment, 3–25.

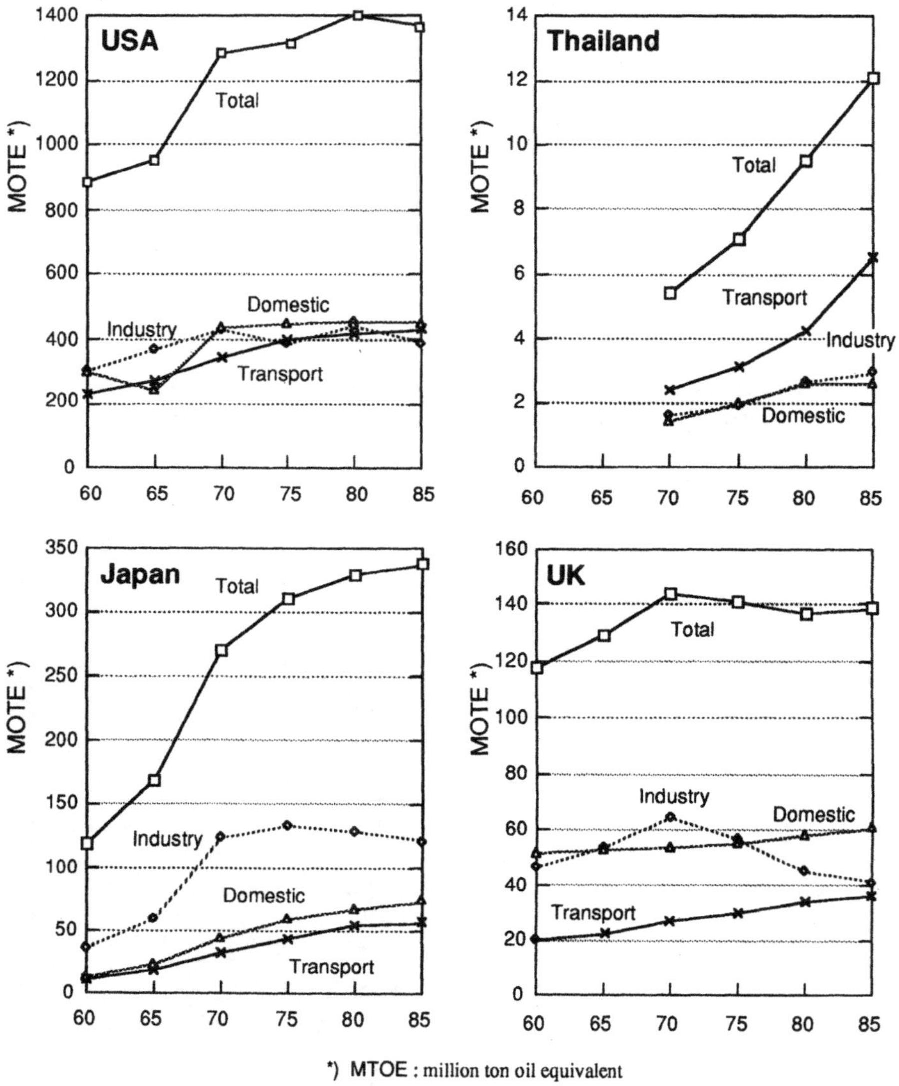

Fig. 1. Changes in Energy Consumption by Sector
Sources: OECD 1989, 1991, 1992.

In major developing countries, transport energy consumption continues to expand in order to cope with their rapid economic growth. For instance, more than 50% of the total energy volume now being consumed in Thailand is from the transport sector. Inevitably, such increasing energy use in the transport sector per se leads to both local and global environmental effects. About 40-80% of NO_X and 70-90% of CO amount in OECD countries as well as 14% of CO_2 amount in the world come from transport emissions (Faiz, 1993).

Apparently in most developing countries, improvement of transport infrastructure and land use regulations cannot keep pace with the speed of economic growth and urbanisation. Realising that the majority of urban population increase will take place in developing countries in the near future, improvement of transport and land use systems at the actual sites is essential so as to ensure a sustainable growth in economic development, while not amplifying the environmental load beyond acceptable limits.

In this study, typical mutual relationships between land use, transport, energy consumption and the environment will be formulated and examined through an international comparison of four metropolitan areas, namely London, Tokyo, Nagoya and Bangkok (Figure 2). These diverse cities are at different stages of their economic development paths, urbanisation and motorisation. Some aspects of financing approaches and land legal systems are also discussed.

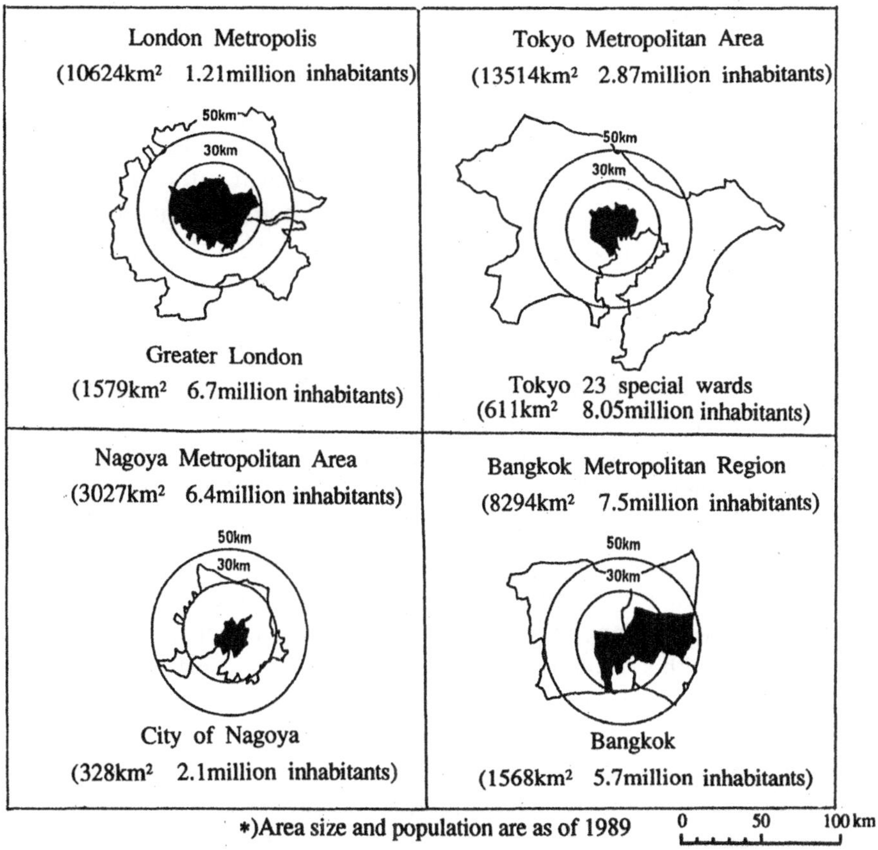

Fig. 2. Study Areas

Source: Central Statistics Office (1991), London Research Centre (1988, 1990), Nagoya City (1990), National Statistics Office (1971,1975, 1980, 1985, 1991), The National Energy Administration (1988).

2. Relationship between Economic Development and Environmental Changes in the Context of Land Use and Transport Interaction

A typical mechanism of increasing energy consumption is illustrated in Figure 3. Economic growth raises freight and person trip demand due to increased production and income growth. At the same time, income growth encourages car ownership, followed by a modal shift from public transport to road oriented transport mode. On the other hand, economic development causes rapid urbanisation and further suburbanisation. Suburbanisation consequently increases trip lengths and, if it appears in a sprawling manner, more difficulty in providing sufficient transport infrastructure. Congestion and increased energy consumption are also the results. Eventually, the output leads to certain degrees of environmental degradation problems which depend upon the scale of environmental loads generated from the actual phenomena.

Figure 4 illustrates the chronological development of built-up areas in London, Tokyo, Nagoya and Bangkok, while Figure 5 illustrates the contemporary spatial distribution of populations in the four metropolises. Both figures are formulated using GIS (Geographical Information System). By classifying these cities in terms of the stages of urbanisation development defined by Klassen (1981), i.e. *(i) urbanisation*, *(ii) suburbanisation*, *(iii) deurbanisation* and *(iv) reurbanisation*, the following points are observed from the two figures. Since the stage of suburbanisation in Tokyo began to develop in the late 1950s, manifest changes can be contrasted with its urbanisation patterns in the period of 1910 ~ 1960. Comparatively, suburbanisation in Bangkok started to develop in the early 1980s as distinctly seen in the period of 1960 ~ 1985 (Hayashi, *et al., 1991, 1993a*). It is noticeable that the built-up area in Bangkok expanded swiftly, about five fold, within this short period of time (Banasopit, 1990). The concentration of urbanisation in Bangkok is clearly depicted in Figure 5. There will be a significant difference in future energy consumption and environmental load in Bangkok depending on whether the currently concentrated population suburbanises with sprawl like Tokyo in the period of 1960 ~ 1985, or forms independent suburban towns with greenbelt barriers like London in the period of 1910 ~ 1960.

Economic development generally brings higher car ownership through increased income. According to Figure 6, before reaching average income level of US$ 5000, the rate of increase in car ownership is even higher than the rate of increase in income. In Bangkok, car ownership rates imitated those of Tokyo before reaching per capita income of US$ 3000 but shifted to

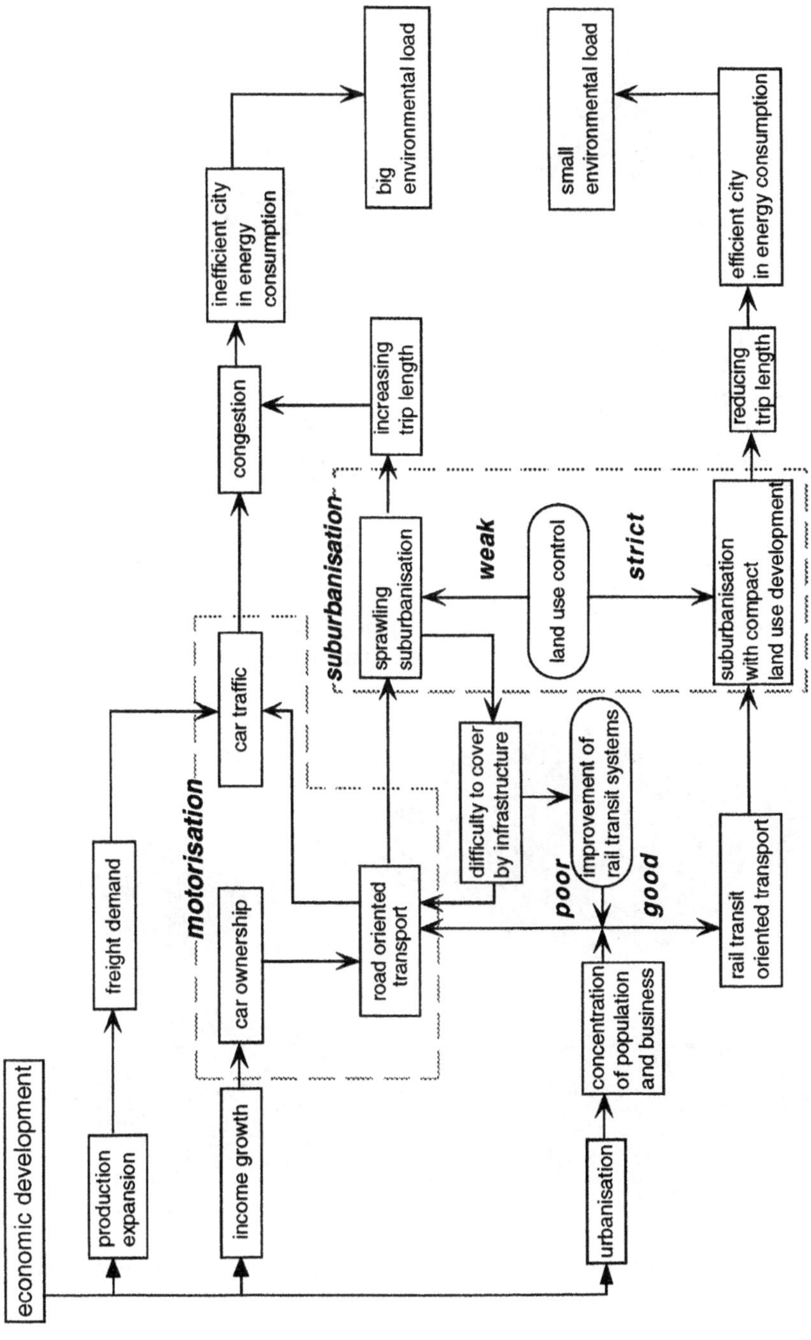

Fig. 3. Mechanism of Economic Development Influences on the Environment
Source: Hayashi (1993a, 1994).

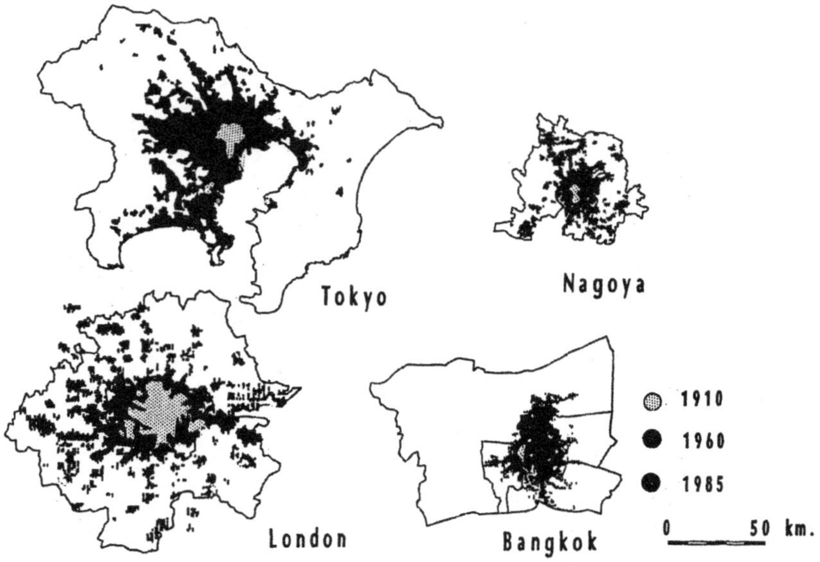

Fig. 4. Comparison of Built-up Areas
Sources: City Planning Institute of Japan (1992), Hall (1982), JICA (1990), Nagoya City (1966), Statistics Bureau (1960, 1985).

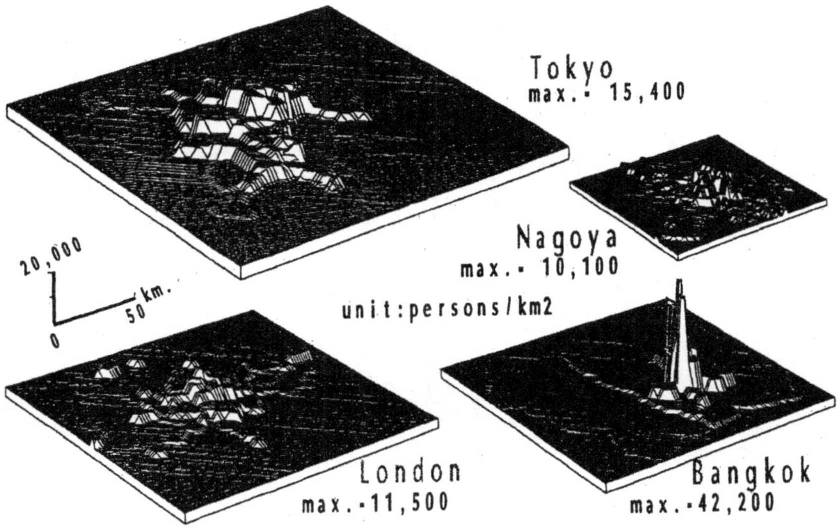

Fig. 5. Comparison of Concentration and Decentralisation of Population (1988)
Sources: Central Statistics Office (1991), London Research Centre (1988, 1990), Nagoya City (1990), National Statistics Office (1971, 1975, 1980, 1985, 1991), The National Energy Administration (1988).

a higher rate of increase like that of Nagoya afterwards. The number of car registrations in Bangkok is increasing at a rate of 200 cars per day (i.e. about 70,000 per year). It is then estimated that car ownership in Bangkok may reach 180 cars per 1,000 inhabitants by the end of 1993.

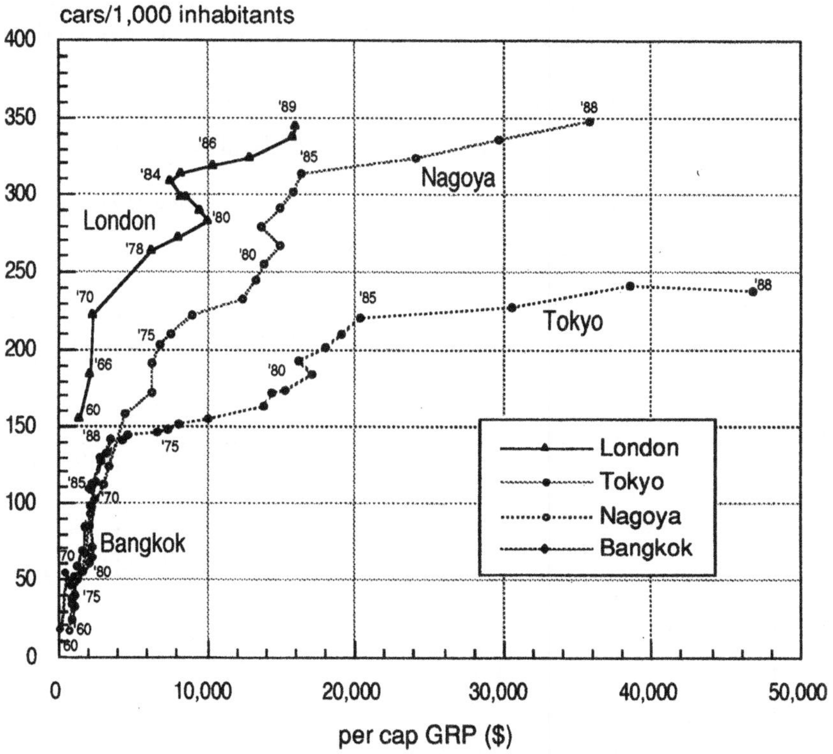

Fig. 6. Economic Growth and Motorisation
Sources: Central Statistics Office (1978-1991), Economic Research Inst. (1976, 1992), IMF (1990), London Research Centre (1988, 1990), Nagoya City (1986, 1990), National Statistical Office (1971, 1975, 1980, 1985, 1991), NESDB (1991), TMG) (1970-1990).

The reason car ownership in Tokyo seems to have become saturated is that its rail network, when compared with Nagoya, is far better than the road network. As there is almost no rail system efficiently functioning in Bangkok, car ownership will increase at the same rate as or even higher than Nagoya if there continues to be no substantial improvement of transport infrastructure, particularly in the rail transport system.

3. The Influence of Fuel Price on Car Use and Suburbanisation

Figure 7 shows the relationship between fuel price normalised by gross regional product (GRP) and car trip length in 1980 in Los Angeles, Tokyo/Nagoya, London and Bangkok. This implies that the normalised fuel price in Bangkok is about ten times higher than that in Los Angeles and three to four times higher than those in London and Tokyo/Nagoya respectively. As shown in the offset in this figure, the average car trip length in Bangkok is only 20% of that in Los Angeles and 60% of that in London. The lower the normalised fuel price, the longer the total car trip length.

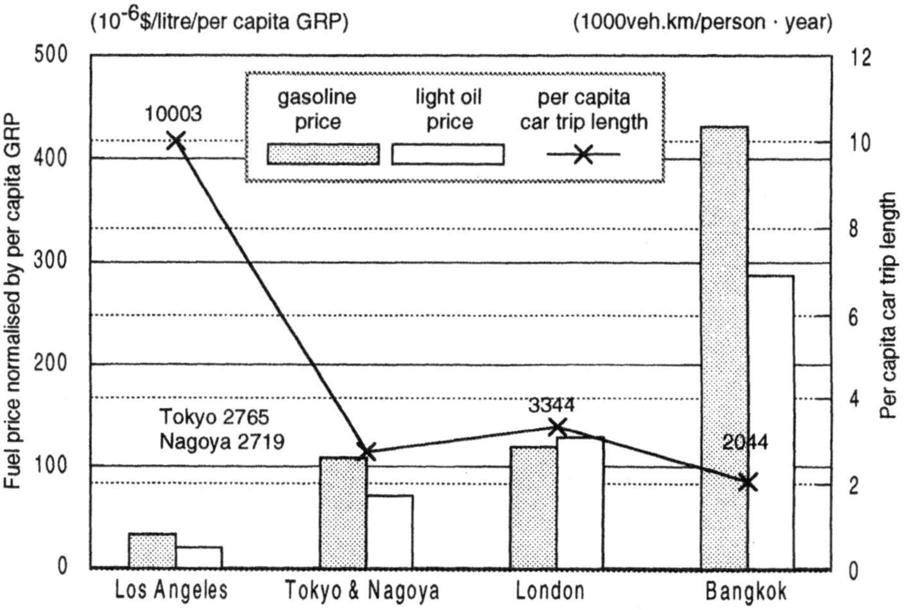

Fig. 7. Comparison of Fuel and per Capita Trip Length (1989)
Sources: EUROSTAT (1990), Nagoya City (1981), Newman (1991), The National Energy Administration (1988).

The figure also shows a significant difference between the price of gasoline and that of light oil. The price ratio of light oil to gasoline in 1980 is 0.68 for Tokyo and Nagoya, which is very low compared with London's ratio of 1.05. This encourages the use of diesel vehicles, which in return, accelerates NO_X and CO emissions.

Figure 8 shows the relationship between normalised fuel price and energy consumption in 32 cities. It clearly shows that even in the United States, which is one of the richest countries, the effect of fuel price elasticity on car energy consumption is extremely high. Talking across the board, a lower fuel price allows more intensive car use and housing location in a remote area.

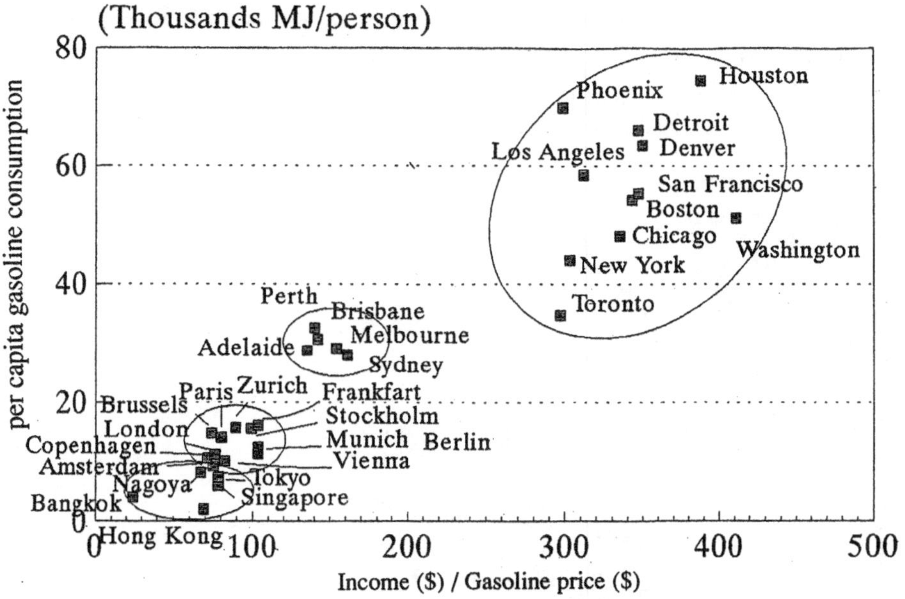

Fig. 8. Income/Gasoline Price Ratio vs. per Capita Gasoline Consumption
Note: Data for Tokyo, Nagoya, and Bangkok are estimated by authors.
Source: Newman (1991).

4. Suburbanisation and Transport Energy Consumption

4.1 Compactness of Cities and Energy Saving

It is naturally assumed that a compact city requires less transport energy consumption because of shorter daily travel distance. Figure 9 shows the relationship between population density and per capita energy consumption in 32 cities. The plots range from consumption of 10 giga-joules per person-year in Asian cities to 80 giga-joules per person-year in American cities. This global comparison shows that cities with a higher population density tend to be more efficient in terms of transport energy consumption. Yet, when income seems the dominant factor in determining geographical boundaries of the cities, some Asian cities do provide exceptions. The per capita energy consumption is slightly higher in European cities than in Asian cities although the population density is much lower in European cities.

The contrast between American and European cities may be explained by the following: a) strict land use control has guided the European cities' compactness, though the land legal system was initially developed to achieve suburban sanitation and amenities early in this century when no one anticipated today's environmental constraints, and b) besides the compactness of cities, better urban rail transit systems and more expensive fuel prices are effective in discouraging car use.

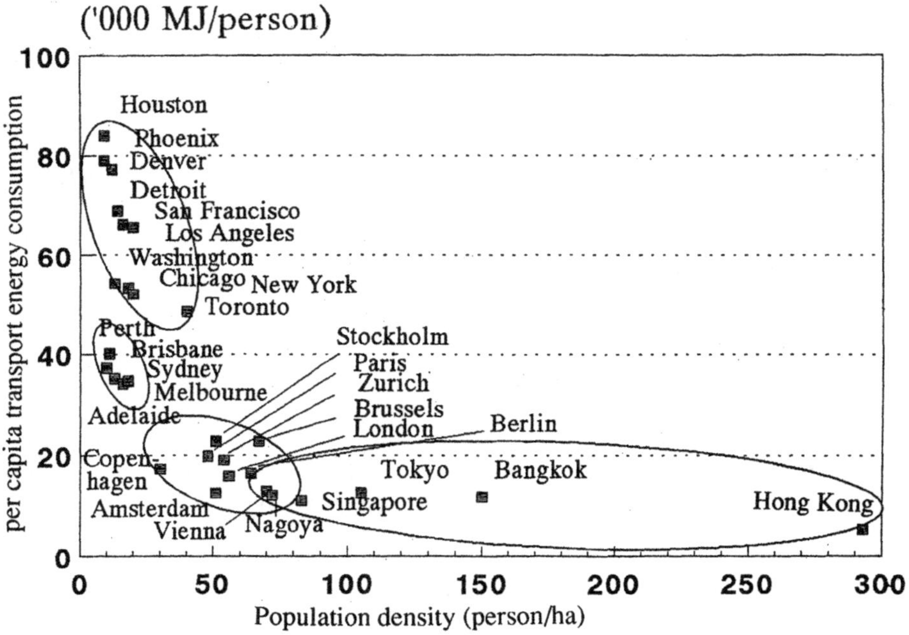

Fig. 9. Population Density versus Transport Energy Consumption
Source: Newman (1991).

4.2 Mutual Enhancement between Suburbanisation and Motorisation

Suburbanisation enlarges the boundaries of human activities by shortening travel times between regions and thus making it possible for people to live in distant suburbs. When houses locate in a sprawling fashion, public transport cannot be provided thoroughly. Consequently, mobility in suburbs heavily depends on cars. This further accelerates motorisation. It is the history of urban development that suburbanisation and motorisation have a mutual multiplier effect, and thus continuous increase in transport energy consumption occurs.

To analyse this mechanism, we apply the indicator *"urban radius index (R)"*, defined as the average radius of the built-up area as specified by the following equation:

$$R = \Sigma a_i r_i / \Sigma r_i \quad (1)$$

where

a_i = the area size of the built-up area in district (I), and
r_i = the distance between district (i) and the city centre.

The value of this indicator implies the spatial compactness of the built-up areas surrounding that city (Hayashi, *et al.*, 1993a).

4.3 Influence of Suburbanisation on Car Trip Length

Figure 10 illustrates how the total car trip length increases due to suburbanisation in London, Tokyo, Nagoya and Bangkok. As also shown in Table 1, the relative ratio between the trip increase rate and the rate of increase in the urban radius index is highest in Bangkok (1.96 between 1972 and 1987), then Tokyo (1.69 between 1960 and 1980), then London (1.35 between 1960 and 1980) and lowest in Nagoya (1.09 between 1971 and 1988).

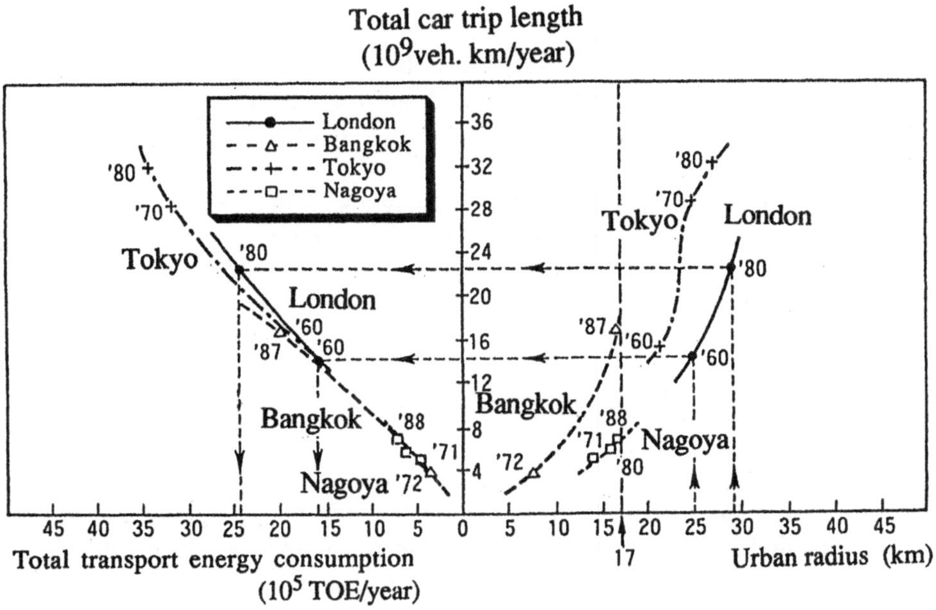

Fig. 10. Change in Urban Radius, Car Trip Length and Energy Consumption
Sources: Kocks (1975), Newman (1991).

This implies that the total car trip length in Bangkok increases at a rate two times greater than in Nagoya. Contrasting Bangkok in 1987 and Nagoya in 1988 the urban radius is equivalent (approx. 17 km), but the total car trip length of 16 billion vehicle-km per year in Bangkok is almost three times greater than the 6 billion in Nagoya. This may be due to the following reasons:

Table 1. Influence of Suburbanisation on Car Travel Characteristics

	increase rate of urban radius (A)	increase rate of total car trip length (B) [B/A]	increase rate of car energy consumption (C) [C/A]
London (1960-1980)	1.16	1.57 [1.35]	1.51 [1.30]
Tokyo (1960-1980)	1.26	2.13 [1.69]	2.12 [1.68]
Nagoya (1971-1988)	1.18	1.29 [1.09]	1.52 [1.29]
Bangkok (1972-1987)	2.27	4.46 [1.96]	4.75 [2.00]

Source: Fig. 10

a) The total number of cars in Bangkok is about 30% higher than in Nagoya.
b) In Bangkok almost all trips depend on road traffic in the extended suburban areas, because urban rail transit systems are inadequate and ineffectively operated.
c) The average road length per car in Bangkok is only 3.9 meters (JICA, 1990), which is even less than half the 8.1 meters available in Nagoya (Nagoya City Report, 1990). Consequently, car traffic in Bangkok is obliged to detour, especially in the suburban areas where the road network is entirely insufficient.

4.4 Influence of Long Car Trips on Transport Energy Consumption

The left hand side of Figure 10 also shows that total transport energy consumption and total car trip length have progressed almost proportionately. Only Bangkok's tangent in the graph has declined, which means that its efficiency has also declined. By tracing the arrow directions in the figure, it is easy to understand the influence of suburbanisation and trip length on transport energy consumption.

As also shown in the brackets in Table 1, the elasticity between the urban radius index and transport energy consumption is 1.29 in Nagoya (a minimum), 1.30 in London, 1.68 in Tokyo and 2.00 in Bangkok (the maximum). This implies that marginal energy consumption to unit suburbanisation is about 60% higher in Bangkok than in Nagoya. The main causes are (i) Bangkok's auto-dependent transport system, (ii) older, higher emitting cars due to loose enforcement of the car-inspection system, and (iii) inefficient fuel use given very serious traffic congestion (Hayashi, *et al.*, 1993a).

5. Traffic Congestion, Ample Fuel Use and Air Pollution

5.1 Traffic Conditions and Transport Energy Consumption

In the context of car travel speed applicable within an inner city territory, Figure 11 implies that a mild speed interval of 40 to 70 km/h provides higher energy efficiency than the others. The slower the travel speed caused by traffic congestion, the more the unnecessary fuel consumption occurs. For instance, the energy efficiency of gasoline cars differs by 1.5 times between the speeds of 10 and 20 km/h.

As a consequence of traffic conditions, transport energy consumption levels in the four cities are different as shown in Figure 12. A 100 km drive requires 10.0 in Nagoya, 11.3 in London, 13.3 in Tokyo and 17.0 liters in Bangkok. These differences depend largely on average travel speeds on the streets within the inner city areas: 21.8 km/h in Nagoya, 18.7 km/h in London, 15.8 km/h in Tokyo and 10.0 km/h in Bangkok. Approximate 60,000 to 70,000 new cars have been registered annually in Bangkok for the five years preceding 1993. Consequently, the average travel speed on the city streets has been drastically reduced and traffic jams have now become extremely serious.

5.2 Causes of Congestion

Road traffic congestion occurs due to the gap between demand for access to private vehicles and the supply level of road infrastructure. Figure 13 summarises the relationship between car ownership per thousand inhabitants and road length per car. The car ownership level of Bangkok in 1988 is about 150 cars per thousand inhabitants which is almost equivalent to that of Nagoya in 1972 or Tokyo in 1976. However, the degree of infrastructure improvement for road length per car in Bangkok is equivalent to only one fourth of that in Nagoya in 1972 or just one third of that in Tokyo in 1976. It is not difficult to imagine that Bangkok will soon face heavier traffic congestion, more serious socio-economic deprivation and associated environmental repercussions.

5.3 Air Pollution: Bangkok's CO Level is 15 Times that of Nagoya

Figure 14 and Figure 15 show the average pollution levels of CO existing along the roadsides and in the background areas. Although the simple comparison of absolute pollution levels across cities may be no longer accurate due to different measurement methods, we can grasp the contributions of traffic congestion to pollution by comparing the discrepancy in pollution levels in the background to those of the roadside.

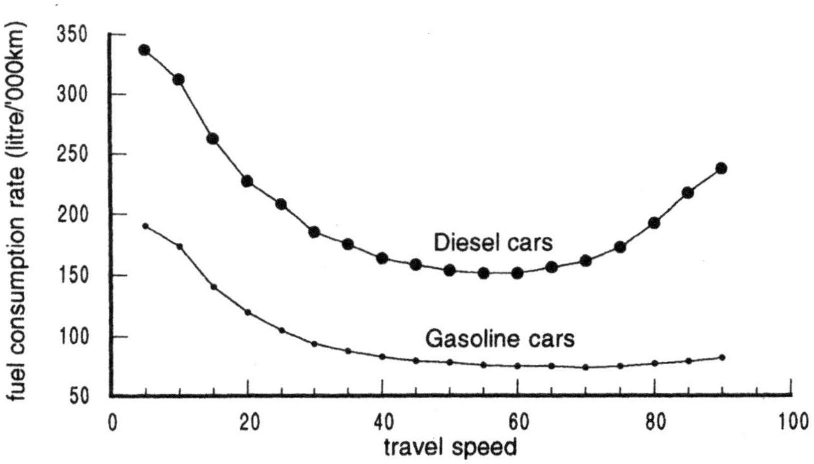

Fig. 11. Efficiency of Fuel Consumption Subject to Travel Speed
Sources: Kanto Engineering Office (1979), Sano (1979).

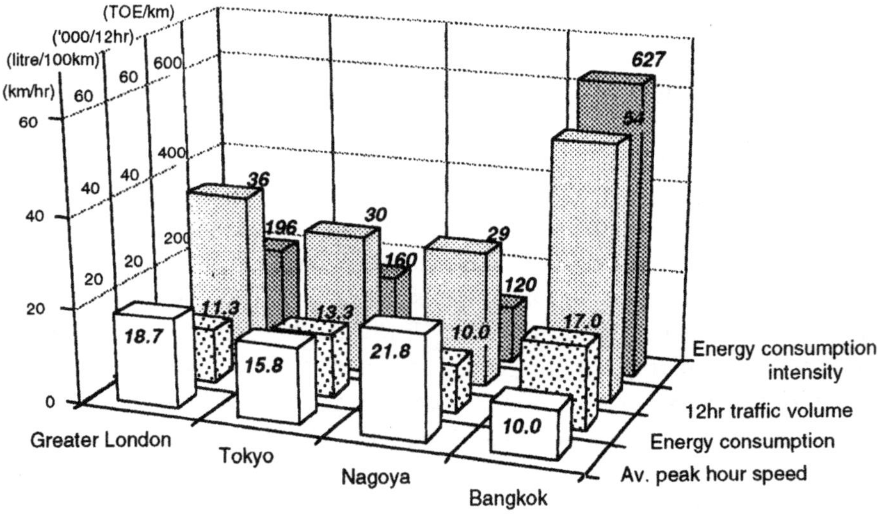

Fig. 12. Average Peak-Hour Car Speed, Traffic Volume and Energy Consumption (1988)
Sources: Department of Transport (1990), JICA) (1990), Nagoya City (1981), NESDB (1990), The National Energy Administration (1988), TMG) (1970-1990).

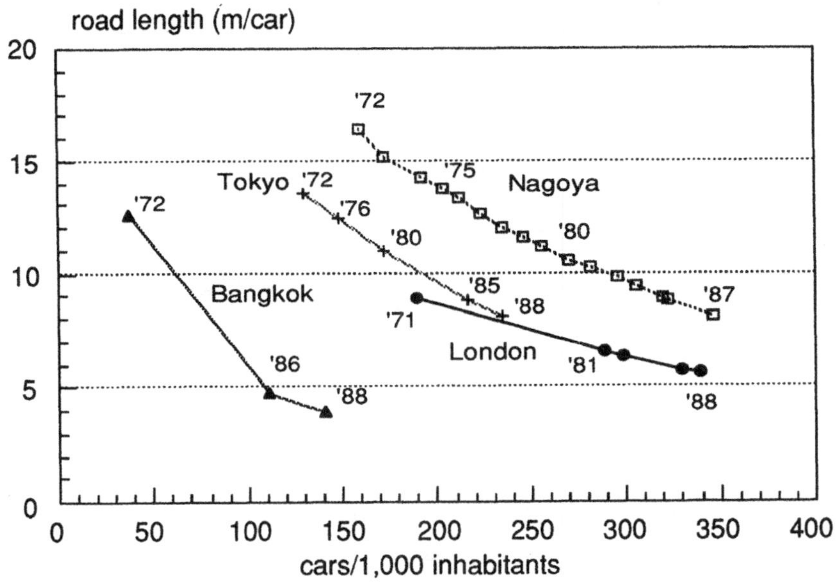

Fig. 13. Motorisation and Road Supply Level
Sources: JICA (1990), Kocks (1975), London Research Centre (1988, 1990), Nagoya City (1986, 1990), TMG (1970-1990).

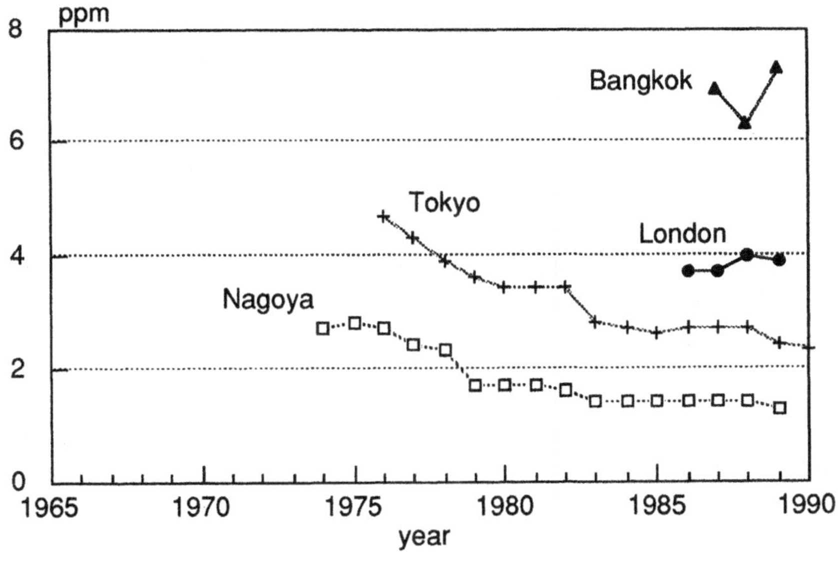

Fig. 14. Changes in Annual Mean Concentration of CO (*Roadside*)
Sources: Bureau of Environmental Protection (1990), LPAC (1990), OECD (1989), ONEB (1989, 1990), Pollution Control Bureau (1980-1990).

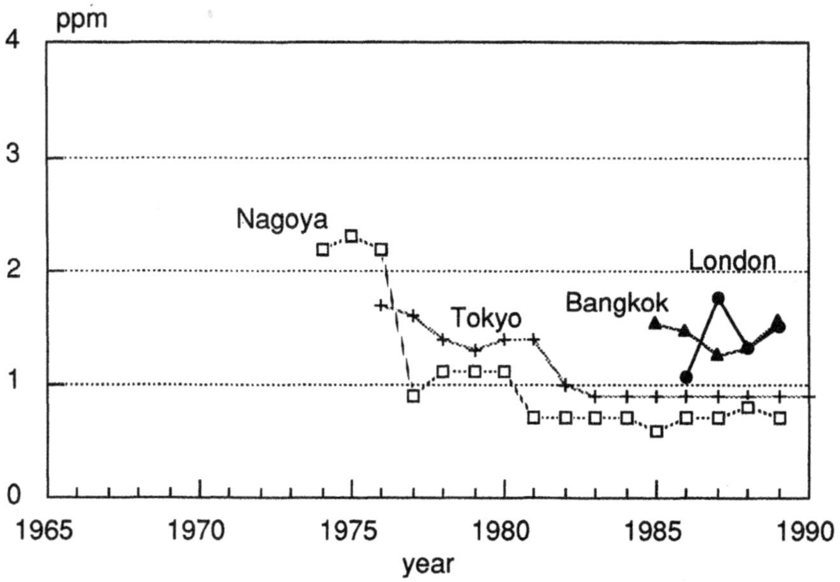

Fig. 15. Changes in Annual Mean Concentration of CO (*Background*)
Sources: Bureau of Environmental Protection (1990), LPAC (1990), OECD (1989), ONEB (1989, 1990), Pollution Control Bureau (1980-1990).

The difference is 0.4 ppm in Nagoya (the minimum), 1.8 ppm in Tokyo, 2.4 ppm in London, and 6.0 ppm in Bangkok (the maximum). Evidently, this means that traffic induced pollution is extremely high in Bangkok. Rationally, this intensity order also matches that of the amount of transport energy consumption per unit drive length: 120 TOE/km (ton oil equivalent per km) in Nagoya (the minimum), 160 TOE/km in Tokyo, 196 TOE/km in London, and 627 TOE/km in Bangkok (the maximum).

NO_X pollution can be analysed in the same fashion. The emission rates of CO and NO_X from diesel vehicles are 5 to 10 times higher than those of gasoline engines.

6. Impact of Infrastructure Investment and Land Use Planning System on the Environment

6.1 Infrastructure Investment Level

Figure 16 shows the time serial changes in the shares of road investment to GNP in the UK, US, Japan and Thailand. The common declining direction on road budget share in the UK and the US is understandable, because these countries have already invested in an ample level of infrastructure stock, and therefore their demands for new construction are not very high. It is well known that the steep decline of infrastructure investment in the US during the '60s and '70s induced its severe transport deterioration, and some roads and bridges had to be closed. We clearly remember the accident on the Brooklyn Bridge and the closure of the West Side Highway in Manhattan, New York City as symbolic events.

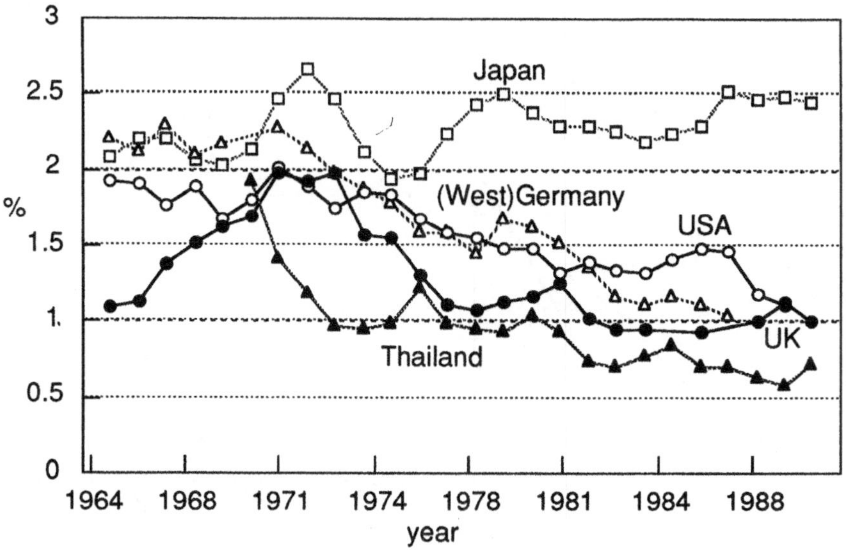

Fig. 16. Ratio of Road Investment to GNP
Source: International Road Federation (1970-1991).

Thailand provided road investment of about 2% of GNP in the late '60s, but it has later declined steeply to reach a level of less than 1% of GNP, which is the lowest among the four countries. It is also clearly seen in Figure 17 that in Bangkok, the rate of increase in investment of infrastructure and in environmental improvements has been much lower than that of GRP, whilst in Tokyo these rates are much closer to each other during its period of rapid economic growth (Hayashi, *et al.*, 1993b). In Bangkok, insufficient infrastructure investment brought about a large gap between car traffic demand and road construction supply as previously mentioned in Figure 13. This leads to a complicated problem with many implications and controversial issues.

On the other hand, Japan has steadily invested in road infrastructure at a rate of 2.0-2.7%, which is almost in proportion to its GNP. The share of new construction investment still remains high at present. This has been accomplished through imposing a fuel tax ear-marked for the construction of road infrastructure at a rate close to that of GNP growth. Since the rate of increase in demand for cars is significantly higher than that of GNP after a threshold level (approx. US$ 1000 per capita annual income), for which Thailand has already passed, a public budget that dominantly consists of income taxes can not follow the rapid increase of construction costs for roads. This implies that certain public budgetary instruments that increase in proportion to car use are necessary. Tokyo and Nagoya have fortunately managed to avoid catastrophic congestion by proportionately generating their budgets for road investment corresponding to the increasing car traffic demand.

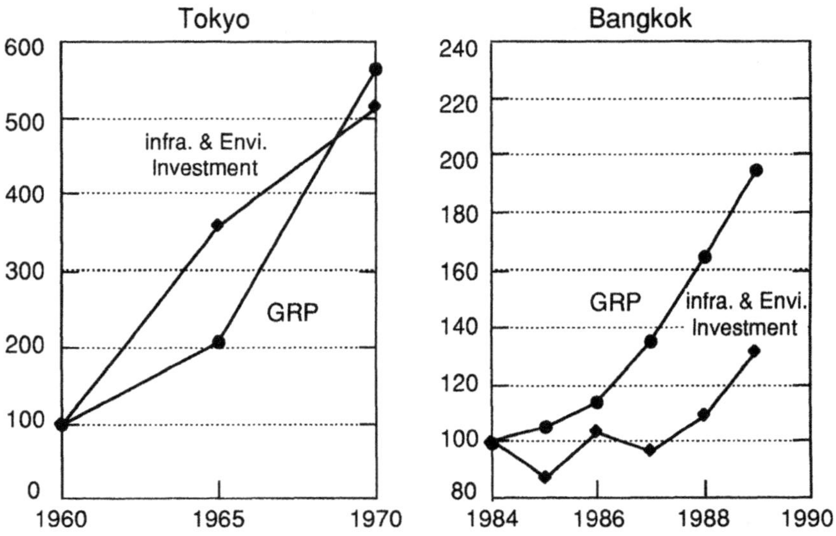

Fig. 17. Economic Growth And Infrastructure Investment (index)
Source: Hayashi (1993b).

6.2 Land Planning System

Lack of urban sprawl control is one of the major causes contributing to broader environmental problems through the interactive mechanism of urbanisation and motorisation. In the U.K., land use control has long been an effective planning tool for protection against rapid urban sprawl and environmental degradation in the suburbs. It has been implemented through several legislative measures, e.g., the Town and Country Planning Act of 1919, the Restriction of Ribbon Development Act of 1935, the Green Belt Act of 1938, and so on. However, legal land use planning through the designation of 'restricted urbanisation area' in Japan has provided less contribution to urban sprawl control and environmental protection in that it does not stop but only reduces the speed of urban sprawl. So far, there have been no concrete or satisfactory countermeasures against rapid urban sprawl in Thailand.

A land use planning system is fundamental to improve the environment for at least two reasons: (i) it is extremely essential in controlling urban sprawl to keep the infrastructure development cost at an acceptably low level, particularly during periods of rapid suburbanisation, and (ii) infrastructure improvement provides various benefits not only to users, the user of related transport modes, lines, and routes; but also the owners, tenants, and developers of land properties in the vicinity. The last benefit portion is imputed to land properties in the form of value increase.

Value capture is one of the contemporary approaches viable not only for financing lofty construction costs of new transport infrastructure, but also applicable to the maintenance or improvement of the deteriorating environment towards a more human-friendly circumstance.

7. Concluding Remarks

7.1 Economic Development Cycle

Intuitively, suburbanisation enlarges the boundaries of human activities due to the shortening of travel times between the regions, thus encouraging people to live in a distant suburb. However, when most houses are located in a sprawling fashion and therefore effective or adequate public transport systems cannot be provided, mobility of the population in suburbs is forced to depend heavily upon car use. This obligation will definitely further accelerate motorisation. It is therefore illustrated by the comparative history of urban development study that the coexisting negative feedback of suburbanisation and motorisation have a mutual or reverse multiplier effect thus booting continuous increase in transport energy consumption. Associated residuals or output discrepancies caused by different characteristics from unique features of individual metropolis are compared and examined as well.

7.2 A Scenario Analysis Method and its Phenomenal Interpretations

By combining the intuitive processes of analysis conducted in this study, Figure 18 proposes a conceptual approach for scenario analysis of environmental influences. In short, the basic process of the scenario analysis method consists of six consecutive components described as follows:

(i) observe and consolidate the past trends to obtain likely scenarios of economic growth,
(ii) examine urbanisation scenarios,
(iii) examine motorisation scenarios,
(iv) examine the road improvement policy scenarios in comparison with the expected increase of traffic demand,

(v) examine the relationship among urban radius index, vehicle-drive distance and the consumed transport energy, and
(vi) forecast the environmental changes.

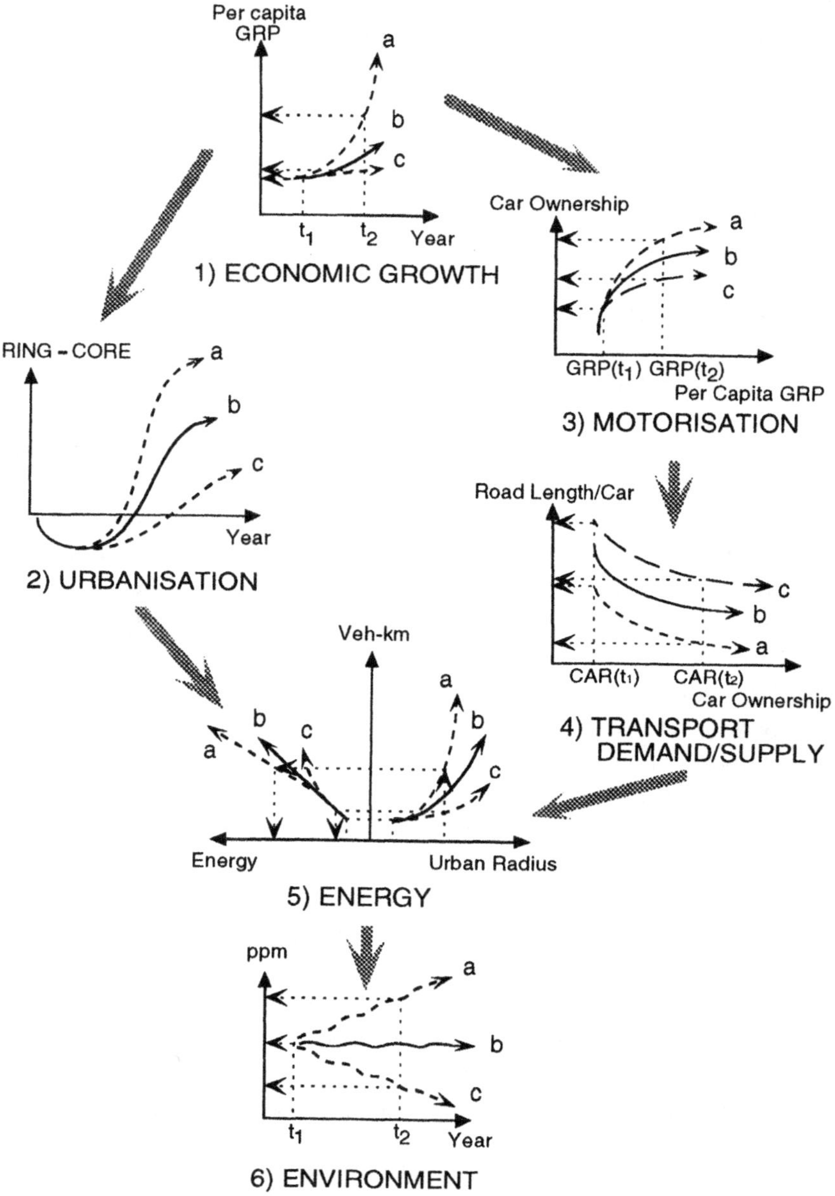

Fig. 18. Basic Process of Scenario Analysis Method
Source: Hayashi (1994).

7.3 Economic Development and the Environment Nexus

At the moment it is still difficult to use empirical trends to forecast across-the-board the future conditions of urbanisation, land use, transport infrastructure and the consequent environment in a dynamic framework. One fundamental reason is that there are many unseen and non-linearly changing factors emerging in the real-world paths of economic development. For instance, these could be some unstable economic growth patterns, inconsistent rates of increase in car ownership and fluctuating traffic demand which is generally higher than that of income, the advanced innovative technology in car engines, the financial viability to supply adequate transport infrastructure, and the shortage of private/public funds to subsidise environmental improvement actions. Such shortcomings propound historical comparisons with another metropolis during the similar stages of economic development. In other words, it would be useful to review the actual past trends and to understand the development mechanisms of elderly aged metropolises when attempting to understand the future scenario of another younger aged metropolis at the same economic development stage.

Fundamentally, the later metropolis per se can incorporate the fact findings and results of analysis from the earlier metropolises into its planning and evaluation process to simultaneously accommodate future sustainable development and a pleasant environment. To a certain extent, this paper has explicitly pointed out the --*common phenomena*-- under similar constraints that could take place repeatedly at the *same stages* of economic development, urbanisation, motorisation and the environment which are presumed to be closely interrelated. Thus, learning from such experiences for future applications, the study try to derive some fundamental guidelines for a late developing metropolis by appropriately shifting and/or accelerating the development cycles of the preceding metropolises.

7.4 Limitations, Coexistence and Final Thought of the Study

Although many attempts have been made mainly to compare and contrast the concurrent dynamic changes of economic development, urbanisation, motorisation, infrastructure and the consequent environmental impact among the metropolises, one should keep in mind that this comparative study is basically based on a *'cross-sectional analogous analysis'* of the outcomes at various stages of development and at different points of time to characterise their common phenomenal mechanisms. It is therefore recommended that further comprehensive research also be carried out through a *'longitudinal (time-serial) analysis'* within the chronological entity and compatible development framework for each metropolis to gain thorough insight into their interactions and actual consequences. As a result of integrating such related research, it would be possible to realise a deeper and broader perspective towards expected sustainability of how and when the comparable dynamic development process continues to change under diverse conditions. Likewise, it would enable us to depict a more reasonable and reliable portrait of our future environment.

References

Bureau of Environmental Protection (1990) The report on air quality monitoring. Tokyo (in Japanese)

Central Statistics Office (1978-1991) Economic trends. HMSO

Central Statistics Office (1991) Regional trends 26. HMSO

City Planning Institute of Japan (1992) *Tokyo Metropolitan Area: Regional Structure, Planning History and Future Perspective*, Shokkokusha. Tokyo (in Japanese)

Department of Transport (1990) Transport statistics report, London traffic monitoring report for 1989. HMSO

Economic Research Institute (1976 and 1992) Annual report on prefectural accounts. Tokyo (in Japanese)

EUROSTAT (1990) Bulletin of energy prices: a survey of import and consumer prices for oil, gas and electricity in the community up to January 1989. European Communities

Hall, P. (1982) *Urban and Regional Planning*. Allen and Unwin

Hayashi, Y., Doi, K., Suparat, R., Sakurai, K. (1991) Comparing the environment between cities in different stages of economic development and urbanisation. Proceedings of the PTRC Summer Annual Meetings, Vol. 19, Seminar A, pp. 95-106

Hayashi, Y., Tomita, Y., Doi, K., Suparat, R. (1993a) An international comparative study on land-use transport planning policies as control measures of urban environment. Selected Proceeding of the Sixth World Conference on Transport Research, pp. 255-266

Hayashi, Y., Suparat, R., Nakazawa, N. (1993b) Influences of economic development and urbanisation on the environment: the case of Tokyo and Bangkok. Paper Presented at the Second International Congress on Asian Planning Schools: Planning for a Better Living Environment in Asia, Hong Kong

Hayashi, Y., Suparat, R., Mackett, R., Kato, H., Anurakkamonkul, A., *et al* (1994) Land use, transport and the environment nexus. *Memoirs of the School of Engineering, Nagoya University*, Vol. 46, pp. 55-98

International Monetary Fund (IMF) (1990) International financial statistics yearbook 1990. Washington DC.

International Road Federation (1970-1991) World road statistics. Washington DC.

Japan International Cooperation Agency (JICA) (1990) The study on medium to long term improvement/management plan of road and road transport in Bangkok. Bangkok

Kanto Engineering Office (1979) Fuel consumption of the vehicle running on

roads: a review on the reports of survey on vehicle fuel consumption. Tokyo
Klaassen, L.H., *et al* (1981) *Dynamics of Urban Development.* Aldershot: Gower, pp. 8-28
Kocks, F.H. KG. *et al* (1975) Bangkok transport study. Bangkok
London Research Centre (1988, 1990) 1986-87, 1988-89 Annual abstract of Greater London statistics. London
LPAC (1990) *Air Pollution Associated with Transport in London.* Romford
Banasopit,M.,*et al* (1990) Urbanisation and environment: managing the conflict. Proceedings of 1990 TDRI Year-End Conference, TDRI, Bangkok
Nagoya City (1966) The outline of city planning for Nagoya '66. Nagoya
Nagoya City (1986) Nagoya centennial: historical statistics data. Nagoya (in Japanese)
Nagoya City (1990) Annual statistics of Nagoya 1990. Nagoya (in Japanese)
Nagoya City (1981) 1980 census of traffic volume in Nagoya. Nagoya (in Japanese)
National Statistical Office (1971, 1975, 1980, 1985, 1991) Statistical year book: Thailand. Office of the Prime Minister, Bangkok
National Economic and Social Development Board (NESDB) (1990) The study on integration of public transport system in Bangkok. Bangkok (in Thai)
National Economic and Social Development Board (NESDB) (1991) Gross regional and provincial product: preliminary series of 1981-1989. Bangkok
Newman, P. and Kenworthy, J. (1991) *Cities and Automobile Dependence: An International Source Book.* Gower
OECD (1989) OECD environmental data. Paris
OECD (1989) World Energy Statistics and Balances 1971-1987, Paris
OECD (1991) Energy balances of OECD countries 1960-1979. Paris
OECD (1991) Energy balances of OECD countries 1980-1989. Paris
OECD (1991) Energy statistics and balances of non-OECD countries 1988-1989. Paris
OECD (1992) Energy balances of OECD countries 1989-1990. Paris
OECD (1992) Energy statistics and balances of non-OECD countries 1989-1990. Paris
Office of the National Environment Board (ONEB) (1989) Roadside air quality in Bangkok (1984-1989). Bangkok (in Thai)
Office of the National Environment Board (ONEB) (1990) Air and noise pollution in Thailand 1989. Bangkok
Pollution Control Bureau (1980-1990) Stage of public nuisance, Nagoya city, 1980-1990. Tokyo (in Japanese)
Sano M. (1979) Fuel consumption on urban streets. *Traffic Engineering,* Vol. 14 No. 2 (in Japanese)
Statistics Bureau (1960, 1985) Population census. Tokyo (in Japanese)
The National Energy Administration (1988) Energy use in road transport. Ministry of Science, Technology and Energy, Bangkok
Tokyo Metropolitan Government (TMG) (1970-1990) Tokyo metropolitan statistics year book 1970-1990. Tokyo (in Japanese)

CHAPTER 2

CONGESTION AND POLLUTION IN A RAPIDLY EXPANDING CITY OF SOUTH-EAST ASIA: THE CASE OF BANGKOK

Kunchit Phiu-nual

1. The Present Transport Situation

1.1 Role of the Transport System

The transport system is one of the main users of land in a city and its task is to serve the movement needs of other land users. But since the other land users depend vitally on transport, the location and density of their activities - and hence their movements - are largely determined by the location and efficiency of the transport system itself. An appraisal of the transport system must therefore include not only its effectiveness in meeting actual traffic demands but, also its impact on the location of activities which gives rise to those traffic demands.

This paper reviews first the performance of the system in meeting the revealed demand for movement, and then its impact in shaping that demand.

1.2 Availability of Transport

Roads

Bangkok has been transformed from a water-based to a road-based city, and many of its problems are related to that change. Its network of main roads (Figure 1) is fairly large, but poorly configured, and it lacks numerous

Y. Hayashi and J. Roy (eds.), Transport, Land-Use and the Environment, 27–45.

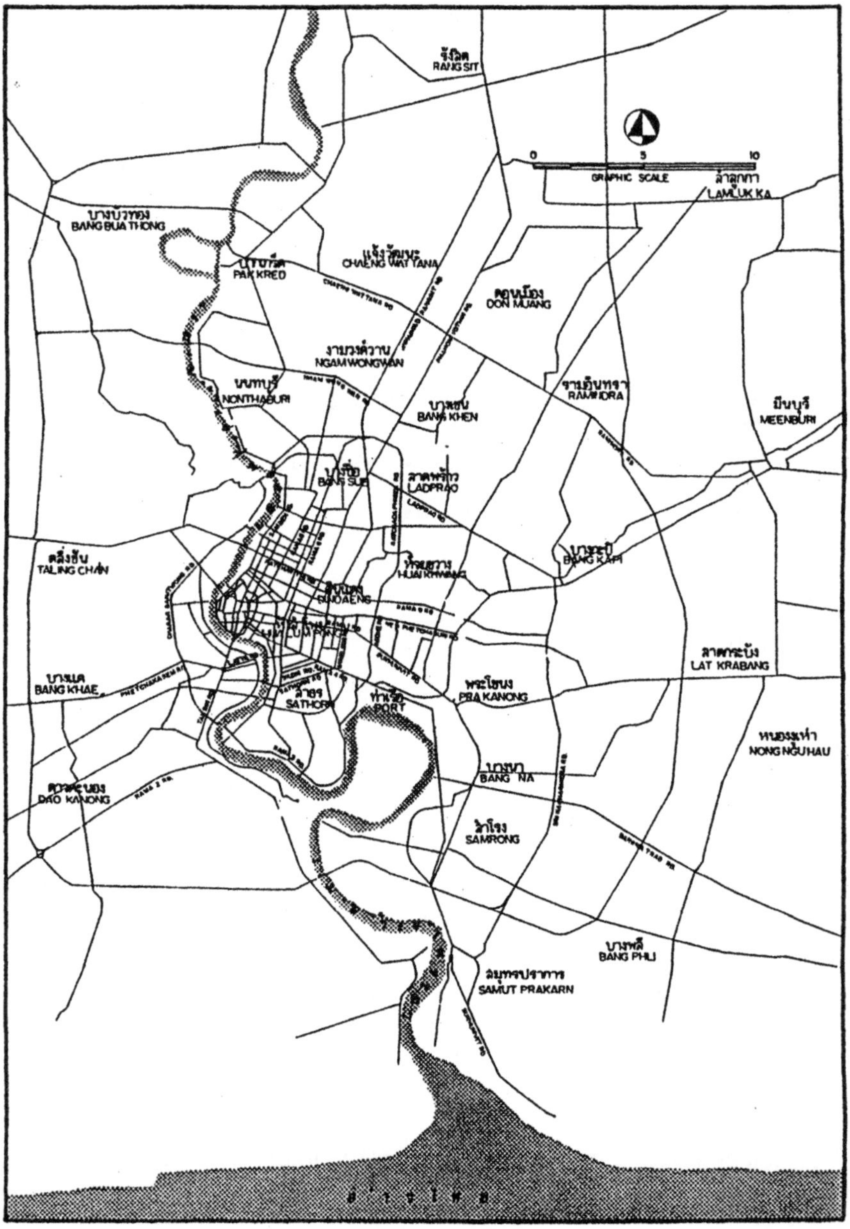

Fig. 1. Road Network in Bangkok

necessary links for better land use, particularly in outer suburbs; these are commonly referred to as "missing links".

The main roads naturally form the boundaries of large blocks which also require internal road systems to provide distribution and access within each block. The construction of these internal roads has been left mainly to private developers, who have built large numbers of access roads to a minimal standard in an incremental, unplanned manner. But, private developers will not build distributor roads, which are needed to connect the access roads to the main roads and also to each other.

Rather than building their own roads, developers naturally prefer to build along the main roads, leaving areas undeveloped in the interior of the blocks. Meanwhile, very few distributor roads have been built. Thus, Bangkok now consists of numerous large blocks containing many little access roads, usually unplanned and unconnected except to the main roads, and lacking distributor roads.

Railways

The railways serve primarily intercity traffic and carry less than 1% of trips in Bangkok. Suburban passenger services are run on four lines and are used mainly by low-income passengers; they are not designed to provide an alternative to the use of the car for longer distance commuters. Some improvements have been made recently, but the capacity for urban services is small and could not be significantly increased without heavy investment in track, signalling, rolling stock and stations.

The Skytrain mass transit system was planned in 1975 and has been approved in principle by Cabinet for many years. The decision in May 1986 to invite the private sector to undertake the project on behalf of the Expressway and Rapid Transit Authority (ETA) has led to years of negotiation, and finally nothing happened. After that, there were three other systems: BTSC, RAMTUFS and MRTA first stage, which were responsible to Bangkok Metropolitan Administration (BMA), State Railway of Thailand (SRT) and Metropolitan Rapid Transit Authority (MRTA) respectively. These projects were agreed to in principle by the cabinet, and two private companies have been selected as the private concessionaires. The first two projects are under construction now, but the MRTA line is still under negotiation with some local companies.

Road transport

Between 1985 and 1992 the number of cars and pick-ups registered in Bangkok more than tripled, and the number of registered motorcycles increased by more than five times. The number of households in the Greater Bangkok Area (GBA) is estimated to have risen from about 1.6 mn. to 2.0 mn. in the same period. Hence, the number of private passenger vehicles registered per household tripled from 0.33 to 0.99. (The registration figures are known to exaggerate the number of vehicles actually kept in Bangkok, but the trend is clear.)

People without private transport are largely dependent on buses, which give a comprehensive coverage of main roads throughout the city. In the lanes where buses seldom operate, there are often minibuses; and for those who can afford them, there are good and plentiful services provided by taxis, motor samlor and motorcycle taxis. Outside the BMA, there are also trishaws, which are not permitted inside.

Urban bus services are managed by the Bangkok Metropolitan Transit Authority (BMTA) subject to policy control by the Ministry of Transport and Communication (MOTC) and Department of Land Transport (DLT). In August 1990, the BMTA owned 4,800 buses and controlled the operation of another 3,200, including 2000 minibuses, by private operators. About 7,200 buses are normally available for service, as compared with about 5,500 in 1985. But in 1985, some 24% of public transport passengers used converted pick-ups (the old style of minibus) which no longer operate in central Bangkok, and the productivity of the buses has suffered from the spread of congestion, both before and during peak hours, and yet there are clearly not enough buses on the road to provide an acceptable service.

1.3 Travel Time and Traffic Congestion

The performance of the transport system is best measured by the door-to-door time it takes to make trips, especially trips to and from work. This, of courses, is the product of travel speed and distances. With the growing sprawl of the city, trips distances are clearly getting longer. Many people are making long trips daily from distant suburbs like Samut Prakarn, Minburi, Rangsit and Nakhon Chaisri. Many children leave home about 5 a.m. to make long journeys to school.

Average traffic speeds in Bangkok are measured periodically by the BMA and are found to vary surprisingly. In general, peak hour speeds inside the Middle Ring Road (but excluding the Expressway) are less than 9 km/hr. This area measures about 90 sq km and includes the central area, about 30 sq km, and the inner suburbs. In other large cities, speeds are typically about 16 km/hr in the central area and 25-35 km/hr in the inner suburbs. Traffic congestion in Bangkok is among the worst in the world. Furthermore, it is not confined to the inner area: heavy congestion occurs regularly in the other suburbs in places such as Bang Kapi, Samrong, Rangsit, Phra Khanong, Ngam Wong Wan Road, Rama II Road and many others. And in Bangkok, few people can avoid the traffic congestion since nearly all trips have to be made by road.

1.4 Traffic Management

Skilled management is needed to obtain the best performance from an urban road network . In Bangkok, about 200 intersections are signal controlled, of which 47 in the old city are linked by a simple computer control system. This is not large by modern standards; other cities employ far more signals and use more sophisticated systems. An Area Traffic Control system, with about 200 traffic - responsive sets of signals, has been recommended by an expert team from Japan and approved by the BMA. This will be supplemented by a Closed Circuit Television system employing 66 camera sites.

The existing signals are well maintained, but their operation by the Traffic Police has been criticised by traffic engineers for many years. In particular, Bangkok has become famed for its long cycle times, which are reputed to be the longest in the world. Experts showed that such long cycle times are inefficient, and fail to maximise junction capacity.

Many main roads are now subject to parking restrictions in order to prevent obstruction to moving traffic, and these restriction are effectively enforced.

One-way and turning restrictions are now widespread. Continuous median barriers, which force drivers to go farther and make U-turns, are a feature of most large roads. Most of these schemes have been introduced by the Traffic Police in order to increase junction capacities. Heavy criticism has been levelled at some of the schemes on the grounds that:

1. they excessively increase journey distances and hence the average volume of traffic, thus absorbing some, if not all, of the extra junction capacity produced by the scheme;
2. they make roads much more difficult for pedestrians to cross;
3. they have adverse environmental effects in terms of noise, air pollution and possible accidents.

Other general defects in the style of traffic management are the lack of channelisation into lanes and the neglect of pedestrians.

Trip distances

Traffic engineers have made an analysis of trip distances in Bangkok. A random sample of 100 trips was selected from the model trip matrix and the distances between origin and destination were measured. A similar sample of trips was selected in London with the same average airline distance, and actual road distance measured. The actual distances were on average 1.34 time airline distance. But in Bangkok, the results showed that:

- actual distances were, on average, 1.55 times airline distance;
- with no one-way system, the average distance would be 1.47 times airline distance;

- with no median barriers, the average distance would be 1.51 times airline distance;
- with neither a one-way system nor median barriers, the average distance would be 1.43 times airline distance.

This means that in Bangkok people drive nearly 16% farther to cover a given airline distance than in London, where the road system is not particularly efficient. About one-half of this excess is due to one-way systems and median restrictions; the rest is due to the poor structure of the road network.

Conclusion
The management of the road system could be improved, to the benefit of all road users, including pedestrians. There is a need for a complete rethinking of traffic management practices and of the way in which these are decided and implemented in Bangkok.

1.5 Road Accidents

The number of road accidents in the BMA is reported to have increased from 15,000 in 1985 to 40,000 in 1988, most of the increase coming in 1988. Doubtless these statistics mask some change in the manner of their collection, but they obviously give cause for concern.

1.6 Transport and Development

Ribbon development
Geographically, in the last five years, Bangkok has exploded: the urban area has expanded rapidly, following the big new road built by the Department of Highways (DOH) and BMA. The Middle Ring Road has attracted enormous development, and all the main radial roads have become lined with ribbon development for between 20 and 50 kms. This process is continuing at an extraordinary pace.

Meanwhile large tracts of land remain undeveloped between and behind the main roads. Figure 2 shows the areas developed in and around Bangkok, as interpreted from aerial photographs. This kind of development is the reverse of good planning; it is inefficient in the use of land, since much potentially valuable land is either not used or used only for small-scale farming; it is highly uneconomic in the use of transport, since it simultaneously causes obstruction of the main roads themselves; and it produces deplorable living conditions for the people who live and work along the roads.

The pattern is now set, because the roads are already built or committed; development has moved in, and there is no prospect of preventing the rapid spread of uncontrolled ribbon development. The huge, unsightly sprawl of shophouses, factories, offices, condominiums, sales yards, etc. can be seen,

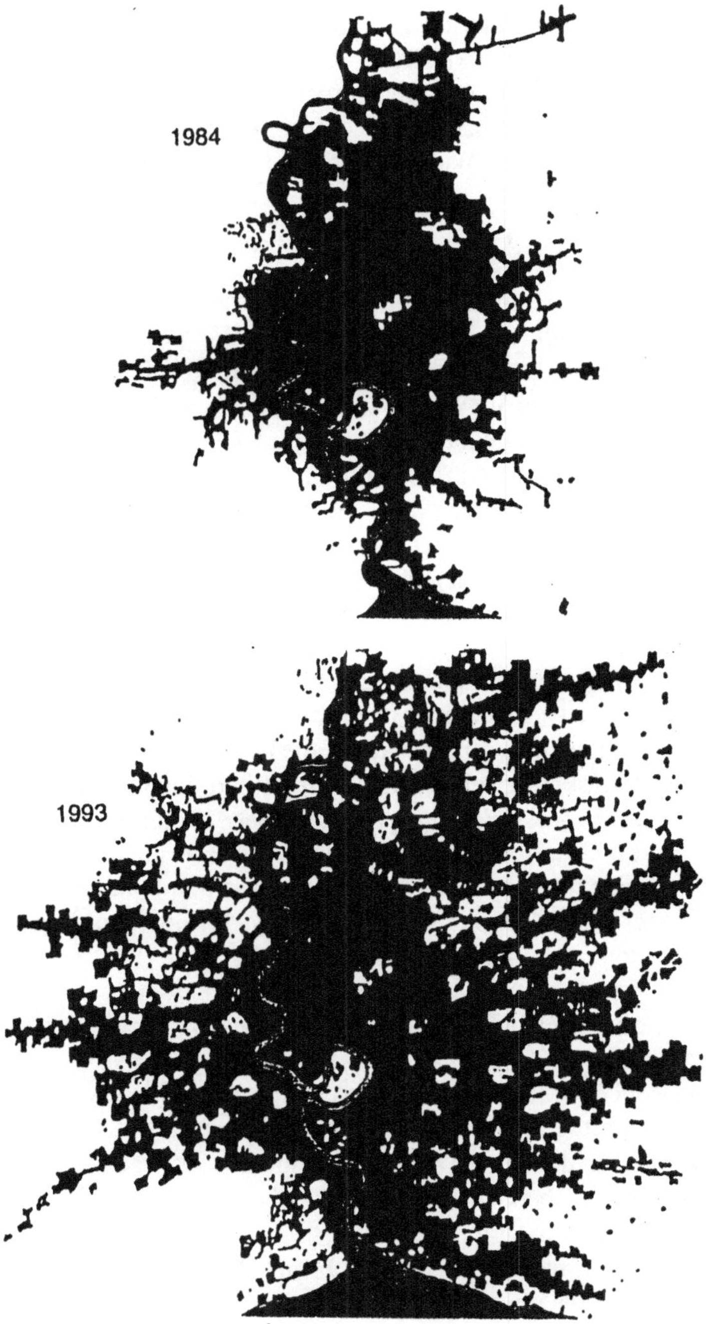

Fig. 2. Built-up Area in Bangkok
Source: Wanisubut (1994). Bangkok Transport: A Way Forward, The Wheel Extended No. 87, A Toyota Quarterly Review, 21

for example, on the Sukhumvit Road to Samut Prakarn, the Bangna Trat Highway, the Rama 2 Road to Pak Thor, the Vibhavadi-Rangsit Highway and Route 4 to Nakhon Pathom. A similar process, but with more emphasis on residential and retail development, is occurring on new big roads like Ngam Wong Wan Road and Chaeng Wattana Road and will doubtless be repeated on the Outer Ring Road.

Managed growth

A key feature of the main transport strategy was the land use policy of "managed growth", which was intended to lie somewhere between the extremes of uncontrolled market forces, which were regarded as unacceptable, and strict planning control, which was regarded as unachievable. With managed growth, it was intended that transport facilities, with some added government incentives and controls, would attract development into the right areas. This has not happened, partly because there has been insufficient government incentive and control, and partly because the main roads have not been supported by the necessary distributor and access roads to encourage development in the right areas off the main roads. There has certainly been growth, but not managed growth.

There is little that transport planning can now do to influence the strategic development of the city, as distinct from the region, apart from encouraging orderly infilling of the empty spaces within the urban area. The big roads have been built and the ribbon development has happened. Planning controls could eventually stop the process and start producing better patterns of development in the right places. Therefore, the main objective of transport investment now should be to provide the supporting networks of distributor and access roads in the right places, and to encourage high density development where highly accessible corridors and nodes are created.

1.7 Transport and the Location of Activities

Transport influences not only land use, ie. the way that land is used, but also who uses it. Given the location of houses, offices, factories, shops, etc., the way that people conduct their activities - homes, jobs, schools, shopping, sports, entertainment - between the available choices is of crucial importance to the transport situation.

Of greatest importance is the home-job link. If everyone lived near their work, there would be no transport problem. Given that there is a severe transport problem, some people in Bangkok go to great trouble and expense to improve their home-job link, by finding a home near to their job or a job near to their home. And similarly with other activities, people avoid going to places where they are likely to get caught in congestion. In some places they will consciously avoid going out of their sois (small access roads or drives) into the main road for fear of getting stuck in the traffic.

In these ways the traffic situation has a restricting effect on people's lives in Bangkok. It is not possible to benefit fully from the opportunities of the city if one cannot travel easily around the city.

2. Existing Environmental Conditions

2.1 Major Environmental Issues in the Bangkok Area

There are a number of agencies responsible for transport and environmental matters in Bangkok. The main ones are listed in Table 1. Some such as the DLT and the Traffic Police have specific interests, whilst others, such as the NEB (Office of the National Environmental Board) and the Institute of Environmental Research at Chulalongkorn University, cover all aspects of the environment. The consensus of these agencies would appear to be that environmental priorities are first air quality and water quality, then noise level, with a lower priority to visual impact.

In a socio-economic survey of 426 households by Chulalongkorn University for the Ekkamai - Ram Inthra Expressway Project, 32% of respondents said they were "highly annoyed" by traffic noise levels of 72 dB(A) or more, 33% considered finding a new residence and 17% wanted better compensation for displacement due to expressway construction. Most people near the alignment preferred at-grade roads to elevated expressways, indicating concern about visual impact. About 77% thought the impact of the expressway would be "most troublesome", but interestingly 50% thought that house prices would go up compared with 33% who thought prices would decrease, and 17% said no change. Also 53% thought that government should give "just and proper compensation", and 38% demanded assistance in finding a new home.

Reports in local newspapers over the last few years have drawn attention to the following issues:

- introduction of exhaust filters on vehicles
- problems for pedestrians on footways and road crossings
- bicycle lanes
- installation of catalytic converters
- policy to reduce production of 2-stroke motorcycles in favour of less polluting 4-strokes
- introduction of lead-free petrol
- noise pollution from expressways
- "Bangkok's killing dust"

Table 1. Thai Agencies and Institutions Concerned with Transport and the Environment

AGENCY	RESPONSIBILITIES
Office of the National Environmental Board (NEB)	Prepare guidelines for Environmental Impact Evaluation (Manual Published in 1979), Monitor Environmental Impact Statements and recommend mitigation measures, Recommend standards to Cabinet, Monitor noise and air quality, Co-ordinate concerned agencies, Prepare national policies and strategies, Environmental education.
National Economic and Social Development Board (NESDB)	Prepare National Economic and Social Development Plans, Co-ordinate Action Plans under Thailand Third Highway Section Loan Review and all investment proposal by government departments and state enterprises.
Department of Land Transport (DLT)	Register all vehicles and inspect vehicles under the Land Transport and Motor Vehicle Acts, with annual inspection of commercial vehicles, once only inspection of private vehicles upon registration.
Traffic Police Division	Enforcement, control traffic during peak hours. Carry out random inspections at roadside using 8 mobile units.
Bangkok Metropolitan Administration (BMA)	Responsible for all environmental problems in Bangkok. Provide pedestrian facilities such as zebra crossings and footbridges and maintain the condition of the footways, eg. policy on vendor. Responsible for most of the highways in Bangkok.
Department of Public Health	Responsible for preventing and controlling effects on the health of the people from all types of hazards, including air pollution.
INSTITUTION	RESPONSIBILITIES
Institute of Environmental Research, Chulalongkorn University	Responds to requests for advice, research and training from other agencies,
Asian Institute of Technology (AIT)	Primary school environmental education programme. Work on vehicle emissions.

- water pollution of khlong (canal) by the Khlong San Saeb elevated road
- traffic accidents
- traffic congestion
- lead pollution

In the opinion of the expert team, based on local information and experience elsewhere, the major environmental issues in the Bangkok region are:

- air pollution
- noise
- visual impact
- severance of communities and agricultural land
- pedestrian/vehicle conflict
- demolition and land-take associated with new infrastructure projects

The following sections indicate the severity and distribution of these environmental problems.

2.2 Air Pollution

Air quality standards and thresholds

The NEB has set air quality standards for pollutants emitted by motor vehicles. Standards set by the US Environmental Protection Agency the (EPA) have been used as guidelines, but appear to have been adapted for local conditions. Thai and EPA standards are shown in Table 2, together with current European Community (EC) standards as a comparison.

Table 2. NEB, EPA and EC Air Quality Standards

Pollutant	Period	NEB	EPA	EC
NO2	Limit	0.32	0.10	0.20
Pb	Limit	0.01	0.0015	0.002
SO2	Year	0.10	---	0.08--0.12
	24 hr.	0.30	0.365	0.25--0.35
Smoke/SPM	Year	0.10	---	0.08
	24 hr.	0.33	0.26	0.25*
CO	1 hr.	50	40	---
	8 hr.	20	10	---

* WHO recommend 0.1 -- 0.15 mg/cu m

Vehicle emission standards

Black Smoke. Two system are specified by the NEB for smoke measurement, the Bosch and the Hartridge, with different levels being specified for each. In practice, the Bosch method is the one that is used for smoke testing at the DLT

Inspection Centre. Two measuring methods are also specified. The first method specifies a standard of 40% for no-load acceleration at 75% of maximum speed, but says that this is to be changed in 1988 to 50% at rapid acceleration under no-load condition to maximum speed. The reading to be taken is the greater of the two. The second method is designed to be used while the engine is running on a test bench at 60% of maximum speed and full load. The average of two measurements is taken as the result.

Carbon Monoxide. This is measured while the vehicle is idling using Non-Dispersed Infra-Red [NDIR] measurement. The average of two measurements is taken, and the standard is 6%, meaning that no more than 6% by volume of the exhaust may be CO.

Application of Standards. Selected vehicles (registered under the Land Transport Act, ie. lorries + buses) are tested for black smoke during their annual inspection. The vehicles selected are those that appear to be emitting excessive black smoke on their approach to the test point. This method catches infrequently only the chronic offenders. Also, a bus was picked at random. Although not emitting visible black smoke, it nevertheless failed the smoke test, registering 51% and 61% on its two tests. The Thai standards are not too lenient, but they need applying more rigorously. They only apply to commercial vehicles. Vehicles registered privately, including large numbers of pick ups and other vehicles fitted with diesel engines, are never tested, and there are no data on their emissions. In routine checks, CO is not normally measured. This is partly due to the fact that most tests are on diesel engine vehicles, which do not have high CO emissions.

It is understood that CO levels may be measured during random checks on the road, as may smoke levels. In view of the sort of NDIR equipment used for CO measurement, it would be surprising if it functioned for long at anything approaching its design accuracy, unless used by competent personnel and calibrated frequently.

Severity of air pollution problems

Data have been obtained in an unpublished form from the NEB for CO, suspended particulate matter (SPM) and lead concentrations. The data are from 16 locations in Bangkok and for the years 1987 , 1988,1989. For the purpose of this paper some of the data, relevant to the air quality standards, are reproduced in Table 3 for the year 1989.

Carbon Monoxide. None of the samples exceeded 63% of the NEB standard. It is not known exactly when or where the data were collected. An NEB Study in 1977 at roadside locations found CO levels equivalent to the 8 hr NEB standard of 20 mg/cu m., which is twice the EPA standard. Variation of the period or position of a roadside test may significantly affect the measurement.

Data from other sources are given in Table 4 and 5.

Table 3. Summary of Air Pollution Data From Nine Sites in Bangkok for 1989 [mg/cu m]

Location	CO	SPM	Lead	
	[max hrly av]	[24 hr av]	[av]	[max]
1. Ratcha Prarop Rd.	15.87	0.81	0.0020	0.0027
2. Yao-Wa-Rat Area	13.76	0.56	0.0023	0.0044
3. Lan Luang Rd.	9.65	0.35	0.0019	---
4. Bamrung Muang Rd.	28.40	0.46	0.0038	---
5. Sukhumvit Rd. (Meteorological Department)	8.12	0.49	0.0017	---
6. Bang Lum Poo area	14.23	0.25	0.0012	0.0018
7. Phahon Yothin Rd.	27.80	0.39	0.0012	0.0017
8. Silom Rd.	31.47	0.58	0.0031	0.0043
9. Si Phraya Rd.	13.92	0.39	0.0028	0.0062
10. Ban Somdet Police Station	18.05	0.38	0.0018	0.0036
11. Somdet Phra Pin Khlao Hospital	7.05	0.19	0.0006	0.0011
12. Chula Hospital	6.21	0.16	0.0009	0.0014
13. Police Department, Rama IV Rd.	10.81	0.34	0.0013	0.0023
14. Ramkamhaeng Rd.	9.96	0.4	0.0017	0.0025
15. Land Development Dept., Bang Kaen	---	0.18	---	---
16. Legal Enforcement Dept.	---	0.22	---	---
NEB Standards	50.00	0.33	0.010 [max]	---

Table 4. Air Quality Near Major Streets in Bangkok and Other Major Cities

Pollutants	Site Measured	Range of Max Values (mg/cu.m)	Standard (mg/cu.m)
Carbon Monoxide (8 hour)	Bangkok	27-37	20
	Chiang Mai	16-18	
	Haad Yai	6.27	
Suspended Particulate Matter (24 hour)	Bangkok	0.23-1.05	0.33
	Chiang Mai	0.41-0.47	
	Haad Yai	0.42-0.45	

Source: NEB (1985)

Table 5. Existing Air Quality Along Proposed Ekkamal-Ram Inthra Expressway Route (Average of 5 Measurements, 1987)

unit mg/cu.m

Measurement Site	SPM	CO	Pb
	24 hr	Max./hr	24 hr
Sacred Heart Convent School	0.187	6.0	0.00115
Military Flat, Sukhumvit Soi 66	0.244	6.3	0.00205
Soi Ruam Rudi 2	0.247	8.1	0.00442
Sukhumvit Bowl	0.206	9.9	0.00261
Lat Phrao Soi 71	0.244	14.7	0.00163
Bangkok Bank, Ram Inthra Road	0.175	5.2	0.00082
NEB Standard	0.33	50	0.01 (max)

Source: Ekkamai - Ram Inthra Expressway Final Report (1)

Suspended Particulate Matter. There are significant problems at almost all the sites. Some sites show daily averages over twice the standard, while one site has an average SPM of nearly three times the limit.

Lead. None of the sites show excessive levels by NEB standards. All but two of the sites, however, exceed the EPA limit (0.0015 mg/cu m), and most are at or above the EC limit (0.0020 mg/cu m). Other studies of streetside locations suggest much higher levels, up to 0.03 mg/m3 for people living and working on the streets, eg. vendors and policemen. The high level of lead in petrol is no doubt primarily responsible for these results. Currently petrol has a limit of 0.415 g/ litre in Thailand, with an unleaded option. It is intended that the lead shall be reduced to 0.15 g/1 by the end of 1993.

Vehicle emissions

There are no published data on emissions from road vehicles in Thailand. Reported observations indicate that the main problems are black smoke from diesel powered vehicles, unburned hydrocarbons from 2-stroke motorcycles, together with the more normal emissions of the car fleet.

Motorcycles form an important proportion of the vehicle fleet in Thailand. Because of the relative unimportance of motorcycles in Europe, there has been little research on their emissions. The difference between two stroke and four stroke engines should be noted. Two stroke engines emit considerably higher unburned hydrocarbons than four stroke, while the situation is reversed for CO. The levels of nitrogen oxides (NOx) from the different types of engine are known. In Thailand, the high numbers of motorcycles and their probable contribution to the total pollution suggest that legislation is needed to limit their emissions.

Conclusions

The NEB has been collecting air pollution data from roadside sites for some years and has published data up to 1993. A new version of the report "Air and Noise Pollution in Thailand" is currently being prepared. From their data, it is apparent that there are severe problems of air pollution in several areas in Bangkok, and possibly in some of the larger towns in the provinces. The worst problem appears to be SPM, but the lack of descriptive information accompanying the data makes it difficult to assess the problems from gaseous pollutants.

The lack of data on vehicle emissions makes quantitative conclusions impossible. There is a need to construct a data base detailing the emissions of the entire vehicle fleet based on sample tests. The data base should include motorcycles. The formulation of realistic emission standards should proceed in parallel.

2.3 Noise

Noise standards

There are no community noise level standards in Thailand. The NEB do, however, refer to the US EPA standard of 70 dB(A) leq 24 hr, as a level when hearing loss occurs. In the UK, indoor noise level standards recommended by the Wilson Report (5) are usually followed:

	Day	Night	Remark
Country area	40	30	Level not to be
Suburban, away	45	35	exceeded for
from main roads			more than 10%
Busy urban areas	50	35	of the time

To estimate outdoor noise standards, 15 dB(A) should be added for the insulation effect of walls and windows. On this basis, levels of up to 65 dB(A) are acceptable in urban areas.

The US Federal Highway Administration prescribes 60 dB(A) for parks and open spaces where quietness is of primary importance, 70 dB(A) for residential areas, hotels, schools, hospitals and 75 dB(A) for other types of development.

Vehicle emissions

The NEB do have vehicle emission standards (Table 6) which are compared below with standards in other countries. In setting emission standards it is important to have a statutory test procedure for measuring noise. The EEC test procedure, Council Directive 81/334/EEC, is the most widely accepted. It simulates the highest noise levels under urban driving conditions and entails 'open throttle' acceleration of the vehicle over a 20 m zone for an approach speed of 50 kph in the second gear. The sound pressure level is measured at 7.5 m from the track centreline. The higher of two measurements is taken. The NEB adopted the stationary test as a measurement method with a standard for all vehicles of 85 dB(A) measured at 7.5 m. from the exhaust and 100 dB(A) at 0.5 m.

Table 6. Vehicle Emission Noise Limits (dB(A))

Vehicle type	1985/86 81/334/EEC	1980/89 84/424/EEC	NEB
Passenger Car	80	77	85
Minibus < 9 seats	81	78-79	85
Bus > 9 seats	82-85	80-83	85
Light truck/ van	81	78-79	85
Medium truck/van	86	81-83	85
Heavy trucks	88	84	85
Motorcycles < 80 cc	77		85
80 - 175 cc	80		85
> 175 cc	82		85

Note: Motorcycle standards are contained in 78/1015/EEC

Although the test procedures are not comparable, if one assumes a measurement distance of 7.5 m, then the NEB standards are lenient compared with EC for all vehicles under 84/424/EEC, the most recent legislation; and similarly, when compared with the US, Japan and Switzerland, where levels range between 70 dB(A) for a moped to 84 dB(A) for heavy vehicles

(Switzerland). The roadside tests carried out by the DLT and the traffic police measure noise at 0.5 m with open throttle. There is no information with which to compare the 100 dB(A) standard, but it seems high, particularly if applied to all vehicles in the same way.

Noise problems

The NEB carries out measurements for 7 continuous days, per year at 14 stations alongside roads in Bangkok and in 5 regional cities. The standard of 70 dB(A) is marked, and all readings lie above the standard. Nevertheless, there appears to be a trend towards lower noise levels. This may be due to the lower speeds on the network now, generating lower noise from vehicles. At least 5 stations have noise levels in excess of 80 dB(A) and all are in excess of 75 dB(A). In comparison with the UK guidelines, Bangkok noise levels are at the top end of the range for arterial and major roads, assuming L10 is roughly equivalent to Leq +3 dB(A).

Noise measurements were carried out at 10 sites within the Ekkamai-Ram Inthra Expressway corridor as part of the expressway study. The results are given in Table 7.

Most sites lie in the range Leq 70-75 dB(A), ie. in excess of the standard. Only 3 sites are below 70 dB(A) and these are where development is less dense.

Table 7. Noise Level Along Proposed Ekkamai-Ran Inthra Expressway

Site No.	Location	Description	L10	Leq
1	Sacred Heart	Traffic flows to port	76.5	66.9
2	Sukhumvit 66/1	30m from Bangna expressway	75.5	70.2
3	Soi Ruam Rudi 2	Next to expressway ramp	81.5	75.9
4	Ekkamai Road	Connects Sukhumvit to New Petchburi, both major roads	81.5	74.6
5	Prakhanong Bridge	Semi-Commercial area	78.0	75.3
6	Ram Kham Haeng/ Rama IX	Commercial centre	81.0	74.7
7	Pracha Uthit Road	Paddy fields mainly	74.0	66.9
8	Lat Phrao 71-73	Housing	79.0	72.8
9	Soi Maiyalap Ram Inthra		64.5	57.1
10	Soi Wat Trirata-naram Ram Inthra	Busy community	82.5	75.5

Source: Ekkamai - Ram Inthra Expressway Final Report.

An example of a severe noise problem was reported in 1988 at Prathomnonsee School, which is located 4.5 m from the edge of the First Stage Expressway System (FES) which is higher than the third floor of the school. Noise from traffic was so high that teachers had to use microphones. Teachers and students were going deaf.

2.4 Visual Impact

Bangkok was until recently a low density, low rise city . Increasingly high-rise buildings have been constructed, giving in some areas of the city a sense of enclosure. At present, the only elevated road is the First and part of the Second Stage Expressway, but its visual impact is high. It is above the height of an average house, can be seen over some distance in certain areas and obstructs the view of people living and working alongside. The junction just south of Rama IV Road, where the expressway is divided, is particularly ugly, and dominates the surrounding area. At present there are no measures to reduce the visual impact.

The rash of huge advertising boards next to the expressway is another eyesore, blocking views of the city and detracting from the adjacent buildings. They are a growing fashion associated with highways, and examples can also be seen alongside the regional highways.

2.5 Severance

Within Bangkok, major roads often sever communities and act as a physical barrier to movement, either due to the number of lanes that need to be crossed, the high traffic volumes at high speeds or the presence of a median divide. Pedestrians have to use a footbridge, if there is one, or cross at the nearest traffic signals. Drivers may have to make a long detour to the next U turn or junction.

Examples of roads with a severance effect are Rama IV, Sathon (median divide), Sukhumvit, Rama I, Phetchaburi (high traffic volumes and speeds) and the First Stage Expressway (FES) where at- grade. Even where the FES is elevated, it still creates a feeling of severance; this is psychological rather than physical, but still real to people.

2.6 Pedestrian / Vehicle Conflict

A good indicator of the degree of pedestrian/vehicle conflict is the set of accident statistics concerning pedestrians. In 1979, 47% of road casualties in Thailand involved pedestrians. Road accidents are now the main cause of death, and are increasing at 13% p.a.. Death from diseases such as tuberculosis and pneumonia is declining (from 20 to 10.3 per 100,000 population, 1970-1985).

Contributing factors are a low standard of road user behaviour, in particular, non-observance of red signals by drivers, issue of driving licenses without passing a test and pedestrians ignoring crossing regulations. According to BMA, most pedestrian accidents occur at zebra crossings. Although zebra crossings are clearly marked with signs and lights installed by the BMA, drivers still tend to disregard them, and do not give pedestrians the right of way. Pedestrians have a false sense of security at zebra crossings, and do not take the same care as at random crossings. BMA have found it dangerous to place zebra crossings midway between an overbridge and signal crossing, because vehicles pick up speed on the section with an overbridge and often do not slow down for the crossing.

As a response to poor driver behaviour, wide and fast roads and high pedestrian flows, the BMA have installed over 200 footbridges and annually install 15 throughout the city. This makes Bangkok one of the highest ranked cities for footbridges. In spite of this, the pedestrian has a hard time. Poor conditions on the footway compound the problem for pedestrians, with vendors and public utilities cluttering the way, as well as pot holes, drainage ditches and poor street lighting. One of the positive benefits of walking in Bangkok is the landscaping and tree planting on many streets, providing a pleasant and shaded environment for the pedestrian.

A growing concern is accidents between motorcyclists and pedestrians in the sois, particularly at night when they are ill-lit. There are many motorcycles now providing a taxi-like service in the sois. Thus the National Safety Council tries to introduce safety measures in the sois such as speed humps, better lighting and footpaths.

References

HFA, PPK, AEC (1980). Seventh Plan Urban and Regional Transport Study, A study conducted by a group of consultants for NESDB (National Economic and Social Development Board), 11-55

Limpaseni, W. and Danich, S. (1987-1988). A Review of the Air Pollution Status in Bangkok During Sixth National Plan, A study for preparing an emission data base for Bangkok, 1-68

Phiu-Nual, K. (1989). Project Implementation and Lessons Learned During the Sixth National Plan, *Engineering Institute of Thailand Journal* (in Thai), Vol. 2, 32-40

CHAPTER 3

BETTER CITIES - BETTER FOR THE ENVIRONMENT

Barry Telford

1. Introduction

As cities grow in numbers and in size awareness increases of the damage being done to the environment through pollution due to storm water and sewage, congestion in urban living and the dominance of the private motor car. The ever increasing costs associated with both hard and soft infrastructure are also becoming unacceptably high.

There are various conferences and reports on the environment and the increasing difficulties being experienced by urban dwellers. The Better Cities Program in Australia, while based on solid urban development principles, is responding to calls that will from a practical approach demonstrate the means of addressing some of the emerging issues.

2. The Better Cities Program

The Better Cities Program is a Federal Government initiative announced in the 1991/92 Federal Budget and provides funds to the States and Territories totalling $816 million over five years. The program involves the three spheres of government, the private sector and the community to ensure a more strategic approach to urban development, and the greatest possible lasting

Y. Hayashi and J. Roy (eds.), Transport, Land-Use and the Environment, 47–54.

benefits for people who live and work in Australian cities. Over the five years of the program around $2.5 billion will be allocated to this program from all sources.

The Better Cities Program promotes an integrated approach to urban development and design by bringing together planning, housing, employment opportunities, transport, environmental improvement and infrastructure investment.

The philosophy of the Better Cities Program is that these problems cannot be seen in isolation. An integrated or holistic view is imperative to ensure the maintenance of the social, economic and environmental viability of an area.

The outcomes that Better Cities is striving to achieve have been related to other elements or components in that area in an integrated manner. Integrated planning is at the core of this innovative national program. It ensures that a complete range of issues that pertain to the development or redevelopment of urban areas are considered at the planning and implementation stage. Environmental considerations feature strongly.

Better Cities seeks to deliver real improvements to urban environments through the development of 25 fully integrated area strategies in major centres across the country.

These area strategies, which are identified in formal agreements between the Federal and State or Territory Governments, consist of a range of urban development activities which promote best practice in the planning, design and delivery of urban environments. They evidence the benefits of a co-operative and co-ordinated approach to urban development by the public and private sector.

A major objective of the Better Cities program is to demonstrate best practices in, among other things, the environmental management in urban areas which is both relevant in Australia and overseas.

The program does not seek to provide one-off solutions to the emerging problems of our cities. Solutions are as diverse as the problems and therefore it is the approach and its transportability that is of major significance.

The five goals of the Better Cities program are:

- economic growth and micro-economic reform;
- improve social justice for the less advantaged;
- reform of inappropriate and outmoded institutional care for people with disabilities and frail aged people (in some cases freeing up land for other uses);
- ecological sustainable development; and
- more livable cities.

These five goals are embodied in each of the 25 area strategies spread across the eight States and Territories in Australia.

The key point being made in this paper is that it is only through an integrated approach to planning and development that an urban form which will be economically sustainable in terms of its own geographic region and those regions which impact upon it can be ensured. The Better Cities program has taken a very practical approach to the problems confronting Australian cities. For too long planning and development has been concerned or focused on the concern of the immediate vicinity rather than take a broader view of the region or district. Better Cities advocates the broader application of planning practices and principles to the urban environment.

This paper will discuss three areas of the program which illustrate this broader approach to environmental management.

3. Water and Sewerage Treatment Programs

Some water and sewerage treatment programs have been developed through the Better Cities programs. As indicated, these programs are designed to take into account the full effect of those activities both within the geographic boundary of the area strategy and adjacent to it.

A good example is the Southern Areas Area Strategy, on the southern fringe of metropolitan Adelaide. The area was, until relatively recently, used for agricultural and farming purposes. Residential and commercial development since the early 1970s has increased pressure on the areas' natural environment and traditional farming industries.

The aquifer currently being drawn on by the farming industry is seriously stressed through heavy usage, threatening the economic viability of this important agricultural area.

The area also has strong environmental appeal and importance, and existing residents and local government are concerned to ensure future development is sensitive to environmental needs.

The Better Cities response to these issues includes proposals for stormwater retention and aquifer recharge, and the piping of treated water several kilometres to the vegetable farming area. The recycling of this water will enable a constant supply of nutrient rich water to this important vegetable growing region of South Australia.

It is estimated that the increase in productivity due to the use of nutrient rich treated water, the reduced requirement of fertilisers and reduction in water usage will result in a 100 per cent increase in productivity in this region.

The environmental restoration program focuses on water resources management, stormwater, drainage, revegetation of creeks and watercourses and environmentally sensitive land-based treatment of effluent. In addition to these measures, the impact of urban development and tourism on agricultural land use and production will be managed to ensure that the environmental and economic assets of the area are not compromised.

Separate, yet related, issues are being addressed in the North West Crescent Area Strategy near the Port Adelaide district in metropolitan Adelaide. While only physically covering an area of 2000 hectares, the area collects 40per cent of all the stormwater runoff from the greater Adelaide metropolitan area. In addition, the Port Adelaide River has, over many decades, received effluence (treated to varying degrees) from local sewerage plants and industries. This volume of contaminated water is doing significant damage to the mangrove environment and the fish nurseries it sustains. Added to these manmade environmental problems is the low rainfall of the area. The evaporation rate is three times that of rainfall, and salty groundwater can be found just 1-2 metres below the surface. It is a tough, relatively dry, often dusty area, making greening difficult.

The impact of all of these factors was taken into consideration in the development of an integrated plan to address the significant environmental problems in the North West Crescent of metropolitan Adelaide. Consideration was also given to the sewerage treatment works adjacent to this area.

A key feature of this plan is the re-use of stormwater and sewerage effluent by local industry to, in turn, minimise their own effluent discharges. Waste water will also be reused by local authorities for improvement of open spaces and recreational facilities.

The treated water from the sewerage work in the area strategy will also be used in the salt mining industry of South Australia which is one of the Australia's largest. This plant accounts for 20 per cent of the use of fresh water in the Adelaide metropolitan area. Again the recycling of treated water to this plant will result in considerable benefits by reducing stress on the limited water resources in this very dry area of South Australia.

This increased sensitivity and understanding of a total management system for water and sewerage has resulted in the better estuarine management in all of the Port Adelaide region and will eventually have significant effects on increasing the mangrove swamps in this area.

This approach to waste management should also result in a reduction in the levels of contaminated water release into the Gulf and will lessen the proliferation of seaweed which is having a damaging effect on the mangrove population. The production of seaweed will hopefully be greatly reduced and eventually this problem will be eliminated. This should allow the mangroves to regenerate and reclaim some of the mudflats currently lying bare.

These mangroves are significant as they are the most southerly located mangroves in the world and could form the basis for future "eco-tourism" ventures.

4. Role of Wetlands

The second major aspect of the Better Cities program is related to the first. It is the need to better understand the role played by wetlands within city environments and how to manage them more effectively.

This is being achieved again in the Adelaide region through the rehabilitation of currently polluted waterways to create a wetlands area. Sensitive management of existing mangroves and the establishment of urban forests will result in repopulation of the area with native aquatic and bird life. The rehabilitation and management of this natural environment will result in significant social and economic benefits for residents of the North West Crescent of metropolitan Adelaide.

Wetlands close to the central business districts of cities such as Perth and Adelaide give increased impetus for the need to give attention to better understanding and management of these areas in an urban context.

"Eco-farms" are laying an important role in the purification and natural treatment of runoff from cities. The attraction of fauna and flora is an additional bonus.

The incorporation of settling ponds in wetland areas as a method for removing contaminants is becoming increasingly understood. In another of the Better Cities sites in the area, the ponds of the wetlands are being constructed with cement bottoms so that they can be drained and cleaned of the heavy metal deposits when necessary. Again this is evidence of the integrated planning process central to Better Cities. It must be ensured that the rest of the wetlands in these areas do not suffer pollution by ensuring that heavy metals are trapped early in the process.

In another area strategy funds have also been provided to undertake projects designed to improve the quality of effluent flowing into the Swan River from the local water mains in Perth.

The pilot scheme in this area is Australian using technology developed by the Commonwealth Scientific and Industrial Research Organisation (CSIRO) and others and involves the establishment of artificial wetlands to strip nutrients from the flow in the drains.

Artificial wetlands will be built through the excavation of sandy material and replacement with suitable soil and the planting of appropriate aquatic vegetation.

The basic principle is that the contaminated water will flow through these areas where nutrients will be stripped by soil and vegetation before the water flows into the river or sea.

5. Use of Private Motor Vehicles and Public Transport

The final area to examine is the role that Better Cities is playing in reducing the use of private motor vehicles.

As stated earlier, one of the major objectives of the Better Cities Program is to stimulate new solutions to urban problems. In nearly all cases the public transport investment component of the program is being incorporated into a wider urban strategy involving significant expenditure on other forms of infrastructure, land reclamation or development, or the provision of housing. It is hoped that this approach will have a major impact on reducing the level of car dependence and the associated environmental degradation.

Approximately half of the Better Cities Program expenditure is to be spent on utility services, with railways and other public transport accounting for over 40 per cent of that, or just over 20 per cent of the total program expenditure.

The significance of public transport as a proportion of total expenditure varies across the States according to the nature of the particular area strategies. While 98 per cent of this expenditure on public transport is concentrated in ten of the twenty-five area strategies, it is estimated that over the five years from 1991-92 to 1995-96 just over half a billion dollars will be spent on public transport projects within Better Cities area strategies.

In many cases this expenditure will enable an acceleration and probable expansion of the conventional public transport infrastructure associated with the developments.

In particular, area strategies in NSW, Victoria and Queensland have been planned so that light rail systems will be in place before residential development is permitted. This together with parking policy that will promote more appropriate parking in the higher density developments will ensure that alternate transport will be available before people become reliant upon a private transport or purchase a second car.

Light-rail is being funded in the northern suburbs of Melbourne so that higher density housing will develop around newly established major transport nodes. Much of the higher destiny housing in this area will be designed for people who are aged or intellectually disabled. These people are now living in group houses as a result deliberate government policy of closing large institutions in these areas.

Another important transport development is the Ultimo-Pyrmont area strategy on the very fringe of Sydney central business district. In this instance light rail is again being used and will be in place ahead of density residential development.

Initial planning for Ultimo/Pyrmont has identified an anticipated high percentage of public transport usage, which will not be able to be met in the long term by bus services. To overcome this, a light rail service is being put in place, utilising an existing goods rail line and linking in with the existing heavy rail services in the city. A parking policy, based on a desired modal split of 65 per cent use of public transport, is included in the development plan for Ultimo/Pyrmont. To facilitate the implementation of this policy, a direct pedestrian and cyclists link from Ultimo/Pyrmont to the Sydney CBD is to be provided.

This strategy is also reliant upon the development of employment and commercial activity around the transport nodes so that people who are living or working in the Ultimo-Pyrmont region will only have to travel a relatively short distance to work and shop and not become reliant on private transport.

Another successful example of how the Better Cities program has attempted to diminish reliance on private transport in cities is in Perth.

While not a new concept, the Better Cities program has funded a "Park 'n Ride" in one of the outer suburbs of Perth. People are offered the option of buses to the suburban surrounding district or high speed trains to the city. The usage of this facility is up by 30 per cent on projected usage levels after only one year of operation.

The Park 'n Ride" project is also developing as an important mode of transport to and from major weekend sporting events in the city thus reducing congestion and reliance on private transport on weekends as well.

The integrated planning for this area strategy ensures the growth of commercial and residential development around this transport node.

6. Conclusion

Environmental considerations are required to be integrated into the overall planning process for Better Cities programs. They are given prime consideration before the commencement of development projects.

Results in the Better Cities programs are very encouraging to date and demonstrate that an integrated and carefully planned approach to urban policy and development will result in cities that not only are better for the environment but better places in which to live.

Better Cities is about demonstrating that environmentally sensitive urban developments are not only viable but economically and socially beneficial and desirable.

Commitment must be maintained to not only environmental issues, but also to those aspects of urban development which inevitably impact on the environment.

References

Population Issues Committee (1991) Population Issues and Australia's Future, National Population Coucal, AGPS, Canberra

Urwin, N. and Searle, G. (1991) Sustainable Development and Urban Development, Urban Futures, Special Issue 4, Nov 1991.

Department of Arts, Sport, the Environment and Territories, (1992) Urban Consolidation: An Environment Perspective, Conference Proceedings, AGPS, Canberra

CHAPTER 4

HIGHER DENSITY HOUSING AND TRANSPORT IN AUSTRALIAN CITIES

John Black

1. Introduction

Similarities of residential density and travel characteristics in the outer suburbs of both Australian and North American urban areas have led to the term "auto dependency". The landscape, originally filled with coherent communities, now has, according to Kunster (1993) in the *Geography of Nowhere*, the unreality of the suburbs and the tyranny of comparative commuting, such that every place is a "no place". Freeway investment encourages low-density development leading to the unsustainability of high car usage, petroleum consumption and air pollution. As referenced fully by Steiner (1994), a few researchers and real estate developers, and an increasing number of policy makers in many US states, and around the world, argue for infill housing, mixed land uses and increased density, especially around public transport services. These urban forms are variously referred to as urban consolidation, urban villages or "neo-traditional development".

With a growing awareness by governments to adopt principles of ecologically sustainable development (see, for example, Commonwealth of Australia, 1991; and Whitelegg, 1993) the type of policy able to contribute more to solving the full spectrum of economic, social and environmental problems is urban consolidation. This was recognised by leaders and representatives attending the October 1990 Special Premiers' Conference in Australia who agreed to an acceleration of efforts to develop a higher density

Y. Hayashi and J. Roy (eds.), Transport, Land-Use and the Environment, 55–80.

housing code. Work on the new code - the Australian Model Code for Urban Housing - commenced in 1991 with a specific term of reference to address urban consolidation and the change from low density, sprawling, cities to higher residential densities in established urban areas. Three principal consultants were given the responsibility by the Commonwealth Department of Health, Housing, and Community Services to prepare the first draft of the code. Technical input and assistance were provided by Gutteridge Haskins & Davey Pty Ltd (1992) in relation to the engineering aspects - urban infrastructure, movement networks, street design and construction, streetscape, environmental management, access and parking, and site facilities.

The more specific tasks on transport were to identify in the literature the traffic and transport aspects of higher density housing, and then to develop a framework for assessing environmental capacity applicable at the site, precinct, neighbourhood and district scales. Advocates of urban consolidation have made assumptions about the relationships between high-density residential development, travel and transport choices, and the resultant environmental impacts, including energy consumption (for example, Newman and Kenworthy, 1989).

This paper addresses these land use, transport and environment linkages and describes how Australian cities are trying to move towards higher density housing through urban consolidation so as to achieve better transport and energy efficiencies. As there is no clear leader in determining the pattern of urban development in Australia, schematic models are presented to indicate the wide ranging responsibilities for co-ordinating land use and transport, which make the rhetoric of urban consolidation convincing but its effective implementation questionable. These models are tested using issues on urban consolidation raised from different perspectives - governments, providers and consumers of housing, environmental groups, and the research sector.

2. Australian Urban Development

Historical factors have helped shape the internal structure of Australian cities. Of these factors, the timing of the emergence of the successive generations of mechanised transport technology in relation to urban development has been an important influence. Australian cities were relatively compact and walking was the dominant mode of personal transport prior to the introduction of tramway and suburban rail services. There has been a steady progression from a compact "walking city" to a sprawling, car-dependent, city at low residential housing densities by world standards. Taking into account semi-urban development of large lot subdivisions, the post-war spread of Australian cities is believed to have exceeded one million hectares (Australian Urban and Regional Development Review, 1993, p. 11). Despite the problems of definition of terms and the relative accuracy of data, Newman and Kenworthy (1987, Table 4, p. 175) have shown that the mean population densities of Australian outer suburbs at 13 persons per hectare are similar to that of US outer suburbs - at 11 persons per hectare. For international comparative purposes they are significantly different to the outer suburbs of European cities: Newman and Kenworthy's analyses indicate they are about one third of the population density of twelve European cities.

2.1 Population Density

A great deal of urban development in Australia has coincided with the increasing availability of motorised road-based transport - first buses then private cars and trucks. Each development in transport has contributed to the low-density, geographically dispersed, car-dependent character of large tracts of contemporary Australian cities. For example, Stuart (1984, Figure 2.3. p. 12) contrasts per capita motor car ownership in Australia, 1921 to 1980, with the changing patterns of municipal population density in Sydney, 1891 to 1971. Population densities have declined whilst car ownership has increased and this trend is likely to continue unless there is stronger government intervention over urban development and transport pricing.

As documented in many cities throughout the world, there has been a flattening over time of the urban population density gradient in Australian cities. The fastest growing regions are Brisbane to the Gold Coast, Brisbane to the Sunshine Coast, the western suburbs of Sydney, the south eastern suburbs of Brisbane, the northern suburbs of Perth and the south eastern suburbs of Melbourne. The pattern of growth in the outer municipalities of Brisbane, Perth, Melbourne and Sydney are expected to remain the fastest growing areas over the next five of six years (Clare, 1991a, p.26). Population decline, through changing demographic structures, has occurred in a number of municipalities between 1976 and 1989, primarily in the inner areas of larger cities (especially Melbourne).

2.2 Energy Consumption

Australia has also one of the highest per capita car ownership rates in the world. Car ownership levels, annual vehicle use, and motor car fuel consumption in selected OECD countries for 1983 are compared in Table 1. Australia, with 0.47 cars per person, was third behind the USA (0.53) and New Zealand (0.50), and, with 15,100 annual km per car on average, marginally second to the UK in terms of vehicle usage. Although fuel consumption figures are included from national statistics, for comparative purposes should be treated with caution as the OECD figures for 1983 by country, and the estimates by the Australian Bureau of Transport and Communications Economics (BTCE), also for 1983, differ substantially. The BTCE figures for the actual on-the-road average fuel consumption in Australia for 1988 is 11.8 litres/100 km - a reduction of about 7 per cent over the previous decade (Walker, 1992, p. 2).

Table 1. Comparison of Car Ownership, Vehicle Use and Fuel Consumption in North America and Selected OECD Countries, 1983

Country	Cars/ Person	Km per Car per Annum (thousands)	Fuel Consumption (litres/100km)	
			BTCE*	OECD**
USA	0.53	14.8	13.7	21.4
Australia	0.47	15.1	12.5	10.5
Canada	0.43	9.7	13.8	31.2
Germany	0.40	13.6	10.9	9.5
Sweden	0.36	12.0	10.8	13.4
Netherlands	0.33	13.9	n/a	7.5
UK	0.29	15.2	n/a	10.4

n/a - not available
*BTCE - Bureau of Transport and Communications Economics, Australia
** OECD - Organisation of Economic Cooperation and Development
Sources: based on OECD, 1986, Table 2.3, p. 21 and Table 2.5, p. 24; and Walker, 1992, Table 2, p. 2

A meaningful comparison of the amount of energy used by different modes in carrying out transport task is fraught with difficulties: the results vary depending on the assumptions about the loads carried as a proportion of the available carrying capacity. The direct energy consumption in passenger kilometres per megajoule and the load factor for trains, buses, cars and motor cycles with varying load factors, typical of Australian conditions, have been calculated. Buses and trains operate at average load factors of about 30 per cent, yielding about 2 passenger kilometres per megajoule, whereas private cars operate with an average load factor of about 25 per cent, yielding about 0.25 passenger kilometres per megajoule (Commonwealth of Australia, 1979, p. 84). The most recent data (for 1987-88) suggest that the average occupancies of passenger cars in urban areas is 1.7 and the average urban rail and bus load levels are down to about 18-19 per cent of the system's full capacity (Commonwealth of Australia, 1991, Table 2.3, p. 14). Despite the inherent imperfections of these aggregate measures of car usage and modal energy consumption it is clear that Australian cities have become increasingly car dependent, their low-density suburbs are extending outwards, and their public transport systems require patronage at higher load factors typical only of higher population density areas.

3. Land Use / Transport Coordination

First, we need to clarify who is responsible for the coordination of the urban land use, transport and environment complex - the "actors" or "stakeholders" involved to make urban consolidation happen. The specific issue of concern here is the relationship between urban density and transport, or, more precisely, the interactions between the density of residential developments and the resultant travel behaviour, together with the derived energy consumption in transport. Issues are defined as topics which are the subject of public and professional discussion, debate and media coverage. Issues are articulated by different groups, or stakeholders, within society, with the joint motive of either changing current policies or programs or of introducing new ones.

In Australian cities, there is no clear leader in determining the pattern of urban development. Urban development does not follow strategic plans and the careful and detailed plans adopted by state and local government. (Canberra, Australia's national capital, could been seen as an exception to this statement.) On the other hand, urban development is not a purely free market process because governmental authorities play an important role as providers of services (for example, transport, water supply and disposal). The Australian public accepts that governments should exercise some control over private development, and this view has been reinforced recently with debates concerning government responsibilities when a greater role for the private sector in infrastructure provision is being contemplated - for example, the appropriate role for the New South Wales Government with build, own, operate, transfer (BOOT) schemes for three private-sector toll roads (M2, M4 and M5) and a tunnel (Eastern Distributor) in Sydney.

In this mixed private-public sector system, decisions made by each of the different "actors" influence the development decisions of others with the actual development outcome being a result of these interactions. Neutze (1977, pp 239-240) suggests there are three main kinds of interaction - market, administrative and political, and illustrates them with flow diagrams the paths through which the activities of each type of decision-maker influence the others. The six groups can be organised into a much simpler classification for the purposes of this research paper.

Based on the implicit model of the two-part economic and social system in relation to energy that was developed by Ironmonger (1982) it has been expanded to include three-groupings: civil society comprising the land use, transport, environment complex: public policy and government agencies; and the research community (Black, *et al*, 1983; Black, 1992). Thus, the private sector of the economic system - here interpreted to mean the more narrow land use/transport/environment sub-system (private developers, households, businesses and lobby groups) - is a "monolithic grouping" to represent the entire civil sphere. The policy sector is also a monolithic grouping, but, in reality, is partitioned into the appropriate levels of government (Federal, State and local government in Australia), together with the various service authorities, such as the transport mode agencies and the planning authorities (State and local government for urban development in Australia). The research grouping encompasses the research and information arms of government, their servicing agencies and planning authorities - where research is undertaken in-house - or contract research by private organisations, and research institutes in the tertiary education sector. This separation of the research grouping is quite deliberate because we will later aim to establish the research contribution to the current debate on urban consolidation in Australian cities, and the relationship between housing density, travel and energy consumption in transport.

Before doing this it should be pointed out that issues are raised by individuals or groups when tensions occur due to a gap between a desirable state and the perceived or actual state of a system. The model is static but can become dynamic in several ways, such as: successful policy instruments can improve system performance and narrow the gap between some desirable (ideal) state of the urban system and its actual state (performance); or growth pressures, and a failure of policies, can cause a deterioration in system performance; or rising community aspirations and a change in the type of issues relevant to society, or more reliable data and information, together with greater public awareness, can also lead to a convergence of the actual system performance and stakeholder perceptions of its performance. Case studies of transport issues based on those appearing in newspapers in Sydney over slices of time - 1930, 1955 and 1975 (Black, *et al*, 1983), and from 1987 to 1992 (Black and Rimmer, 1987; Black and Rimmer, 1992) - provide examples of the dynamic application of this conceptual model. Although the historical analysis of urban consolidation and transport over time, especially following the first global "oil shock" of 1973/74, has yet to be written the next section applies the model for this task.

4. Conceptual Model And Urban Consolidation

The three sector conceptual model is applied to the issue of urban density and transport, with particular reference to contemporary Australian cities. A descriptive analysis is provided of different perspectives on urban consolidation - by Federal and State Governments, by consumers of housing, by real-estate developers, by environmental groups, and, importantly, by the research sector. Such a context is essential to avoid criticism of over-simplified relationships between urban form, travel and energy consumption.

4.1 Federal Government

Historically, the Australian Federal Government has not played a dominant role in urban affairs leaving the matter to State governments. Those who framed the Australian constitution did not anticipate that the Commonwealth would participate directly in the development of Australian cities. Section 51 of the Constitution makes it plain that the prime responsibility for the provision of essential services and facilities to Australia's cities and regions remain with the states. Clare (1991b) documents briefly the essence of Federal Government involvement with urban policies. A full account is provided by Lloyd and Troy (1981). More recent Commonwealth initiatives in urban affairs, including major policies and programs, are summarised by Bunker and Minnery (1992). One over-arching policy initiative is ecologically sustainable development.

Although the idea of ecologically sustainable development (ESD) has been around since the United Nations Stockholm Conference in 1972, it came to prominence in Australia with the establishment of nine Working Groups by the Prime Minister to consider the implementation of ESD principles in sectors of the economy with major impacts on the environment. The final report from the Transport Working Group was released in late 1991 (Commonwealth of Australia, 1991), and it contained 40 recommendations. Those of specific relevance to the urban transport sector are synthesised from the report and summarised in Table 2 in a form consistent with the main steps of the systems approach to land-use and transport planning. It should be noted that the recommendations are numerous in terms of establishing goals, objectives and values and in offering solutions to urban problems.

Recommendations 13 through to 16 of the final report are measures specifically directed towards urban form. Recommendation 13 is for a comprehensive program to increase understanding amongst planners, managers and the community of the ways in which the benefits of higher urban densities can be achieved. Recommendation 14 is that Australian cities should be developed more in the present urban areas than on the fringe, and with a range of housing types and densities. The thirteen

Table 2. Ecologically Sustainable Development Recommendations Organised by the Urban Land Use/Transport Planning Process

Phase of the Process	Summary of ESD Recommendations
Establishing goals, objectives and values	Rec. (1) comprehensive information, education and consultation programs; Rec. (6(d)) review air quality objectives; Rec. (13) comprehensive education program for planners/managers/community on benefits of higher densities and alternative urban forms; Rec. (22) car parking policies; Rec. (24) traffic calming prominence and funding; Rec. (25(a)) include cycling in transport planning and decision making; Rec. (26) technical advice on demand management, traffic calming and bicycle facilities; Rec. (29(a)) measures to encourage public transport patronage; Rec. (36) coordination between spheres of government and planning agencies; and Rec. (37) review existing planning arrangements to achieve better integration.
Data, research	Rec. (3) study how to incorporate full economic social and environmental costs into energy prices; Rec. (6(a),(c)) air quality studies and monitoring; Rec. (8(c)) in-service vehicle emissions; and Rec. (12) research into transport system in line with ESD; and Rec. (30) investigation of freight terminals.
Modelling	No specific recommendations, although implicit in several recommendations.
Solutions	Rec. (14) urban consolidation with range of housing types and densities; Rec. (15) location of affordable housing; Rec. (16) suburban employment at public transport nodes; Rec. (18) locations for reduced travel demand and traffic calming measures; Rec. (19) programs for car and van pooling; Rec. (20) road pricing mechanisms; Rec. (21) route advisory systems; Rec. (23) transit and HOV lane priority; Rec. (25(c)) dual mode facilities for cyclists; Rec. (28) urban public transport investment; and Rec. (34) re-vegetation of arterial roads.
Evaluation	Rec. (2) ESD considerations in decision making; and Rec. (38) apply environmental impact assessment to policies, programs and projects.
Implementation	Rec. (40) develop appropriate mechanisms to monitor implementation of recommendations and provide information.

Source: based on Commonwealth of Australia, 1991, pp. 131-175

additional measures to achieve urban consolidation under Recommendation 14 are summarised in with a range of housing types and densities. The thirteen additional measures to achieve urban consolidation under Recommendation 14 are summarised in Table 3. Recommendation 15 is for policies to be implemented that ensure affordable housing for all housing groups, but especially low-income households, be located in places accessible to public transport, employment, and other essential community facilities so as to lessen the need for travel. Finally, Recommendation 16 encourages new suburban employment to focus as far as possible on public transport nodes (especially rail) in suburban centres with access to medium-density affordable housing.

Table 3. ESD Transport Working Group Recommendations on Urban Consolidation for Australian Cities

Recommendation	Measure
14(a)	full cost pricing of urban development and phasing out of subsidies on greenfield suburbanisation
14(b)	remove inappropriate regulations and processes in building industry preventing re-urbanisation
14(c)	re-orienting land banking and land availability to redevelopment sites
14(d)	conduct detailed study of infrastructure capacity to assess capability for redevelopment
14(e)	integration of local government desires for new development with metropolitan-scale plans that have clear zoning and priorities specified
14(f)	redevelopment of land around key public transport nodes to include dense housing and some commercial activity so that travel is minimised
14(g)	local government develops community consultative mechanisms for guidelines on appropriate urban design
14(h)	National Housing Strategy affordable housing be directed to locations that are close to sub-centres, employment and good public transport
14(i)	review of government taxation systems as they apply to housing
14(j)	develop planning processes that provide for precinct development (for example, urban villages) rather than for individual housing lots
14(k)	demonstrations of higher density housing systems and urban villages that can minimise or reduce travel
14(l)	where urban fringe development is necessary, apply principles of more efficient land utilisation and locations near to public transport and employment
14(m)	where large-scale redevelopment is not appropriate, facilitate dual occupancy and other small-scale additions to the building stock

Source: based on Commonwealth of Australia, 1991, pp. 151-155

Other Federal Government Departments are also issuing related policy statements. Long-term planning policies should aim, through urban consolidation, to increase the populations along public transport routes and concentrate developments around public transport nodes (Australia, Department of Primary Industries and Energy, 1991, p 26).

4.2 New South Wales State Government

The way that state governments are responding to challenges of ecologically sustainable development and urban consolidation can be illustrated with reference to the Sydney, Newcastle and Wollongong metropolitan region. In October 1993 *Sydney's Future: A Discussion Paper on Planning the Greater Metropolitan Region* (New South Wales Department of Planning, 1993) and the *Integrated Transport Strategy for* Greater Sydney: A First Release for Public Discussion October 1993 were launched by the Premier of New South Wales. The vision for the future of the region is supported by four goals: equity, efficiency, environmental quality, and livability. To achieve its vision and goals, the strategy adopts four key directions - all directly of relevance to urban consolidation:

> " • a more compact city .... getting more out of new and existing infrastructure, with improved transport links, and bringing jobs, housing and facilities closer together
> • a better environment ... more public transport use
> • a more equitable and efficient city, with improved services provision and accessibility, better location of jobs relative to housing ...
> • effective implementation of the strategy ..."
> (New South Wales Department of Planning, 1993. p. vi)

To achieve these key directions, the strategy includes specific proposals for policies and actions (New South Wales Department of Planning, 1993, pp.25-29). Of specific relevance to urban consolidation are the linkages between housing, employment and transport.

- Housing - ensure better planning and development of new housing, including improved public transport accessibility and reduced environmental impact; and progressively increase the proportion of new housing constructed annually in multi-unit form from the current 42 per cent to 65 per cent in 2011.
- Transport - the housing proposals specifically support the development and use of public transport; providing new transport facilities and networks, including public transport links, to support land-use patterns

that deliver efficiency, equity and environmental benefits; providing transport choices and assisting public transport are integral to meeting transport demand.

- Employment - the containment and consolidation of the Sydney region and the close integration of transport and land use will close the distance between housing and jobs.

These proposals should be read in association with the actual experience to date of urban consolidation in Sydney (Bunker, 1983; Searle, 1991).

The *Integrated Transport Strategy* takes the fundamental principles and the preferred land-use scenario set down in the metropolitan strategy as its primary building block. Its efficiency, environmental (including energy efficiency), and social objectives are outlined in Figure 1, together with the performance measures and expected outcomes sought from the integrated transport strategy.

The "environmental" objective is interpreted in the *Integrated Transport Strategy for Greater Sydney* as "minimising the environmental impact of transport" - a primary goal of the move towards travel demand management (p 59). The sub-set of strategic actions to support this objective are reproduced as Table 4.

Table 4. Policies to Minimise the Environmental Impacts of Transport

1	Fuel Efficient and Emission Reducing Buses - Continue the introduction of CNG-powered buses and other fuel efficient and emission reducing vehicles into the Sydney bus fleet
2	Electric Bus - Continue to monitor and assess electric bus and alternative fuel technology
3	Emission Control Strategies - Develop a range of vehicle emission control strategies including the introduction of a program of emission testing of motor vehicles at registration
4	Bicycle Route Network - Continue development of a bicycle route network for the Greater Sydney Region to improve accessibility and safety and to encourage use of non-motorised forms of transport (walking and cycling)
5	Storage of Bicycles - provide facilities for the secure storage of bicycles at railway stations, ferry wharves and other transport nodes where feasible
6	Car Pooling - Facilitate car pooling through the extension of transit lanes for High Occupancy Vehicles (HOVs) and by identifying and removing legislative and institutional impediments.

Source: *Integrated Transport Strategy for Greater Sydney*, 1993, p. 60

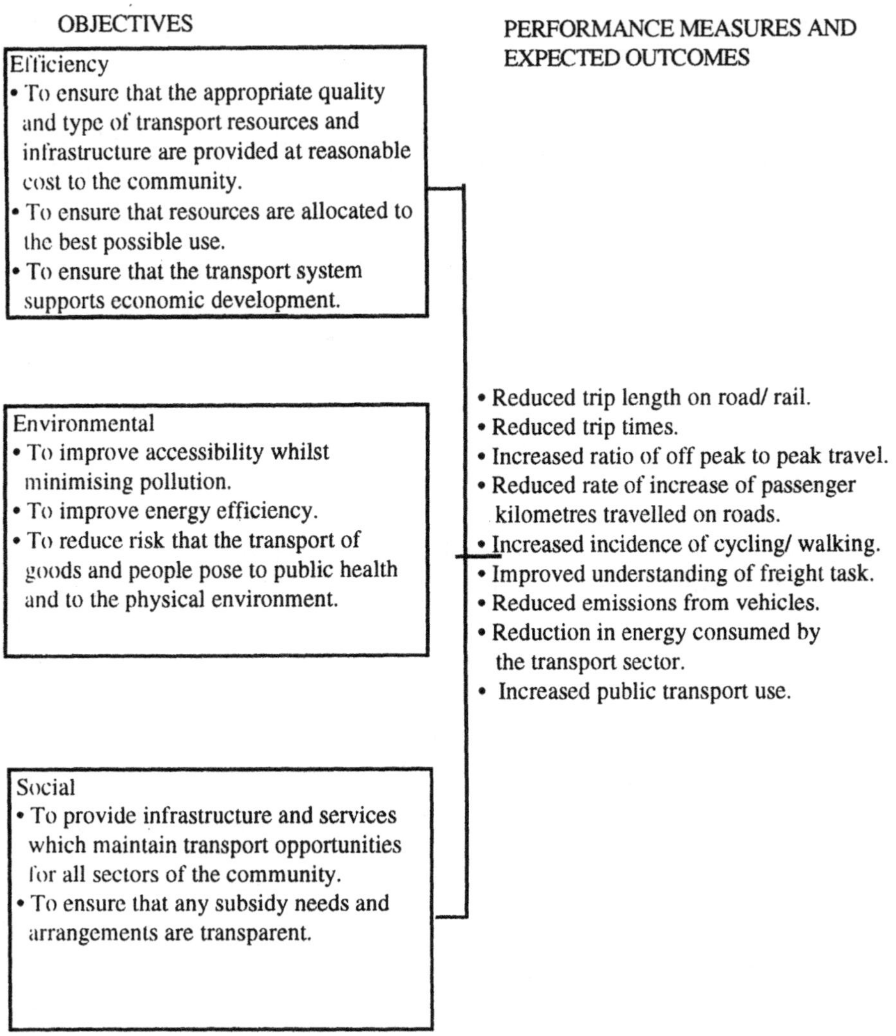

Fig. 1. Objectives for Transport and Performance Measures and Expected Outcomes as Part of the Integrated Transport Strategy for Sydney, Newcastle and Wollongong
Source: based on *Integrated Transport Strategy*, 1993, Figure 2, p. 26.

4.3 Community Perspective

The two things that communities oppose most are sprawl and density. Clearly, consumer acceptability of higher density types is required. Surveys in Australian cities would confirm that most people prefer detached dwellings on separate blocks, with private open space. Much of Australia's national wealth has been invested in the search for a particular suburban lifestyle. The Housing Preference Survey undertaken for the Adelaide Planning Review, for example, demonstrated this clearly. There is a view within the building industry that state governments need to promote and sell urban consolidation to the public, local government, home buyers, land agents, builders, developers, bankers, and designers. Sabino (1991) suggests that the secret of success of the Delfin Property Group with higher density housing is that they base their designs on market segments, their characteristics, their needs and what can be afforded. The chairperson of the National Housing Sector of the Master Builder Construction and Housing Association of Australia also sees design as a crucial issue if Australian cities are to achieve significant increases in housing density and diversity (Dening, 1991, p. 14).

Of the environmental groups, Greenpeace Australia probably has been the most active in promoting urban consolidation, urban villages and a visionary approach to metropolitan development. The Greenpeace Urban Strategy for Sydney (11.12.92) - an urban consolidation strategy in conjunction with improvements to public transport, vehicle performance and renewable energy use - suggests that of the 145 railway stations in Sydney, 90 could be suitable for urban consolidation, accommodating some quarter of a million people. The development of urban villages on the metropolitan fringe are seen as alternatives to the conventional subdivision of the "quarter-acre block". A well-developed concept for an urban village is illustrated by Canberra Land Pty Ltd (*et al*, 1991) for the development of Gungahlin in Canberra.

4.4 Summary of Conceptual Model

To summarise the points arising from two of the three parts of the conceptual model discussed so far we can suggest the following. The thrust of government policy both at the Federal and State levels in Australia is on urban consolidation. There is strong support from environmental lobby groups in the hope that governments will embrace higher densities and better public transport systems more enthusiastically than in the past. There is scepticism in the development industry about the market for higher density housing but a recognition in some parts of the housing industry that design aspects are a crucial element in achieving higher residential densities. There is also scepticism as to how medium density housing at selected nodes will have much of an impact on overall urban population densities.

When we now turn to the third part of the model we would expect to see a strong research base relating density, transport and energy, but this is not the case. *Urban Consolidation and Related Subjects: A Selective Bibliography* by the New South Wales Department of Environment and Planning (Harris and Shearer, n.d.) makes no reference to such methodological issues in 549 entries. On such basic information as the traffic generating characteristics of conventional and higher density housing in Australia there is little quantitative information readily available. A scan of the bibliography being compiled on housing in Australia at the Ian Buchan Fell Institute, University of Sydney, found not one reference amongst about 400 entries, and the former Director of the New South Wales State Transport Study Group knew of no such work undertaken in Sydney.[1] With this enormous gap in the Australian information base in mind, we summarise the literature review on the traffic and transport aspects of the Australian higher density housing code.

5. Model Code for Urban Housing - Transport and Energy

Advocates of higher residential density development and public transport oriented urban consolidation have made assumptions about the relationship between higher density residential development, transport choices and the resultant environmental impacts. They assume that people will make fewer and shorter journeys by motor car and walk, or use public transport more often compared with residents living at lower densities. Underlying these assertions is a largely untested assumption that people will be willing to move into these new, higher density, urban forms, and when they do, they will change their travel behaviour as outlined above. The expected outcome of urban consolidation is reduced vehicular emissions and a lower consumption of petroleum. If, however, people living at higher densities continue to take the same number of trips, at the same average distances, as people with similar socio-economic and demographic characteristics who live in lower density residential areas, petrol consumption could increase because of greater road traffic congestion.

A literature search strategy was devised using key words such as "residential density" "housing", "traffic" and "transport". Three computer-based information systems were searched: European Space Agency (ESA) in Italy; TRIS in North America; and the Australian Road Research Board (ARRB). Whilst a growing body of literature is recently emerging from the USA, Australian research on the subject remains inadequate for policy formulation.

5.1 Urban Density and Transport

Some general observations on the results of this work are as follows. There is much work at the highly aggregate level relating traffic and transport to urban densities (for example, Smith, 1984; Newman and Hogan, 1987).

---

1 Dr J.P. Lea and Mr D. Graham, pers. comm.

There are several international reports on design for higher density housing with sections on parking, streets and public transport, but they make little reference to the wider context of location and transport. There is a considerable amount of information and data on housing type and density and trip generation/modal split relationships (for example, Pushkarev and Zupan, 1977) and on vehicular trip rates by housing type (Institute of Transportation Engineers, 1991) but there are no such comparable Australian studies. Finally, there is renewed interest in urban density and travel in the USA (see, for example, Pisarski, 1991) but, in Australia, the only active centre researching this topic is the Institute of Urban Policy at Murdoch University.

The most aggregate scale at which the relationship between urban density and public transport use can be viewed is comparing entire urban areas, and this is a convenient starting point. Data from 105 US cities in 1960 and 1970 suggest that the percentage of workers using public transport increases linearly with higher gross population densities. A decline in the level of public transport use from 1960 to 1970 is evident (Pushkarev and Zupan, 1977, pp. 24-26). There is a scatter of points around the regression line to the extent that average density does not say much about public transport use. The authors note: that 103 of the cities fall into a density range of between 1500 and 5300 persons per square mile; that city centre office space concentration is a determinant of public transport use; and that presence of rail or bus on its exclusive rights of way are also influences of public transport use.

A recent, unpublished, study gives the results of an analysis of the 1990 National Personal Transportation Survey in the USA and provides insights into aggregate relationships. For urban areas (SMSA) of more than 1 million people, the relationship between annual public transport ridership per capita and average gross population density is positive, although there is a wide scatter of data points at the same urban density. Simple linear regression analysis yields a correlation coefficient (r^2) = 0.26 (Dunphy and Fisher, 1994).

5.2 Density and Public Transport

Newman and Hogan (1987) have collected data for 62 cities and have plotted the relationship between public transport use and residential density. The overall picture suggested is: (a) US and Australian cities with low density and motor-vehicle dominance; (b) European and Japanese cities with medium density and a substantial public transport base; and (c) Third World cities with high density and walking as the dominant mode. The authors state that although there is a very large scatter, "the bell shaped curve best represents the suggested relationship between public transport and density" (p. 12).

Table 5 shows the mode of transport used for daily personal travel in US urban areas at different population density levels. The density data have been combined at the lower ends, followed by the mid-point density, as defined by Dunphy and Fisher (1994). About 60 per cent of the American urbanised population lives at gross residential densities below 2000 persons per mile, and about three quarters live below 4000 persons per square mile (6 persons per acre). If we examine the travel behaviour of the remaining one quarter of urban American (and noting that their travel behaviour at 4500 persons per square mile is very similar to that at the lowest densities) a number of trends are apparent.

Table 5. Person Daily Trip Rates by Mode, USA Standard Metropolitan Statistical Areas, 1990

Density**	Trip Rates or Range of Trip Rates by Density				
	Private Vehicle	Public Transport	Walk/Cycle	Taxi	Total*
<2000	3.28-3.53	0.02-0.08	0.21-0.33	0.0	3.77-3.96
2000-4000	3.34-3.46	0.07-0.08	0.28-0.29	0.0	3.81-3.92
4500	3.51	0.06	0.30	0.0	3.95
6250	3.29	0.11	0.36	0.01	3.83
8750	2.92	0.16	0.45	0.02	3.62
30000	1.90	0.50	0.95	0.03	3.42
60000	0.59	1.03	1.55	0.16	3.40

* Not shown in the table are the "other" modes, which account for 0.16 daily personal trips at 50 persons per square mile to 0.04 trips at 30000 persons per square mile

** density information in the National Personal Transportation Survey was calculated for each household's zip (postal) code

Source: adapted from Dunphy and Fisher, 1994, Table 5, p. 31

First, higher density living does result in lower person daily trip rate from about 4 (the "suburbia" mean) to 3.4 - a reduction of about 15 per cent - at densities of 30,000 persons per square mile and above. Secondly, the daily number of private vehicle trips, including drivers and passengers, declines consistently with density from 3.5 to 0.6 at the very highest density. Thirdly, public transport (bus and train) trips rise sharply above densities of 10,000 per square mile from less than 0.16 to 0.50, with a further doubling at the very highest density level. Fourthly, numerically, walking and cycling is more important than public transport at all density ranges with about 0.3 daily trips per person being the "suburbia" norm, which increases rapidly to

1 at 30,000 persons per square mile, and then to 1.6 with another doubling of population density. Finally, taxi travel, with 0.16 daily trips per person is only a significant means of travel in the very highest density range shown in Table 5.

The conclusions drawn by Dunphy and Fisher (1994, p. 32) are that public transport trips increase sharply above densities of 10000 persons per square mile, and rail only plays a role at density levels of around 50000, where most US rail systems are concentrated. Only about 10 per cent of urban America is living at gross residential densities above 7500 persons per square mile.

When the relationship between density *within* an urban area and public transport use is examined a similar trend emerges. Based on the Tri-State Regional Planning Commission Study of the New York region in 1963, both the variation in urban density is very large (from 0.8 dwellings per acre to 200 dwellings per acre) and the difference in travel demand (total weekday trips per person by all vehicular modes and the weekday trips per person by public transport) is substantial . Densities of 7 to 30 dwellings per acre appear necessary to sustain significant public transport use and cause a reduction in travel by private transport (Pushkarev and Zupan, 1977, p. 30). In relating mass transport for high-rise, high-density living, Smith (1984) also reproduces the same data. Recently, Frank (1994) has analysed the Puget Sound Transportation Panel data for 1989 at the census tract level and has found a non-linear relationship between shopping mode choice (car, bus, walk) and gross population density per acre. Densities need to exceed to 13 persons per acre before significant modal shift occurs away from single occupancy car driver trip.

Less attention has been paid to the question of whether good public transport starts attracting ridership by suppressing car ownership in the vicinity of their good services. Pushkarev and Zupan (1977, pp. 39-41) found that at suburban densities of less than 10 dwellings per acre, the presence of suburban rail reduces the number of cars per household by about 5 per cent in the upper income areas (there were no reductions in middle and low income areas close to suburban railway stations). On the other hand, mass rapid transit was found to have an influence on all income groups: "the effect of a rapid transit station on auto ownership in adjoining census tracts can be equivalent to that of more than a tenfold increase in residential density" (p. 41). However, data extracted for Sydney from 1976 to 1986 found no evidence of reduced car ownership for increased population density at three locations near railway stations (Gutteridge Haskins & Davey, 1992, Appendix 4). The 1990 US National Personal Transportation Survey reveals that there is only a noticeable reduction in the average number of car per adult above a gross populated density of about 7.5 persons per acre - where rates drop in a log linear way from about 1 to 0.7 at 150 persons per acre (Frank, 1994, Fig. 8, p. 22).

There is an intervening variable in that car ownership is also related to urban density and proximity to good public transport services. Household car ownership is a function of household size (the number of persons of driving age, the number of workers in a household), the income of the household, and (possibly) the residential density of the area in which the household is located. The various social factors influencing whether an household owns a car or not were analysed by Johnson and Black (1977) in a study of 965 households in four Melbourne suburbs, using a technique called automatic interaction detector. For instance, whether there are any workers in the household or not is the most important explanatory variable (79 per cent of working households had a car in the early 1970s). Further classifying this group, two or more people in the household (83%) the head of the household with a (relatively) high income (91%) and owner-occupier (95%) are factors increasing the probability of whether a car is owned. On the other hand, households with no one working (20% ownership), and living in other than their own house have the lowest proportion of households with a car (7%).

6. Australian Urban Housing Code and Transport

McGlynn (*et al*, 1991) have suggested that the link between land use and transport is absent in Australia from urban consolidation policies, and this was confirmed by the work by this author for the Australian Model Code for Urban Housing. Consequently, it was decided to determine an analytical procedure from first principles to determine the social, economic and environmental costs and benefits of alternative plans and courses of action with regard to housing density. This is much along the lines of recommendations in Table 3. Such a procedure is sketched out below, with emphasis on the steps involved rather than on the details that would be required in transport engineering or planning practice.

The first step is a sieving process to determine the areas potentially most suitable for urban consolidation within a metropolitan region. This will establish the obvious locations as well as excluding clearly inappropriate areas. Those marginal areas will require further investigations. The outline of such an assessment procedure is set out in Table 6.

The second step recognises that any intensification of residential land use around suburban commercial/retail nodes and transport interchanges where urban consolidation is most likely will mean additional pedestrian and passenger traffic and demand on existing infrastructure. Traffic impact analysis procedures are well developed in traffic engineering practice (for example, Traffic Authority of New South Wales, 1984; Fricker and Tsay, 1985) but these traditional approaches require augmentation to include environmental capacity considerations to ensure information presented to decision makers conforms to the requirements of ecologically sustainable development and the State government's metropolitan strategies. Similarly, there needs to be studies undertaken of spare infrastructure capacity, especially on public transport. Such a framework for assessing the traffic impacts of higher density housing is provided in Figure 2.

Table 6. Identifying Suitable Locations for Urban Consolidation Within Existing Urban Areas

(i) Identify locational factors supportive of higher density housing (market survey, user preferences, developer experience). These may vary from city to city.
(ii) Distinguish in (i) above any market segments within each metropolitan area.
(iii) Use factors in (i) and (ii) to undertake a spatial search (see Massam 1980) to define the spaces which satisfy the locational factors.
(iv) Assuming that proximity to good public transport is a positive locational attribute (potential substitution of public transport for private transport on the journey-to-work and travel to other major activities) undertake an analysis of spare capacity.
(v) Assuming that high accessibility to activities is also a positive locational attribute (potential substitution of walking and cycling for private transport as an access mode) undertake an analysis of accessibility of residential areas to community facilities.
(vi) Combine analyses such as (iv) and (v) together with any others (iii) to produce strategic maps for definite higher density locations and possible higher density locations. The maps may be stratified to indicate different market segments in (ii).

It is important at this step to establish the travel patterns. When additional housing is proposed the likely person and vehicular traffic generating characteristics of these households requires careful consideration in terms of amount, mode, whether walking and public transport is substituted for car travel, distances travelled, and energy consumption (see the previous section). In addition, there are well-defined analytical methods and land use/transport interaction models to apply to this task. One important parameter is on vehicular trip generation rates of housing at different densities, and, in the absence of any authentic data for Australian cities, Table 7 reproduces data compiled from a large number of studies in the USA conducted in different years for daily weekday average vehicular trip ends per dwelling type. Because of the importance of the peak period in traffic engineering practice,

Table 8 has been included as well to summarise the same data but for the morning peak two hours.

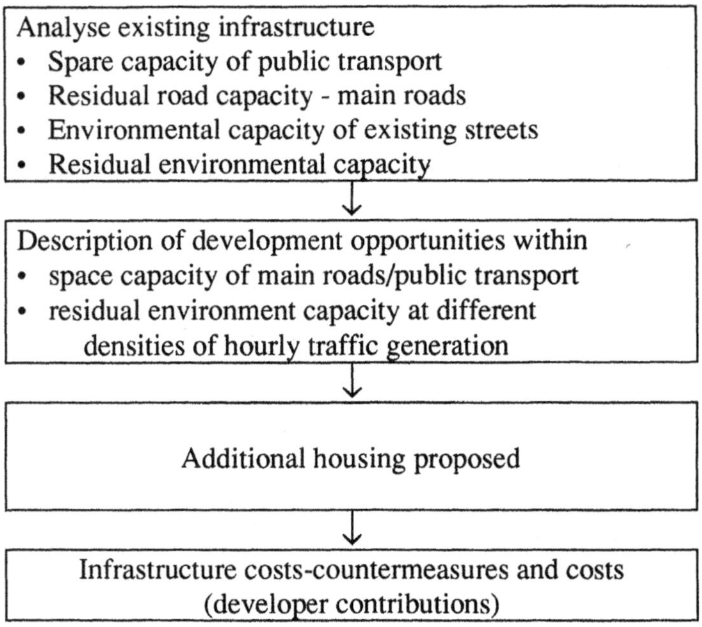

Fig. 2. Framework for Assessing the Traffic Impacts of Higher Density Housing

Table 7. Daily Weekday Average Vehicle Trip Ends Per Dwelling Type, USA, Various Years

Dwelling Type	Mean	Standard Deviation	Range	Number
Family detached	9.6	3.7	4.3-21.9	348
Apartment (rental pre 1973)	6.5	2.9	2.0-11.9	109
Apartment (rental post 1973)	6.3	2.8	2.0-11.8	33
Low rise	6.6	2.8	5.1-9.2	22
High rise (>10 storeys)	4.2	2.3	3.0-6.5	9
Townhouse/condominium (Owner)	5.9	3.1	1.8-11.8	53

Source: Institute of Transportation Engineers, 1991, various pages

Table 8. Daily Average Vehicle Trip Ends Per Dwelling Units Per Hour for Maximum Adjacent Street Traffic (Morning Peak 7.00 am - 9.00 am), USA, Various Years

Dwelling Type	Mean	Standard Deviation	Range	Number
Family detached	0.74	0.9	0.3-2.3	280
Apartment (rental pre-1973)	0.51	0.7	0.1-1.0	87
Apartment (rental post 1973)	0.44	0.7	0.2-0.7	16
Low rise	0.47	0.7	0.3-0.9	26
High rise (>10 storeys)	0.30	0.6	0.2-0.5	17
Townhouse/condominium (owner)	0.44	0.7	0.2-1.6	54

Source: Institute of Transportation Engineers, 1991, various pages

Despite the value of these US data for preliminary strategic analyses, surveys are urgently required in Australian cities to establish four main factors. One, to estimate the person and vehicle traffic generation rates by time of day and day of the week for different housing types (and socio-economic and demographic compositions). Two, to investigate the degree of substitution of walking and cycling for private transport at higher densities, as in the US studies cited. Three, to investigate the degree of substitution of public transport for private transport in locations well served by public transport. Four, to decide whether good public transport and high accessibility to facilities at both the home and destination ends of travel influence household car ownership rates.

The third step in the proposed methodology is to put energy considerations into a broader assessment framework of ecologically sustainable development. As previously noted in Table 2 under the evaluation of land use/transport infrastructure proposals, urban consolidation requires a consideration of a broad range of economic, social and environmental factors. Energy consumption of transport modes is only part of this evaluation process. The Australian Model Code for Urban Housing recognised this and placed considerable importance on environmental capacity concepts - traffic noise, vehicular pollution and pedestrian safety, and associated accident risk (Gutteridge Haskins & Davey Pty Ltd, 1992, pp. 55-60).

Whilst careful analysis will assist in the broad determination of the costs and benefits of urban consolidation and so indicate the efficacy of this policy, there are other considerations. As noted by a number of people, the detailed designs (both of dwelling units and the surrounding micro environment) will play a major part in determining consumer acceptability. There are issues of housing affordability and the quality of the residential environments at higher densities. This leads onto the final step of Figure 2 - the countermeasures. The environmental adaptation of roads and land-use frontages in the proximity of railway stations or bus interchanges will help create a more pedestrian-friendly and safer urban form that is supportive of walking and public transport usage. The design measures are detailed in *Sharing the Main Street* (Roads and Traffic Authority of New South Wales and Federal Office of Road Safety, 1993).

7. Conclusions

The evidence that increasing motorisation is unsustainable is compelling (Whitelegg, 1993). The direction of Australian urban development, with their low densities of development and high levels of private vehicle ownership, means that cities are becoming increasingly unsustainable in economic, social and environmental terms. This has been recognised at a Special Premiers' Conference, where urban consolidation - the change from low density, sprawling cities to higher residential densities in established areas - is seen as an important policy instrument. The Australian Model Code for Urban Housing is one Commonwealth Government initiative in this area.

This paper has outlined two conceptual models of the "actors" or "stakeholders" involved in urban development, with particular reference to transport and urban consolidation. One of these models was applied to the issue in Australia, in general, and in Sydney, in particular, of urban density and transport. The results of this analysis were presented from different perspectives: Federal Government; the State Government of New South Wales; consumers of housing; and the research sector. The important concept of ecologically sustainable development considerations in decision making was stressed. Whilst there is strong support for urban consolidation amongst these "actors" the information provided by the research sector was found to be deficient on land use and transport issues.

Research undertaken for the development of the Australian Mode Code for Urban Housing confirmed enormous gaps in the Australian information base. There is a large literature at the urban macro-scale on housing density, transport and energy, but little at the micro-scale, suitable for detailed studies of urban consolidation projects. The paper sketched out analytical procedures to determine the social, economic and environmental costs (including energy considerations) of alternative urban plans and policies, but more detailed modelling work is clearly required.

The methodological requirements can be summarised as: one, a technical procedure to identify areas and locations within metropolitan regions that are suitable for urban consolidation; two, traffic impact assessment procedures for higher density housing; and; three, a synthesis of this material into an appropriate assessment framework that recognises energy considerations are part of a broader context of social, economic and environmental factors. The real challenge for researchers is to devise appropriate analytical techniques which embrace the principles set out for ecologically sustainable development.

Acknowledgements

The author was a technical adviser to the Australian Council of Social Service (ACOSS) who were represented on the Transport Working Group of the Ecologically Sustainable Development Working Groups, although the interpretations of the ESD recommendations are his responsibility. He also advised Gutteridge Haskins & Davey who prepared the Australian Model Code for Urban Housing on transport matters and Dennis Smith and his group provided valuable discussions. The Chair of the Residential Reference Group responsible for drafting the code, Emeritus Professor Hans Westerman, also provided helpful comments to the author in the development of assessment procedures Tony Cargnelutti, Physical Sciences Library, University of New South Wales, advised on the strategy for the literature search, and Kevin Horgan, Department of Transport Engineering, School of Civil Engineering, assisted by collecting the bibliographic material.

References

Australia, Department of Primary Industries and Energy (1991) *Energy and Land Transport: Discussion Paper.* Australian Government Publishing Service, Canberra

Australian Urban and Regional Development Review (1993) Australian cities and regions: A national approach. *Information Paper No. 1.* Australian Urban and Regional Development Review, Melbourne

Black, J. (1992) Policy measures for land use and transport demand management and their implications in managing rapidly growing Asian metropolises. *Regional Development Dialogue,* 13 (3). 3-25

Black, J. and Rimmer, P. (1987) Sydney's transport problems. *New South Wales Parliamentary Library Background Paper, 1987/4.* The Library for the Information of Members of the New South Wales Parliament, Sydney

Black, J.A. and Rimmer, P. (1992) Sydney's transport: contemporary issues, 1988-1992. *New South Wales Parliamentary Library Current Issues.* New South Wales Parliamentary Library, Sydney

Black, J., Kuranami, C. and Rimmer, P.J. (1983) Transport - land issues, problems and policy implications: Sydney since the thirties. *8th Australian Transport Research Forum: Forum Papers*, Vol. 1. 92-118

Bunker, R.(1983) Urban consolidation - the experience of Sydney, Melbourne and Adelaide. *AIUS Publication*, No. 111. Australian Institute of Urban Studies, Canberra

Bunker, R. and Minnery, J. (1992) Recent Commonwealth initiatives in urban affairs. *Australian Institute of Urban Studies, Urban Studies National Monograph*, No 1. Australian Institute of Urban Studies, South Australian Division, Adelaide

Canberra Land Pty. Ltd., Bryan R. Dowling & Associates Pty. Ltd., W.P. Brown and Partners Pty. Ltd., Paul Whalan and Associates, and McCann and Associates (1991) *Canberra - A New Vision: Gungahlin Urban Village*. City Graphics, Canberra

Clare, R. (1991a) Factors in regional trends. In *Background Papers on Urban and Regional Issues - Studies Prepared for the Office of EPAC, Background Paper No. 10*. Australian Government Publishing Service, Canberra. 13-35

Clare, R. (1991b) A short history of regional and urban policy. In *Background Papers on Urban and Regional Issues - Studies Prepared for the Office of EPAC, Background Paper No. 10*. Australian Government Publishing Service, Canberra. 1-10

Commonwealth of Australia (1979) *Transport and Energy Overview: Prepared to the Request of the Australian Transport Advisory Council*. Australian Government Publishing Service, Canberra

Commonwealth of Australia (1991) *Ecologically Sustainable Development Working Groups, Final Report - Transport*. Australian Government Publishing Service, Canberra

Cracknell, J. (1993) Issue arenas, pressure groups and environmental agendas. In A. Hansen (ed.). *The Mass Media and Environmental Issues*. Leicester University Press, Leicester. 3-21

Dening, K. (1991) Design and residential environments: the need for change. *Urban Futures: Issues for Australian Cities*, Special Issue 3, November. 10-14

Dunphy, R.T. and Fisher, K.M. (1994) Transportation, congestion, and density: new insights. Paper No 940904 presented at the 73rd Annual Meeting, Transportation Research Board, January, 9-13

Frank, L.D. (1994) The impacts of mixed use and density on the utilisation of travel: the single occupant vehicle, transit, and walking. Paper No 940425 presented at the 73rd Annual Meeting, Transportation Research Board, January, 9-13

Fricker, J.D. and Tsay H-S. (1985) Urban transportation impacts of tall buildings. *Journal of Transportation Engineering* 111(4). 395-409

Gutteridge, Haskins and Davey Pty Ltd (1992) *Australian Model Code for Urban Housing Engineering Aspects - Research Study*. Sydney: Gutteridge Haskins & Davey Pty Ltd.

Harris, J.A. and Shearer, L.J. (n.d.) *Urban Consolidation and Related Subjects: A Selective Bibliography*. New South Wales Department of Environment and Planning, Sydney

Institute of Transportation Engineers (1991) *Trip Generation, 5th Edition.* Institute of Transportation Engineers, Washington, DC.

Integrated Transport Strategy for Greater Sydney: A First Release for Public Discussion, October 1993 (1993)

Ironmonger, D.C. (1982) The interaction between economic and social research and policy in relation to energy. In J. Black (ed.). *Liquid Fuels in Australia: A Social Science Research Perspective.* Pergamon Press, Sydney. 82-86

Hanly, S. (1991) Neo-traditional development: the debate. *Berkeley Planning Journal* 6. 135-144

Johnson, K. and Black, J. (1977) An aid to understanding car ownership: a case study of households in four Melbourne suburbs. *International Journal of Transport Economics* 4. 147-167

Kunster, J.H. (1993) *The Geography of Nowhere: The Rise and Decline of America's Man-Made Landscape.* Simon and Schuster, New York

Lloyd, C.J. and Troy, P.N. (1981) *Innovation and Reaction: The Life and Death of the Federal Department of Urban and Regional Development.* George Allen & Unwin, Sydney

Massam, B.H. (1980) *Spatial Search: Applications to Planning Problems in the Public Sector.* Pergamon Press, Oxford

Neutze, M. (1977) *Urban Development in Australia - A Descriptive Analysis.* George Allen & Unwin, Sydney

Newman, P.W.G. and Hogan, T.S.F. (1987) Urban density and transport: a simple model based on three city types . *Transport Research Paper,* 1/87, Environmental Science, Murdoch University

Newman, P.W.G. and Kenworthy, J.R. (1987) Transport and urban form in thirty-two of the world's principal cities. In W. Young (ed.). *Proceedings of the International Symposium on Transport, Communication and Urban Form: 2 Analytical Techniques and Case Studies.* Department of Civil Engineering, Monash University, Clayton, Victoria. 169-194

Newman, P.W.G. and Kenworthy, J.R. (1989) *Cities and Automobile Dependence: An International Sourcebook.* Gower Publishing, Aldershot, Hants

New South Wales Department of Planning (1993) *Sydney's Future: A Discussion Paper on Planning the Greater Metropolitan Region.* New South Wales Government Department of Planning, Sydney

Pisarski, A.E. (1991) Overview, *Special Report 231, Transportation, Urban Form and the Environment: Proceedings of a Conference, Bechman Centre, Irvine, California, December 9-12, 1990.* Transportation Research Board, National Research Council, Washington D.C. 3-10

Pushkarev, B.S. and Zupan, J.M. (1977) *Public Transportation and Land Use Policy.* Indiana University Press, Bloomingdale, Indiana

Roads and Traffic Authority of New South Wales and Federal Office of Road Safety (1993) *Sharing the Main Street: Practitioners' Guide to Managing the Road Environment of Traffic Routes Through Commercial Centres.* Federal Office of Road Safety, Canberra

Sabino, T. (1991) Urban consolidation - is urban design the key? *Urban Futures: Issues for Australian Cities,* Special Issue 3, November. 29-36

Searle, G. (1991) Successes and failures of urban consolidation in Sydney. *Urban Futures: Issues for Australian Cities*, Special Issue 1, July. 22-30

Smith, M.J., Black, J. and Gilchrist, J. (1993) Prototype transport and urban development concepts for the Australian multi-function polis. *Lyon '92 Selected Proceedings of the Sixth World Conference on Transport Research: Volume III Transport Policies*. L'Imprimerie Charat , St-Just-La-Pendue. 2129-2140

Smith, W.S. (1984) Mass transport for high-rise high-density living. *Journal of Transportation Engineering* 110 (3). 521-535

Steiner, R.L. (1994) Residential density and travel patterns: a review of the literature and methodological approach. Paper No 940890 presented at the 73rd Annual Meeting, Transportation Research Board, January, 9-13

Stuart, R. (1984) *Liquid Fuels, Transport and Urban Structure: A Social Science Research Review*. Academy of the Social Sciences in Australia, Canberra

Traffic Authority of New South Wales (1984) *Policies Guidelines and Procedures for Traffic Generating Developments*. Traffic Authority of New South Wales, Rosebery

Walker, I. (1992) International developments on transport and environment policy. *Energy Consumption in Transport - Where to Now: Proceedings of Workshop held on 19 February 1992 at the ACA Building, Canberra*. Energy Research and Development Corporation, Canberra.. n.p.

Whitelegg, J. (1993) *Transport for a Sustainable Future: The Case of Europe*. Belhaven Press, London

CHAPTER 5

URBAN PRODUCTIVITY AND SUSTAINABILITY — IMPACTS OF TECHNOLOGICAL CHANGE

J.F. Brotchie, M. Anderson, P.G. Gipps and C. McNamara

1. Cities in Competition

Both the developed and developing worlds are becoming increasingly urbanised. Population and economic growth is located largely in cities. In the developed world, up to 90% of population and GNP is now based in urban areas. Competition among cities for this growth is increasing. Cities are the new engines of the economy.

Telecommunications and fast transport networks are expanding to become increasingly continental and global in scale. Telecommunication bandwidths are also expanding, increasing their capacity to meet new industry needs, including multi-media transmission. Fast transport networks in Europe and Japan include developing networks for high speed rail in addition to air transport. These electronic and fast transport networks are creating continental and global markets for financial and securities transactions, services, and goods, and for location of new industries; and are increasing the spatial scale of organisation of activities. At this larger scale of organisation, production of goods and services is being increasingly linked to telecommunications, fast transport, and computing. Cities at major nodes on these networks are developing as global and regional centres, as locations for higher level services and as market places. Regional alliances and trading blocs are diminishing the significance of national borders, further increasing

Y. Hayashi and J. Roy (eds.), Transport, Land-Use and the Environment, 81–99.

the importance of cities as competing entities at the global and regional level. They are competing for primacy in the region for provision of goods and services, for the attraction and nurturing of new industries, and as principal centres for corporate headquarters and for production and distribution in that region.

2. Transition to an Information Economy

A range of technological, economic, social and political forces are operating to influence the development of urban centres.

The global expansion of networks and markets, the confluence of information and communication technologies and their linking with production and transport is changing profoundly the nature of work activities. There is a shift from the rigid work practices of the industrial era to more flexible arrangements and hours, from manual, material and energy based processes to increasingly information and knowledge based; from predominantly male jobs to activities and environments suited to females and males; and from mass production to flexible specialisation and personalisation of products.

These changes are creating another industrial revolution and a transition to an information economy. Phases of this transition are the globalisation or regionalisation of networks, markets and organisations, the confluence of information and communication technologies, and their integration with production and transport as outlined above. Flexible specialisation linked with automation has enabled a whole range of options to be incorporated in a product. This range of options, however, could create enormous inventory problems — because of the number of combinations required for stock — leading to re-engineering of the production process — compressing it dramatically in time — to enable production and distribution on demand, thereby eliminating inventories and personalising products to client needs. Examples are the reduction of production time for a pocket pager in the US from several weeks to one hour, and a house or car in Japan to a few days or less. Time is now replacing mass as the factor causing producers and suppliers to select locations in close proximity — in time, if not in space.

The inclusion of artificial intelligence in these processes will enable further specialisation and quality enhancement, and provide another phase in the transition to an information economy.

3. Winners and Losers

Global and regional cities at network nodes are increasing their share of corporate control activity, specialist and producer services and income. However the greater spatial scale of organisations is apparently distributing this income beyond the physical boundaries of the city — to larger 'virtual' cities including other cities in the region. As a consequence, per capita income differentials between cities of different size are reducing, and in Australia are only half what they were 15 years earlier. National income redistribution policy may be contributing to this effect. A similar levelling has occurred in the US.

On the other hand, variations in urban operating costs, particularly for the transport task, have also reduced due to dispersal of industry beyond the central city to be closer to markets and suppliers. Jobs have followed people into the suburbs — into urban metro centres, suburban centres, edge cities and special use zones or are even more dispersed. The net result is that both economies of scale and diseconomies of scale have decreased, but net income per capita still increases with city size throughout the range of city sizes in the developed world.

The variations in income within these cities however appear to have increased as the information revolution has created further winners and losers at the individual level. Many routine tasks have been automated, eliminating jobs. Other tasks have been deskilled — while demand has increased for specialist services — facilitated by new technologies and by increasing market size.

4. Metropolitan Dispersal

In major metropolitan centres in the US, UK and Australia, only about one third of jobs remain in the central core city. The remainder are in sub-centres or even more dispersed. Commuting patterns have changed accordingly. Most work trip destinations are in the suburbs and are not well served by public transport, which handles about 5% or less of suburban commuting. Work trip distances are increasing but at a lesser rate than they would if employment were not dispersing.

Travel times for commuting trips are remaining essentially constant via two sets of changes:

(1) substitution of a suburban destination for a central city one, and a less congested route for a more congested one, and

(2) substitution of a faster transport mode or technology for a slower one — e.g. drive alone by car is substituted for almost all other modes — including public transport and car pooling.

Public transport remains a major mode for the remaining central city jobs. These jobs are increasingly specialised, e.g. in specialist or producer services or corporate headquarters — employing white collar workers, professionals and managers. Blue collar workers, retail staff and part time staff have largely suburban destinations which are less well served by transit.

5. Changes to the Urban System

In pre-industrial cities, employment was dispersed in cottage industries and the surrounding fields. The journey to work was primarily by foot. The industrial revolution saw a reversal of this trend with new mechanised industries concentrating in the central city. Radial rail networks catered for commuting, enabling longer but faster work trips, and a physically larger city as a consequence. The information revolution is creating another reversal of urban form with the major share of industry and commerce dispersing to the suburbs and beyond. The principal commuting mode in this case is the private car, enabling still longer but faster trips, and a still larger city as a consequence. Fast rail transport in Europe and Japan is enabling even longer and faster commuting trips, and the extension of cities along its corridors. The merging of cities into Mega conurbations can occur in this way.

These changes in transport technology and urban form have allowed travel times to be stabilised (Pisarski, 1992; Gordon *et al.*, 1990) by switching to a faster mode or closer (e.g. dispersed) location, while urban population and employment capacities have grown.

This behaviour is supported by the view (Marchetti 1992) that there are anthropological invariants in human travel behaviour, which influence these commuting characteristics and urban scale; that people have territorial instincts and budget a sensibly constant portion of their day to movement about that territory, and that this travel time (about one hour per day) is essentially independent of the transport technology used. Normally much of that time would now be devoted to the journey to and from work. Pre-historic man used this time to venture from his cave in search of food. Pre-industrial cities were apparently limited in size by the return distance that could be covered by foot in that time. Horse drawn vehicles extended that distance. Mass rail transit allowed an expansion of that scale by almost one order. The motor car has allowed a further increase assisted by employment dispersal, and the potential population and employment capacity has increased accordingly. Fast rail is allowing further increases in speed and development along its corridors. Telecommuting would allow still further increases creating larger virtual cities.

Transport provides a force for clustering of interacting activities, e.g. homes and jobs, producers and suppliers and markets. On the other hand broad band telecommunications allows their dispersal. Back office functions based on the processing of documents can separate from front office functions which involve face-to-face contact for negotiation, transaction and joint decision making. The front office may stay in the central city while the back office may move to the suburbs. There can be further concentration of activities by the movement of front office functions to larger cities, and the centralisation of back office functions on a regional or national basis at a smaller urban location with low factor costs. (At a global level, telecommunications has allowed a concentration of producer service activities in global cities — by increasing their locational options.)

Models which enable these post industrial activities and transitions to be simulated, analysed and visualised are needed to guide further urban development towards competitiveness and sustainability.

6. A Simple Model of the Urban System

The simplest model of urban land use — transport interactions has only two parameters (Figure 1, and Brotchie 1984, *et al.* 1991). The *land use parameter* measures dispersal of employment relative to housing and is expressed as the ratio, x, of (1) average distance from the city centre of all employment locations to (2) the average distance of all workers' houses from the city centre. This ratio, x, would be zero if all jobs were at the city centre and 1.0 if there was one job next to (or in) each house.

The *transport parameter* measures the ratio, y, of (3) the average work trip distance to (2) what it would be if all jobs were at the city centre. This trip length ratio, y, would be 1.0 if all jobs were at the city centre (the point A), and would vary from 0 to about 1.5 if there was a job next to each house (the line BC).

Thus the triangle ABC defines the feasible limits to this system. A is a single centred city, BC is complete dispersal, AC is the most efficient trip length over the range of employment dispersal where each worker chooses the closest job. AB is the trip length if jobs are chosen without regard to trip distance. D represents a particular city (Melbourne, 1991).

The line AC represents the trip distribution equation (Appendix and Figure 2) with trip decay coefficient $b=\infty$. The line AB represents decay coefficient $b=0$. AC also represents infinite transport cost. AB represents zero transport cost. The line ADC^1 represents a constant finite value of β and of transport cost.

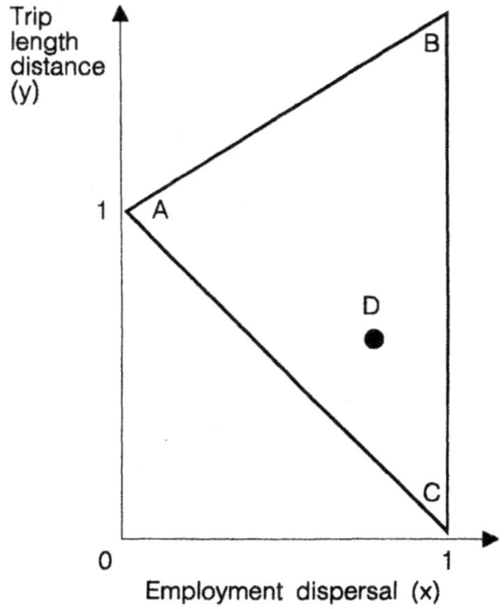

Fig. 1. Employment Dispersal Versus Work Trip Length and Feasible Limits ABC — Normalised by Average Distance of Housing from the City Centre. ABC is the 'Urban Triangle'. D is Melbourne, 1991

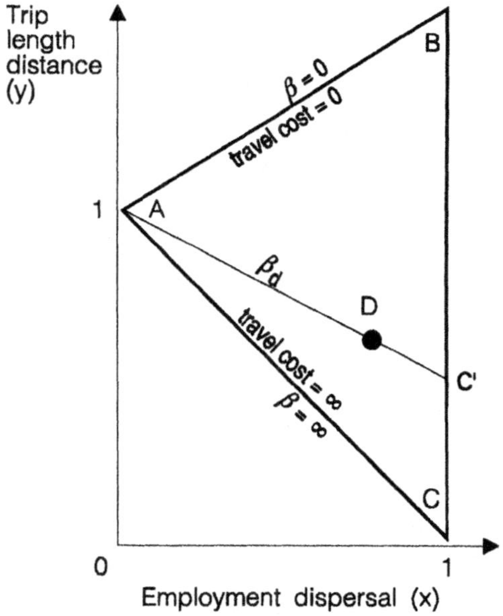

Fig. 2. Showing Constant Transport Cost/Constant Travel Behaviour β_d Along Radial ADC[1]

7. Consequences

Further employment dispersal under present transport costs and behaviour, will mean a move from D towards the point C' (Figure 3) and a relative reduction in trip length, although if the city is expanding, the absolute trip length may still increase. In the latter case, travel times may remain stable by a switch to a less congested suburban route and/or a faster mode.

Major reductions in trip length would require a substantial increase in travel cost through road pricing, fuel price rise or traffic congestion, or a constraint on travel. Increased travel cost will reduce trip length to a lower radial, e.g. AC" and this too can be evaluated — from the triangle of Figure 3.

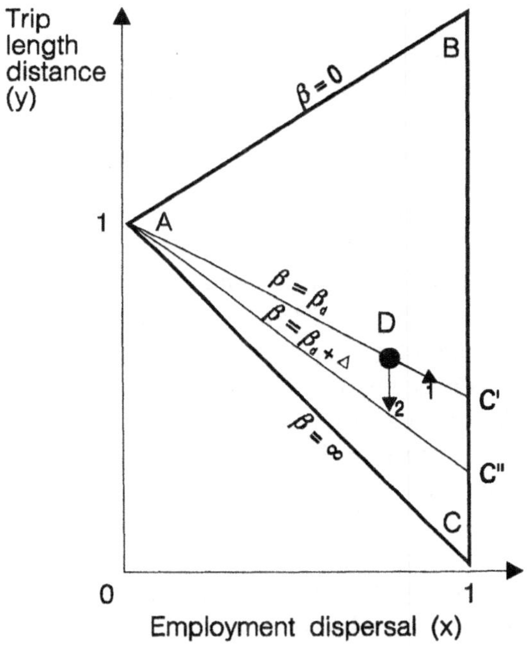

Fig. 3. Trip Length Reduction by
(1) Further Dispersal of Jobs
(2) Increased Transport Cost or Increased β

The triangle model (Figure 4) indicates the reduction which has already occurred in the transport task (for Melbourne 1991 — compared with a single-centred city) of nearly 40%, potential for further reduction within the existing land use pattern — of about 35% (or over 50% of the existing level) and with further dispersal of jobs — of another 25%. Thus the urban triangle indicates the strategies needed for energy savings and the potential savings which can occur for a sustainable future. Only under constraints on travel behaviour or large travel cost rises, however, would these savings be realised.

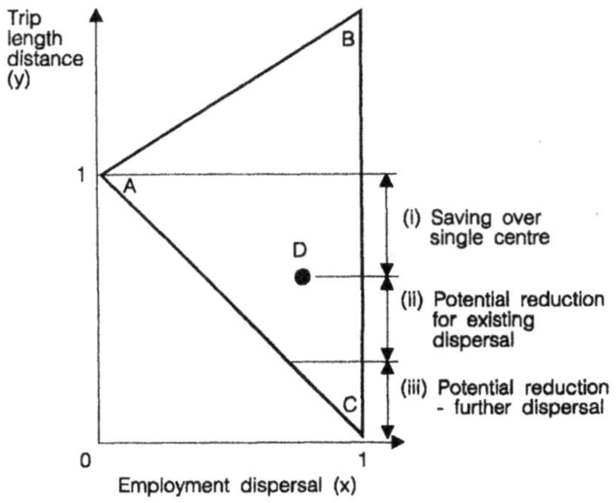

Fig. 4. Urban Triangle Showing
(1) Saving in Trip Length over Single-Centred City
(2) Potential Reduction with Existing Dispersal
(3) Potential Reduction with Further Dispersal

The type of network required for the city also varies with location on the urban triangle (Figure 5). At point C, representing pre-industrial cities, networks were minimal. Point A represents the industrial city with radial rail networks for mass transit. Point B represents the post-industrial telecommunications (and largely private transport) city with high connectivity grids and links of increasingly broad band width.

A notional trajectory of urban development through these ages is indicated in Figure 5 showing a fork in the trajectory for dispersal of post-industrial employment — with travel moving to point D and telecommunications notionally to point D' — as a result of the separation of information from the movement of goods and people enabled by the development of telecommunications networks. Telecommunications costs are relatively low and 'speed' of message transmission is increasing rapidly with increased band widths. Hence the average metropolitan telephone trip length can be expected to be closer to the line AB — and as dispersal continues, closer to the corner B. There is a link however between telecommunication and transport interactions (as noted earlier) which will reduce the distance between the points D and D' of Figure 5. The extent of present dispersal, D, means only a small proportion of trips are on radial rail networks. Most jobs are in the suburbs and the large majority of commuting trips are not served by public transport routes.

A comparison of the major Australian cities, each a State capital, is shown in Figure 6 and Table 1. A feature of this comparison is the consistency of their land use (employment) dispersal ratios, x, and transport commuting

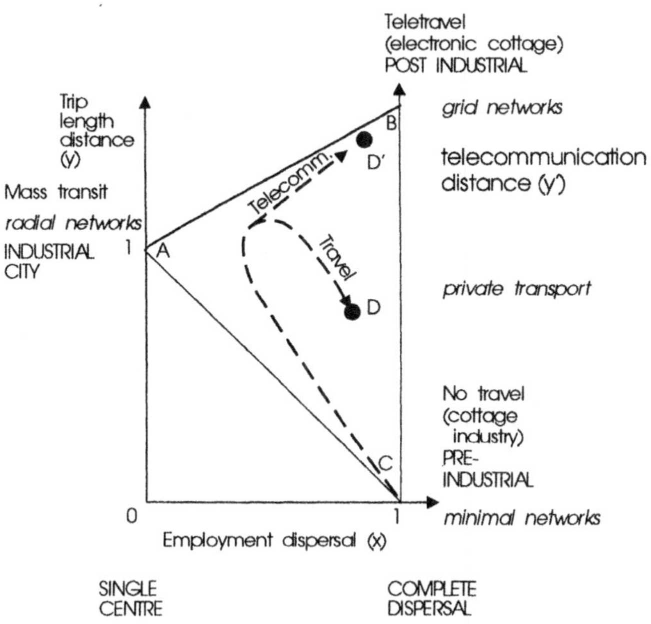

Fig. 5. Notional Urban Development Trajectory Showing Transition from Preindustrial Cottage Industry at C with Minimal Networks, Towards a Single-Centred Industrial City at A with Radial Mass Transit Networks, Turning Back as Employment Decentralises with Commuting D Moving Towards Point C and Diverging from Telecommuting D^1 which is (Notionally) Moving Towards the Postindustrial Electronic Cottage at B with Extensive Electronic Networks (and Smart Highway Grid)

ratios y. The three eastern metropolises, Sydney, Melbourne and Brisbane, are the most dispersed (x = 0.73-0.77 at 1991), with the shortest trip length ratios (y = 0.58-0.63). Adelaide and Perth are less dispersed (x = 0.69-0.71) with longer trip length ratios (y = 0.75-0.79). The consistency of location on the triangle of the three eastern cities (and two other cities) contrasts with their differences in population size, latitude, State and city governments and planning regimes. Is there 'one nation' after all?

In this range of dispersal and trip length ratios, private transport is essentially as efficient in energy use and greenhouse emissions as a single-centred city with predominantly public transport use — assuming existing vehicle loading levels for each (ESD Transport Working Group, 1991). Further trip shortening (via energy cost rises or constraints) will add to ecological sustainability of this form of development and encourage the use of lower energy consuming modes, including teleworking.

A comparison of selected US, UK and Australian cities is shown in Figure 7. The level of employment dispersal is relatively consistent across all these cities. US cities are the most dispersed at x = 0.87 average. The major UK

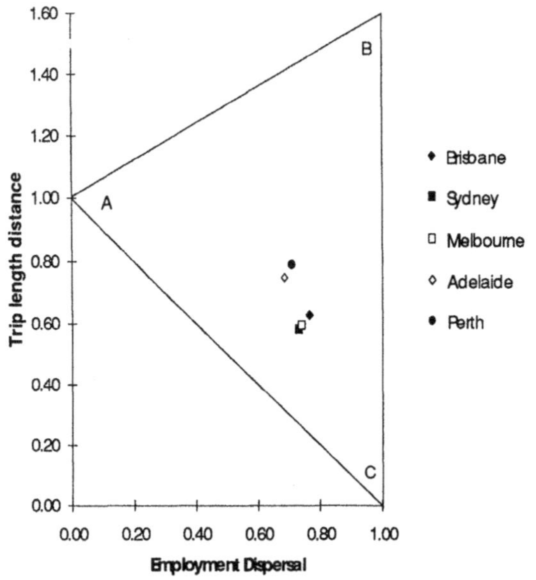

Fig. 6. Comparison of Major Australian Cities, 1991 (see also Table 1 — showing consistency of land use and transport parameters for Sydney, Melbourne and Brisbane, and for Adelaide and Perth)

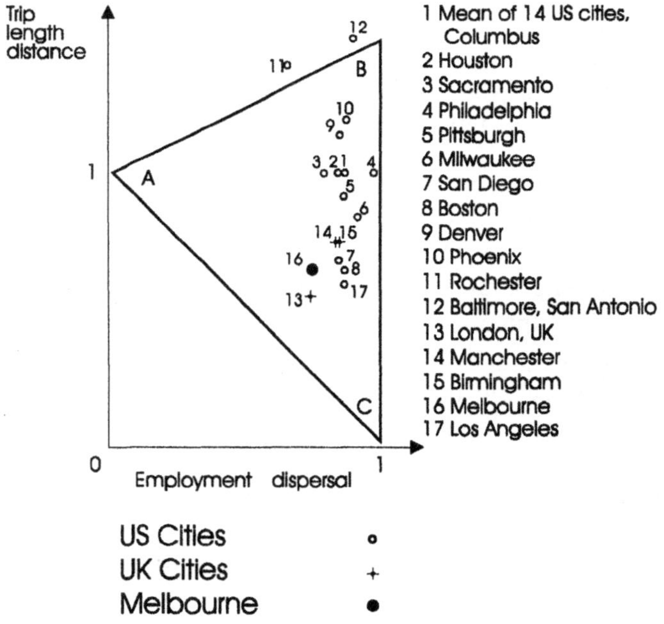

Fig. 7. Comparison of Selected US, UK and Australian Cities Based Largely on Data by Hamilon (1982), Small and Song (1992) and Spence (personal communication)

Table 1. Comparison of Major Australian Cities (1981-1991) Urban Employment Dispersal and Commuting Parameters

Distance of houses from city centre (km)						
	Brisbane	Sydney	Melbourne	Adelaide	Perth	Average
1981	15.94	22.29	20.51	15.97	13.74	19.53
1986	20.1	23.43	21.39	16.48	14.41	20.75
1991	21.58	24.55	22.44	17.37	15.88	21.83
Average	19.55	23.47	21.48	16.64	14.78	20.77
Distance of jobs from city centre (km)						
	Brisbane	Sydney	Melbourne	Adelaide	Perth	Average
1981	12.25	16.34	15.19	11.29	10.24	14.4
1986	15.11	17.13	15.67	11.39	10.34	15.14
1991	16.54	18.41	16.88	11.95	11.75	16.31
Average	14.9	17.34	15.94	11.56	10.86	15.34
Distance of journey to work (km)						
	Brisbane	Sydney	Melbourne	Adelaide	Perth	Average
1981	13.8	13.3	12.3	11.5	11.3	12.6
1986	13.2	14.1	13.0	12.3	12.0	13.3
1991	13.6	14.3	13.3	13.0	12.5	13.6
Average	13.5	13.9	12.9	12.3	12.0	13.2
Trip length ratio y						
	Brisbane	Sydney	Melbourne	Adelaide	Perth	Average
1981	0.87	0.60	0.60	0.72	0.82	0.65
1986	0.66	0.60	0.61	0.75	0.83	0.64
1991	0.63	0.58	0.59	0.75	0.79	0.62
Average	0.69	0.59	0.60	0.74	0.81	0.63
Employment dispersal x						
	Brisbane	Sydney	Melbourne	Adelaide	Perth	Average
1981	0.77	0.73	0.74	0.71	0.75	0.74
1986	0.75	0.73	0.73	0.69	0.72	0.73
1991	0.77	0.75	0.75	0.69	0.74	0.75
Average	0.76	0.74	0.74	0.69	0.73	0.74

Source: ABS Journey to Work census data, and straight line distances.

cities are less dispersed at x = 0.74–0.84 and Australian cities even less at x = 0.69-0.77. However the average trip length ratio, y, varies substantially with US cities averaging 1.0 but varying from 0.6 to 1.5 and UK and Australian cities substantially less at 0.55 to 0.85, reflecting higher transport costs, and possibly lower quality infrastructure or more congested routes. (City size may be a further factor as later discussed.)

A further interpretation of the triangle follows.

8. Further Interpretation

An urban system can be considered as a set of elements with interactions between them. It shares these properties with many other systems such as statistical mechanics systems. The mathematical formulation for this class of system (Brotchie and Lesse, 1979 and Appendix) is developed in terms of total utility U, base utility R, entropy S and diversity 1/l to give

$$U = R + \frac{S}{l} \tag{1}$$

which may be interpreted in terms of the land use transport model of the previous section and Figure 8.

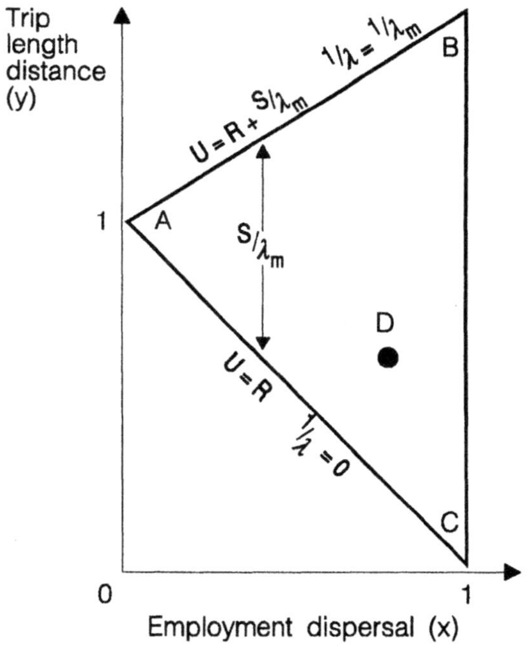

Fig. 8. Total Utility (U = R + S/λ, R = base utility, S = entropy, 1/λ = spatial diversity of opportunities over city.)

The line AC represents the base utility R of the system — the most efficient solution in terms of transport cost or energy use. The line AB represents the total utility U for a particular value $1/l_m$ of spatial diversity of employment (land use), and a particular formulation of entropy S associated with the land use pattern.

Entropy S is here defined as the log of the number, n, of possible commuting patterns for the given ratio x of employment dispersal. At A, (x = 0), all employment is at the city centre, there is one commuting pattern only and entropy is zero. The number of patterns increases with number of centres and is assumed to increase exponentially with x. (Entropy is maximised when all centres are of equal size.)

Entropy is a measure of destination choice offered by the particular dispersal x. Spatial diversity 1/l is the qualitative range of that choice (e.g. the range of occupations and their spatial distribution).

S/l is a measure of (the value of that choice and of) robustness of the urban system under a range of possible futures, i.e. of the number of ways the commuting pattern can be reconfigured by changing origins or destinations or both to provide lower energy uses, and the effectiveness of these solutions in reducing energy use. S/l_m is the distance between the lines AB and AC for a particular value of diversity, $1/l_m$.

The line ADC' in Figure 9 represents a constant value of diversity, 1/l (where $0 \leq 1/l \leq 1/l_m$, and a constant value, b, of trip decay behaviour — Figures 2 and 3).

In statistical mechanics or thermo dynamics (Brotchie *et al.*, 1979), Eqn 1 relates total energy (–U), free energy (–R), entropy (S) and temperature (t = 1/l). The equation applies to weakly interacting particle systems. (Eqn 1 also represents the formulation of the TOPAZ model for zero diversity and residential location fixed, eliminating the quadratic term).

Inclusion of the quadratic term in Eqn 1, gives the equation (Eqn 2, Appendix) for strongly interacting systems which reduces to the TOPAZ formulation when diversity is zero. Inclusion of the entropy term in TOPAZ allows behavioural solutions to be predicted and explored (see Appendix).

9. Implications and Discussion

The diversity 1/l of employment distribution across the city is another factor causing relatively long trips — particularly for workers with specialist skills. Diversity of shopping opportunities can result in long shopping trips also. Provision of the same goods and services in each suburban centre reduces this spatial diversity and shopping trip lengths. Reducing the costs of residential relocation for home owners would enable commuting distances to be reduced.

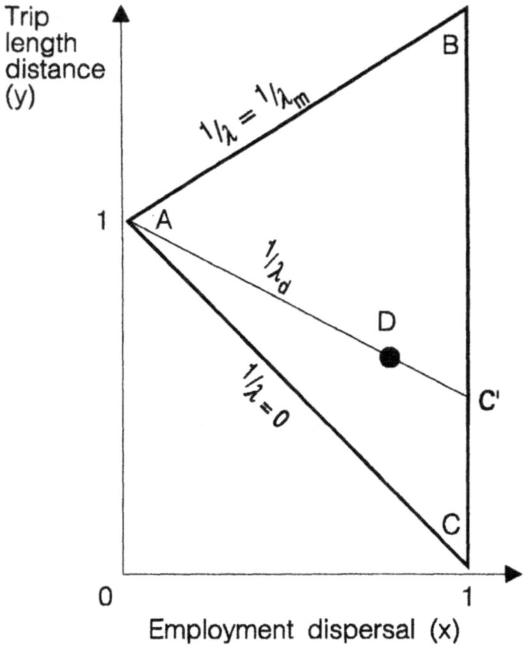

Fig. 9. Urban Triangle with Spatial Diversity of Employment (or shopping) Opportunities $1/\lambda = 0$ (line AC), $1/\lambda_d$ (line ADC^1) and $1/\lambda_m$ (line AB)

Provision of a similar range of residential choices in each suburb would assist this process. Reductions in travel cost or increases in spatial diversity of opportunities across the city would cause increases in transport trip distance. Faster travel would again cause increases in trip length.

The distance of D above the line AC is termed "wasteful commuting" by Hamilton (1982), who assumes for his minimum solution that each job and home pair is on the one radius with jobs closer to the city. A series of further papers, (e.g. Small and Song 1992, Suh 1990, White 1988) have since disputed the term 'wasteful' and provided explanations for the excess travel distances. In terms of the present discussion, this distance D represents existing behaviour with existing transport networks and costs, and existing spatial diversity of job opportunities. It represents freedom of choice, to select the most appropriate job available in whatever direction.

It also represents robustness of the city to operate at a range of commuting patterns with lower energy uses, i.e. the robustness to adapt to lower energy use futures by changing trip origins or destinations to a closer one in order to transform the city from largely multi-centred to essentially a cluster of self sufficient single centred cities with much lower energy use. If all trips were confined to a single zone and there were sixteen equal zones, trip length ratio would decrease to 0.25 or less and with one hundred zones, to 0.1 or less. As noted earlier this containment of travel would require major changes in travel behaviour or in travel cost.

Industry dispersal and trip length varies with the type of industry and occupation of worker. Corporate headquarters and specialist services tend to remain in the central city, and are served there by fixed rail transport. In Melbourne, their managers and professionals tend to live closer to the city, exchanging increased house prices for shorter travel distances. Average travel times, however, appear to be relatively constant for all origin distances from the central city (TRC, 1993). Retail activities are more likely to be in suburban centres. Part time workers (largely female) appear to have shorter travel distances and times — principally to suburban jobs.

In the inner suburbs of Melbourne the city tends to behave as largely single centred with less dispersal of trip destinations; in the middle suburbs it behaves more as if multi-centred, with more dispersal and more inter suburban trips, and the outer suburbs with still more dispersal behave to a slightly greater degree as a cluster of self contained centres.

Transport infrastructure has an undoubted influence on urban behaviour. A peripheral ring freeway would increase transport interactions among outer suburbs, allowing further development of specialist industries and services along it, including re-engineered industrial processes, and would speed the flow of components from suppliers around the outer suburbs — or from regional cities on inter city highways intersected by this ring road. Interestingly, Boston, with its radial tollway and two ring freeways, and Los Angeles with an extensive freeway grid, have the lowest trip length ratios of the US cities of Figure 7. (There does seem to be at least a weak inverse relationship between trip length ratio y and city size in Figures 6 and 7 which is consistent with the concept of constant travel time budgets. It is consistent also with reduced diseconomies of urban scale.)

Concentration of employment in a few key suburban centres, and of housing around these centres — and along a few key transport routes linking these centres with each other and the central city — would increase viability of mass transit on these routes (see also Cervero, 1994; Newman and Kenworthy, 1989) providing another path towards sustainable development. Broadband telecommunications could reinforce this key centre development and provide suburban work centres for telecommuting.

It is interesting to note that the line ADC^1 of existing behaviour and costs (Figure 2) is relatively flat and that movements towards further dispersal or concentration will have only a small effect on trip length and transport energy use. This result was also noted in a previous study of Melbourne (Brotchie *et al.*, 1980) where changes in residential density had only small impact; and in a recent study of deconcentration of urban activities at a national level over the last 30 years in the UK by Breheny (1994) where the net effect of complete prevention of this deconcentration (by containment) would have been a net energy use decrease of only 2.2 per cent. Changes in travel behaviour, e.g. via energy use constraints or substantial changes in travel cost, appear to be necessary if major reductions in energy use are to be achieved.

Telecommuting (Moss and Carey, 1994) in which employees work from home (or from a suburban centre) using information and communication technologies — for part or all of the week — is likely to have further impacts on employment dispersal. Employees can live further from work on the basis that they make the trip less often — thereby further increasing their locational choices. For full time teleworking from home, the job effectively moves to their home, thereby further increasing dispersal, moving them towards the point C (in Figures 1 and 5) from the viewpoint of vehicle traffic — and towards the point B — in terms of telecommunications traffic. The functional city would change from a physical entity to a virtual system which could encompass parts of the surrounding countryside, and neighbouring cities or towns.

10. Conclusion

Cities are the new entities competing in global markets, making international transactions and forming international alliances. Telecommunications are the new international trade routes. Fast transport however is still an integral part of this new system. Information is travelling increasingly by electronic means, but expertise or knowledge is still travelling in peoples heads. Hence cities linked by telecommunications and fast transport and at the nodes of these new networks are the engines of the information economy.

Many of the agglomeration economies of cities with dispersed activities are provided by telecommunications interactions. Models of post industrial cities must consider land use/transport/communication/computing interactions and environmental impacts. The new TOPAZ model is being developed on this basis. The SUSTAIN model (Roy and Marquez) provides a level of planning which points the way to subsequent, more detailed studies. It provides a preliminary phase to the New TOPAZ. The urban triangle is the simplest land use-transport interaction model and provides a still earlier phase of analysis. The urban triangle model is a simple tool for broader brush and comparative studies of cities and their sustainability; and for initial evaluation of potential savings in transport energy, of environmental impacts and of land use/transport/communication policies, to guide development towards a sustainable future.

References

Breheny, M. (1994). 'Counter urbanisation and sustainable urban forms', in *Cities in Competition*, John Brotchie, Michael Batty, Ed Blakely, Peter Hall and Peter Newton, eds, Longman Cheshire (in press).

Brotchie, J.F. (1984). 'Technological change and urban form', *Environment and Planning A*, Vol. 16, pp. 583-596.

Brotchie, J.F. (1987). 'A generalised design approach to the solution of the non-convex quadratic programming problem', *Applied Mathematical Modelling* pp. 291-295, Vol. 11.

Brotchie, J.F., Anderson, M. and McNamara, C. (1994). 'Changing metropolitan commuting patterns', in *Cities in Competition*, John Brotchie, Michael Batty, Ed Blakely, Peter Hall and Peter Newton, eds, Longman Cheshire (in press).

Brotchie, J.F., Lesse, P.F. and Roy, J.R. (1979). 'Entropy, utility and planning models', *Sistemi Urbani*, Vol. 1, No. 3, pp. 33-53.

Brotchie, J.F., Batty, M., Hall, P. and Newton, P. (eds) (1991). *Cities of the 21st Century* Longman Cheshire, Melbourne.

Brotchie, J.F., Dickey, J.W. and Sharpe, R. (1980). *TOPAZ*, Springer Verlag, Berlin.

Cervero, R. (1994). 'Changing live-work spatial relationships, implications for metropolitan spatial structure and mobility', in *Cities in Competition* , John Brotchie, Michael Batty, Ed Blakely, Peter Hall and Peter Newton, eds, Longman Cheshire (in press).

ESD - Transport Working Group (1991). Final Report - Transport, AGPS, Canberra.

Gordon, P., Richardson, H.W. and Jun M.J. (1990). 'The Commuting Paradox — Evidence from the Top Twenty', Working Paper, School of Urban and Regional Planning, USC Los Angeles.

Hamilton, B. (1982). 'Wasteful commuting', *Journal of Political Economy* V90, pp. 1035-53.

Hopfield, J.J. and Tank, D.W. (1985). 'Neural computation of decisions in optimisation problems', *Biological Cybernetics*, Vol. 52, p. 141-152.

Marchetti, C. (1992). *Anthropological Invariants in Travel Behaviour*, International Institute of Applied Systems Analysis, Laxenburg, Austria.

Moss, M. and Carey, I. (1994). 'Information technologies, telecommuting and cities', in *Cities in Competition*, John Brotchie, Michael Batty, Ed Blakely, Peter Hall and Peter Newton, eds, Longman Cheshire (in press).

Newman, P. and Kenworthy, J. (1989). 'Gasoline consumption and cities', *Journal of the American Planning Association*, Vol. 55, No. 1, pp. 20-37.

Pisarski (1992). *New Perspectives on Commuting*, Federal Highway Administration Washington DC.

Small, K. and Song, S. (1992). 'Wasteful commuting: a resolution', *Journal of Political Economy* Vol. 100, No. 4, pp. 888-98.

Suh, S. (1990). 'Wasteful commuting an alternative spproach', *Journal of Urban Economics*, Vol. 28, pp. 276-86.

TRC (1993). 'Travel time budgets', *TRC News*, Melbourne University, Vol. No. 1, p. 5.

White, M. (1988). 'Urban journeys are not wasteful', *Journal of Political Economy*, Vol. 96, pp. 1097-110.

APPENDIX

A general equation for weakly interacting systems of elements or particles takes the form (Brotchie *et al.*, 1979)

$$U = R + S/\lambda \tag{1}$$

$$R = \Sigma\, a_{ij}\, x'_{ij}$$

where U is the total utility or effectiveness of the system,

- R is a base utility representing efficiency — the most efficient solution
- S is entropy, defined here as the log of the number n of possible micro states — or of origin and destination combinations or the number of possible commuting patterns (or job centres) for a given level of dispersal x, i.e. S = log n. If n increases exponentially with dispersal x then $n = e^{kx}$ where k is a constant and S = kx
- $1/\lambda$ is the diversity of energy states or the spatial diversity of destinations, i.e. the spatial distribution of job opportunities (occupations) over the city
- x is dispersal
- x'_{ij} is allocation (e.g. activity i to zone j), so that $x = f(x'_{ij})$
- a_{ij} is a constant for each activity i in each zone j.

These equations are derived statistically in Brotchie *et al.* (1979) for three constraint conditions on (origin) zone capacity yielding three forms of entropy — and for distinguishable and indistinguisable elements.

For a single-centred city only one commuting pattern exists. The number of patterns (and centres) increases exponentially with dispersal x.

R is linear in x and the most efficient solution, AC in Figure 8, corresponding to $1/\lambda = 0$.

S is also linear in x and increases from 0 at x = 0 to (a maximum at) x = 1. For a particular value $1/\lambda_m$ of spatial diversity $1/\lambda$, in Eqn. 1, $U = R + S/\lambda$ corresponds to the line AB in Figure 8 and S/λ_m corresponds to the distance between the lines AB and AC. All values of $1/\lambda$ between 0 and $1/\lambda_m$ are radials from A and intersect BC. In statistical mechanics/thermo dynamics:

-U is total energy of the system
-R is free energy
S is entropy, and
$1/\lambda$ is absolute temperature, t (or diversity of energy states of particles).

For strong interactions between particles, the corresponding equation (Brotchie, 1987) is

$$U = \Sigma a_{ij} x'_{ij} + \Sigma b_{ijkl} x'_{ij} x'_{kl} + S/\lambda \qquad (2)$$
$$= R_1 + S/\lambda$$

S represents choice, and uncertainty of outcome of that choice. It represents also the number of ways the pattern of commuting may be changed to achieve different (including lower) energy states or lower transport costs or urban operating costs.

$1/\lambda$ indicates the range or spatial variation of values, costs or energies and hence the potential effectiveness of selecting other patterns.

S/λ is a measure of the value of the choices made and at the same time, the robustness of the system in adapting to a range of possible energy use futures.

b_{ijkl} is net benefit of interaction between element ij and element kl, e.g. of trips between activities i and k in zones j and l respectively.

Trip distribution T_{ijkl} is given by

$$T_{ijkl} \propto x'_{ij} x'_{kl} e^{-\beta_{ik} c_{jl}} \qquad (3)$$

where c_{jl} is the cost of travel between zones j and l, and β_{ik} is the decay constant for trips between activities i and k (and $b_{ijkl} = T_{ijkl} c_{jl}$).

With $1/\lambda = 0$, Eqn. 2 reduces to the TOPAZ model. (It also reduces to the energy equation for a neural network (Hopfield and tank, 1985).)

Including the robustness term in Eqn. 2 allows behavioural patterns or solutions to be analysed and the behavioural response to policies for sustainable development to be evaluated. (It also allows for uncertainty of that behaviour in an urban system — or in a neural network.)

Part B.
Policy Implications in Modelling

CHAPTER 6

REDUCTION OF CO_2 EMISSIONS OF TRANSPORT BY REORGANISATION OF URBAN ACTIVITIES

Michael Wegener

1. Introduction

It is generally believed that the private automobile has been the primary cause of the expansion of cities over wider and wider areas. However suburbanisation was not caused by the car but has been the consequence of the same changes in the socio-economic context of urban life that were also responsible for the growth in car ownership: increase in income, more working women, smaller households, shorter work hours and a consequential change in lifestyles and housing preferences towards quality of life, leisure and recreation. Under these conditions, the car and low fuel prices brought low-density suburban living within the reach of not only the rich, with the result that for the last thirty years the growth of cities has occurred primarily in the suburbs. Offices, light industry, services and retail started to decentralise later following either their employees or their markets or both taking advantage of attractive suburban locations with good accessibility, ample parking and lower land prices.

However, while this deconcentration process clearly reflects the preferences of the majority of the population, its negative side effects are more and more becoming apparent: longer work and shopping trips, increasing rush-hour congestion and less and less acceptable levels of noise, air pollution and traffic accidents. In particular the high energy consumption of transport in low-density cities has become an issue of growing concern. The fear of diminishing fossil fuels and the threat of long-term climate changes due to greenhouse gases have

Y. Hayashi and J. Roy (eds.), Transport, Land-Use and the Environment, 103–124.

sharpened the awareness that present energy prices do not nearly cover the environmental and social costs of energy use and that the level of energy consumption in affluent countries represents a gross unfairness against developing countries which can never be allowed to rise to the same standards. At the United Nations conference on the global environment in Rio de Janeiro in 1992 many governments pledged to substantially reduce their use of fossil energy and emissions of carbon dioxide (CO_2). The German government promised to reduce CO_2 emissions from all sources by 30 percent compared with 1987 by 2005. As transport represents a major share of primary energy consumption, serious attempts to lower the energy use of urban transport are necessary to achieve this goal.

There have been numerous proposals to respond to this challenge. The majority of them follow the hypothesis that energy use of urban transport is a direct function of settlement density and suggest a return to mixed-use, compact land-use patterns. The most frequently quoted support of this hypothesis is the study by Newman and Kenworthy (1989), who analysed 32 cities in four continents and found a significant negative statistical correlation between residential density and transport-related energy consumption per capita (see Figure 1, top). In many European countries the reliance on this hypothesis has led to policy recommendations such as the following:

"[Local governments are advised] to apply a policy of short distances which reduces the length of trips between residences, workplaces and public and private facilities in order to avoid car traffic and increases the attractiveness of public transport, cycling and walking."

German Council of Cities: Ten Points to Improve Urban Transport (1989)

"The first fundament of sustainable mobility is a policy of concentrated development of residential and industrial areas and public facilities. In a concentrated spatial structure the distances to be travelled are shortest; the fastest transport mode is the bicycle, and good public transport can be provided."

Fourth Note (Extra) on Spatial Development in the Netherlands (1990)

"The strict zoning policies of the past decades which have led to the separation of land use and the subsequent development of extensive residential suburbs have in turn stimulated commuter traffic, which is at the heart of many of the environmental problems currently facing urban areas. We therefore need a fundamental review of the principles on which town planning practice has been based. Strategies which emphasize mixed use and denser development are more likely to result in people living closer to

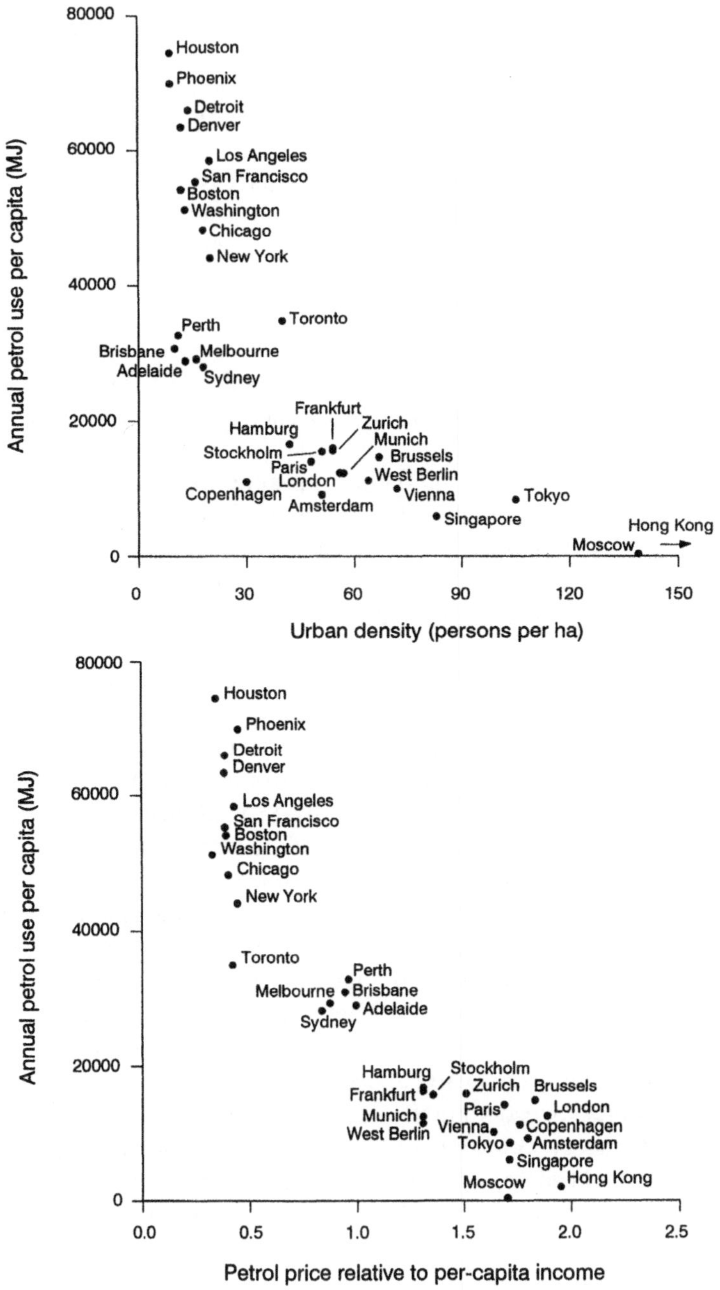

Fig. 1. Petrol Use v. Urban Density (top) and Petrol Price (bottom), 1980
Source: Newman and Kenworthy (1989)

work places and the services they require for everyday life. The car can then become an option rather than a necessity."

Green Paper on the Urban Environment
of the European Communities (1990)

The problem with these policy prescriptions is that there has been no evidence so far that under today's conditions, i.e. with an unconstrained transport market and present travel costs, a return to higher densities would lead to a reduction of energy consumption of urban transport. In fact there have been several studies contradicting this hypothesis:

- Rickaby (1987) found by model simulations that 'decentralised concentration' (urban areas with medium-sized secondary centres) is the most energy-efficient settlement structure.
- Banister (1992) showed that petrol consumption per capita in England declines with city size, but is higher in London than in other large cities.
- Breheny (1995) demonstrated that, if the population of England and Wales had *not* suburbanised between 1961 and 1991, total energy savings would have been less than three percent.

Moreover, even the data presented by Newman and Kenworthy can be interpreted in a different way which sheds doubt on the simple relationship between density and energy use. For instance, if one plots transport energy consumption not against urban density but against the petrol price data contained in the study, one finds the same, but even stronger, inverse relationship. In Figure 1 (bottom) annual petrol use per capita is plotted against petrol price relative to per-capita income, where 1.0 indicates the average relative petrol price of all 32 cities. Now it becomes plausible why petrol consumption in Australian cities is much lower than in cities of the United States, although Australian cities are no less dispersed than cities in the United States: because petrol is twice as expensive in Australia. One might hypothesise that urban density is only an intermediate variable and that the real cause behind a high level of transport energy consumption is the availability of cheap transport energy.

More evidence questioning the importance of urban density as a determinant of transport energy consumption is contained in Breheny (1992). However, beyond that doubt there is not much agreement about what the energy-efficient city of the 21st century should look like. The situation is characterised by Schmitz (1991):

"A settlement structure which is 'ideal' with respect to transport is not known today. Planning paradigms such as small-scale mixed land use, promotion of inner cities through higher densities, decentralised concentration and development axes in regional planning or the development of balanced functional urban regions still have the character of catchwords. They require first to be

specified in more concrete terms and second to be assessed with respect to their efficiency and feasibility."

In this situation computer simulation models of urban land use and transport may have a new role to play. There is a long history of ambitious efforts to introduce computer simulation techniques into planning since the appearance of the first digital computers in the late 1950s. However, with very few exceptions none of these models has made a permanent impact on the practice of planning and only few of them have survived as research tools in university departments (see Wegener, 1994). Today the environmental debate poses questions which are not likely to be handled by incrementalist, piecemeal approaches but require a fundamental review of the way cities are organised. This requires once again a comprehensive view of cities as complex systems. However, this time the models will not be used, as previously, to forecast the direction of urban *growth*, but to guide the spatial *reorganisation* of metropolitan areas towards environmental sustainability.

2. Do We Need to Rebuild Dortmund?

This chapter reports on a project in which a land-use transport model was used to explore the impacts of strategies to reduce transport-related CO_2 emissions by transport demand management in the metropolitan area of Dortmund in Germany.

2.1 The Study Area

The study area was the metropolitan area of Dortmund in Germany. Dortmund (population 615,000) is the most eastern of the cities of the Ruhr Area, the largest industrial region in Germany. It used to be one of the major centres of coal mining and steel manufacturing in Germany, but with the decline of the mining and steel industries it has been reduced to being the administrative, service and retail centre for a large metropolitan area (see Figure 2).

The region represented in the model is the commuter catchment area of Dortmund containing Dortmund itself and eighteen neighbouring communities. The region is relatively compact; most of its settlements lie within the 30-minute travel-time isochrone by car from central Dortmund. The municipalities in the hatched area are exclusively oriented towards Dortmund; the dotted areas are larger self-contained cities or communities oriented towards more than one centre. The study area has a population of approximately 2.3 million.

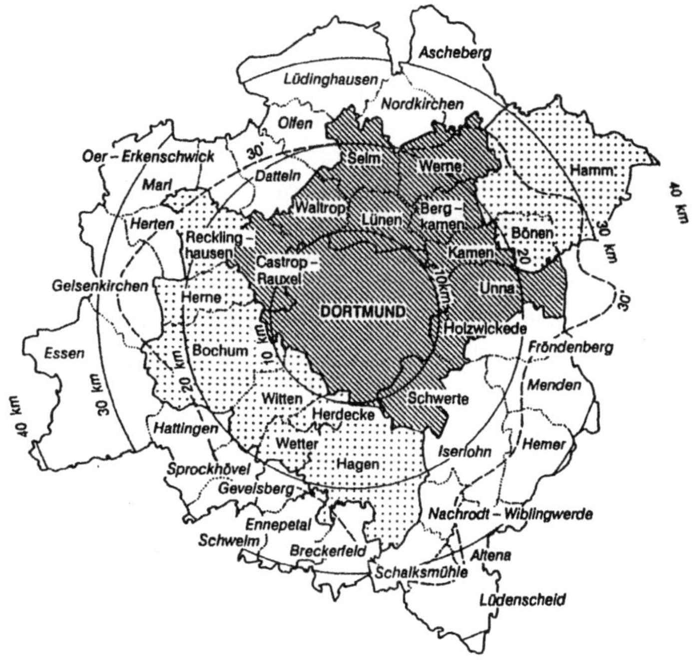

Fig. 2. The Metropolitan Area of Dortmund

2.2 The Model

The Dortmund model is a model of intraregional location and mobility decisions in a metropolitan area (Wegener, 1983; 1985; 1996). It receives its spatial dimension by the subdivision of the study area in thirty *zones* connected with each other by transport networks containing the most important links of the public transport and road networks coded as an integrated, multimodal network including walking and cycling and all past and future network changes. It receives its *temporal* dimension by the subdivision of time into fifteen time periods of three years' duration.

Figure 3 is a schematic diagram of the major subsystems considered in the model and the interactions between them and of the most important policy instruments. The four square boxes in the corners of the diagram show the major stock variables of the model: *population*, *employment*, *residential buildings* (housing) and *non-residential buildings* (industrial and commercial workplaces and public facilities). The actors representing these stocks are *individuals or households*, *workers*, *housing investors* and *firms*. These actors interact on five *submarkets* of urban development. The submarkets and the market transactions occurring on them are:

- the *labour market*: new jobs and redundancies,
- the *market for non-residential buildings*: new firms and firm relocations,
- the *housing market*: immigration, outmigration, new households and moves,

- the *land and construction market*: changes of land use through new construction, modernisation or demolition.
- the *transport market*: trips and their consequences: changes of accessibility, road accidents, traffic noise, air pollution and energy use.

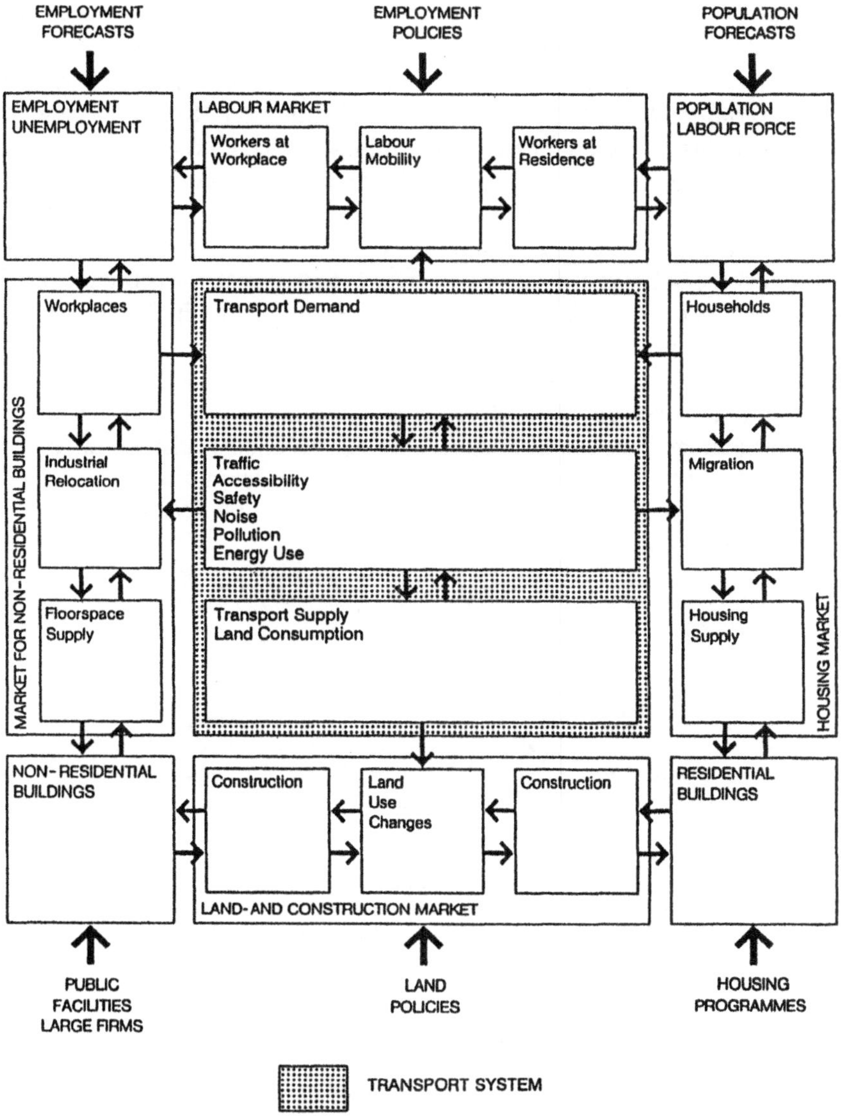

Fig. 3. Major Subsystems of the Dortmund Model

For each submarket, the diagram shows supply and demand and the resulting market transactions. Choice in the submarkets is constrained by supply (jobs, vacant industrial or commercial floorspace, vacant housing, vacant land, network capacity) and guided by attractiveness, which in general terms is an actor-specific aggregate of *quality* and *price*. The large arrows in the diagram indicate exogenous inputs: these are either *forecasts* of regional employment and population subject to long-term economic and demographic trends or *policies* in the fields of industrial development, housing, public facilities and transport.

With this model structure the Dortmund model is one of the few operational urban models in which the two-way interaction between land use and transport in urban areas is explicitly modelled (Webster et al, 1988; Wegener, 1994). Figure 4 summarises these interactions: The distribution of *land uses*, such as residential, industrial or commercial determines the location of households and firms and so of human *activities* such as living, working, shopping, education or leisure. The distribution of *activities* requires spatial interactions or trips to overcome the distance between them. These trips occur in the *transport system* in a sequence of choices: decisions to own a car, to make a trip and to select destination, mode and route. These decisions result in flows in the networks and congestion and increases in travel times, distances and costs. Travel times, distances and costs create opportunities for spatial interactions and can be measured as *accessibility*. The distribution of *accessibility* in space, together with other attractiveness indicators, determines location decisions of investors and so results in new construction, modernisation or demolition, i.e. changes to the *land-use* system, which in turn determine the location decision of users and hence the location of activities.

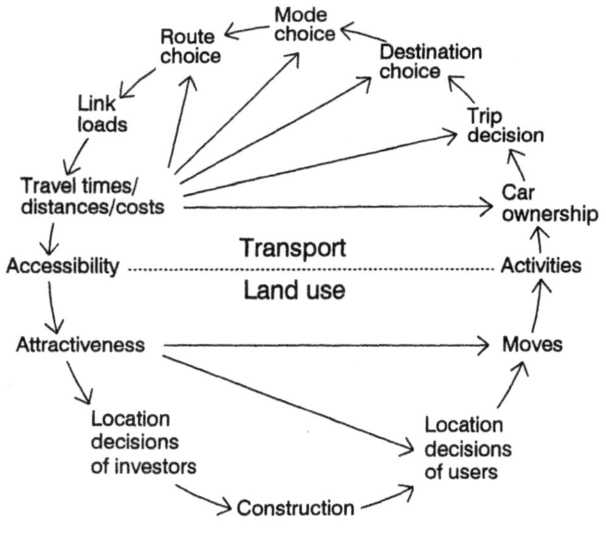

Fig. 4. The 'Land-Use Transport Feedback Cycle'

Figure 5 portrays the same interrelationships in a format proposed by Brotchie (1984). The 'Brotchie Triangle' represents the universe of possible constellations of spatial interaction and spatial structure in an urban area. Spatial structure is represented on the horizontal axis as spatial dispersal (for instance, mean travel distance of employment from the centre of the region), spatial interaction on the vertical axis as some measure of total travel such as mean travel distance to work. Any city will lie between three hypothetical points in the diagram: point A represents a situation in which all jobs are at the centre, i.e. dispersal is zero. Both points B and C represent regions in which all jobs are as dispersed as the population. Point B represents a situation in which workers choose their residence without regard of distance, point C a situation in which they walk to work. The model answers the question in which direction the real city, point D, will shift: a shift up or down indicates reorganisation or *moves*, a shift to the left or right indicates *construction*, or rebuilding the city.

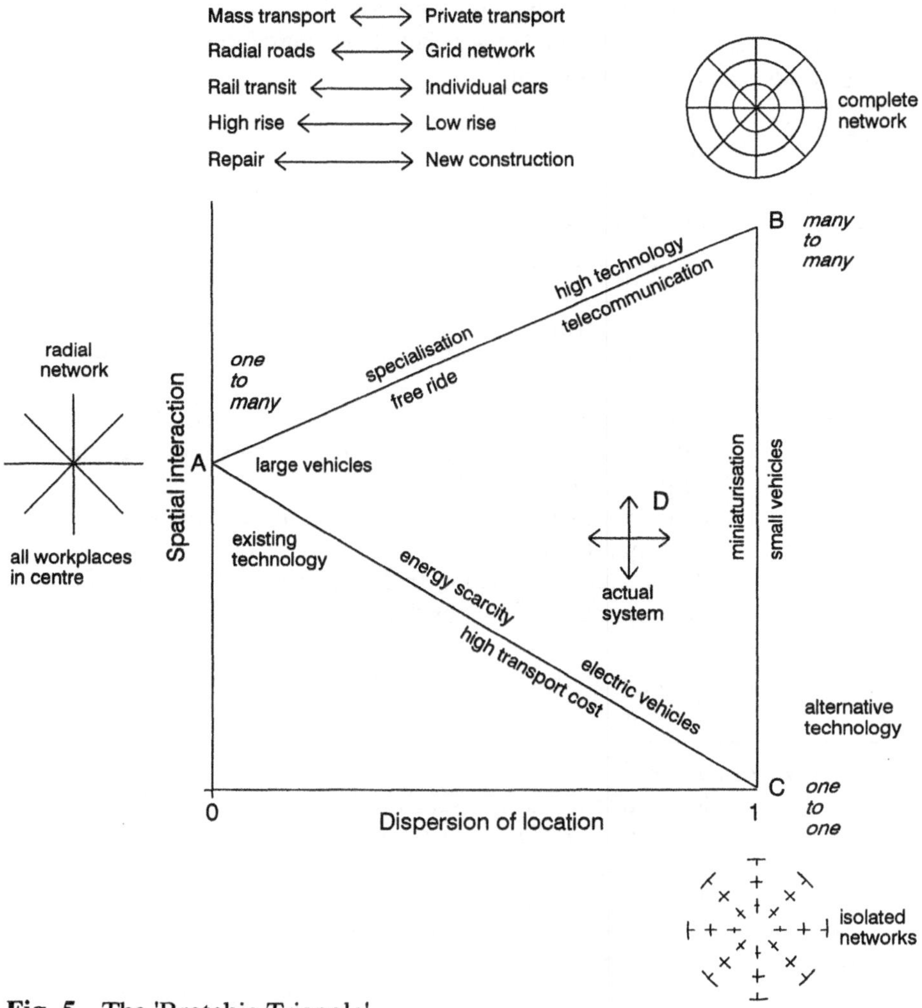

Fig. 5. The 'Brotchie Triangle'
Source: Brotchie, 1984.

2.3 The Scenarios

Three types of scenarios were simulated: scenarios of travel cost changes, scenarios of travel speed changes, and scenarios in which changes of both travel costs and travel speeds were combined. Table 1 shows the list of scenarios and their specification. The first two groups of scenarios are similar to the policy tests conducted by the International Study Group on Land-Use Transport Interaction (ISGLUTI) (Webster et al, 1988); the combination scenarios go beyond ISGLUTI. Scenarios 30 and 40 were defined differently from their ISGLUTI counterparts. In Scenario 30 a much faster increase of petrol price was assumed (all price increases include inflation), but the scenario was made more realistic by assuming that car manufacturers would respond to significant increases of fuel price by offering more energy-efficient cars. In Scenario 40 public transport was not only made faster but also having more trains and busses to accommodate additional ridership.

Table 1. Scenarios

Scenario	*Specification*
Base scenario:	
00	Base scenario.
Travel cost scenarios:	
30	Increase petrol price incrementally to 12 DM/l by 2015 and reduce average petrol consumption of cars incrementally to 5 l per 100 km by 2015.
32	Increase inner-city parking charges incrementally, after 2000 quintupled.
33	Reduce public transport fares incrementally, after 2000 free.
35	Increase public transport fares incrementally, after 2000 doubled.
37	Increase all transport costs incrementally, after 2000 doubled.
Travel speed scenarios:	
40	Make public transport faster (25 %) and reduce headways (50 %) and make cars slower (40 %).
46	Make public transport and cars faster (25 %).
47	Make public transport and cars slower (40 %).
Combination scenarios:	
53	'Promotion of public transport': scenarios 30+32+40.
54	'Reduction of mobility': scenarios 30+32+35+47.

The results of the simulations are summarised in Figures 6 to 8. In each of them the evolution of the urban system between 1970 and 2015 is represented by trajectories of one variable for each of the simulated scenarios. Until the mid-1990s, all scenarios coincide because the policies specified in Table 1 are introduced after 1993; this serves to visualise the development in the past. The trajectory of each scenario is indicated by its number as in Table 1; Scenario 00 is the 'base scenario' defined as the trend scenario without policy changes.

Figure 6 (top) shows the effect of the various policies on average trip length. It can be seen that in the base scenario average trip length increases from 8 to 13 kilometres between 1970 and 2015, and that policies to reduce travel cost (Scenario 33) or increase travel speed (Scenario 46) result in longer trips. Increasing travel costs (Scenarios 30, 37) and making travel slower (Scenarios 40, 47) result in shorter trips, but this effect is diluted after 2000 by growing affluence and greater fuel efficiency of cars. The reduction effect is strongest in the combination scenario which penalises mobility altogether (Scenario 54), whereas in the combination scenario which promotes public transport (Scenario 53) the loss of mobility is much smaller. Figure 6 (bottom) shows that the impact of the combined policies is even stronger if only car travel is considered. Here Scenario 53 shows its superiority because it results in a much stronger reduction of car-km travelled than Scenario 54, in which no attractive travel alternatives by public transport exist. Indeed the total distance travelled by car in the region is more than halved in Scenario 53.

Figure 7 (top) shows that this is due to the substantial modal shift occurring in the region in that scenario. It can be seen that public transport has declined from 30 percent of all trips in 1970 to less than 20 percent today. One can see that neither massive investment in public transport at the expense of car traffic (Scenario 40) nor making public transport free (Scenario 33) will result in substantial increases in ridership. Nor will an increase in the out-of-pocket cost of car travel (Scenario 30) help to revitalise public transport; in combination with increased fares (Scenario 54) it will even discourage public transport use. However, if the improvement of public transport is combined with monetary disincentives to car travel (Scenario 53), the effect is a dramatic rise in public transport use to over forty percent of all trips.

All this translates into significant savings in energy use and CO_2 emissions as shown in Figure 7 (bottom). The diagram shows the savings in energy use and CO_2 emissions by all transport, including the additional busses and trains necessary for the growing number of passengers. Despite the growth in car ownership and travel distances, CO_2 emissions per capita are likely to decrease after 2000 because of greater energy efficiency of cars. However, without intervention the goal to reduce CO_2 emissions by 30 percent compared with 1987 cannot be achieved. None of the policies meets this target except those in which car travel is made significantly more expensive. Of these Scenario 53 implies the smallest sacrifice in mobility.

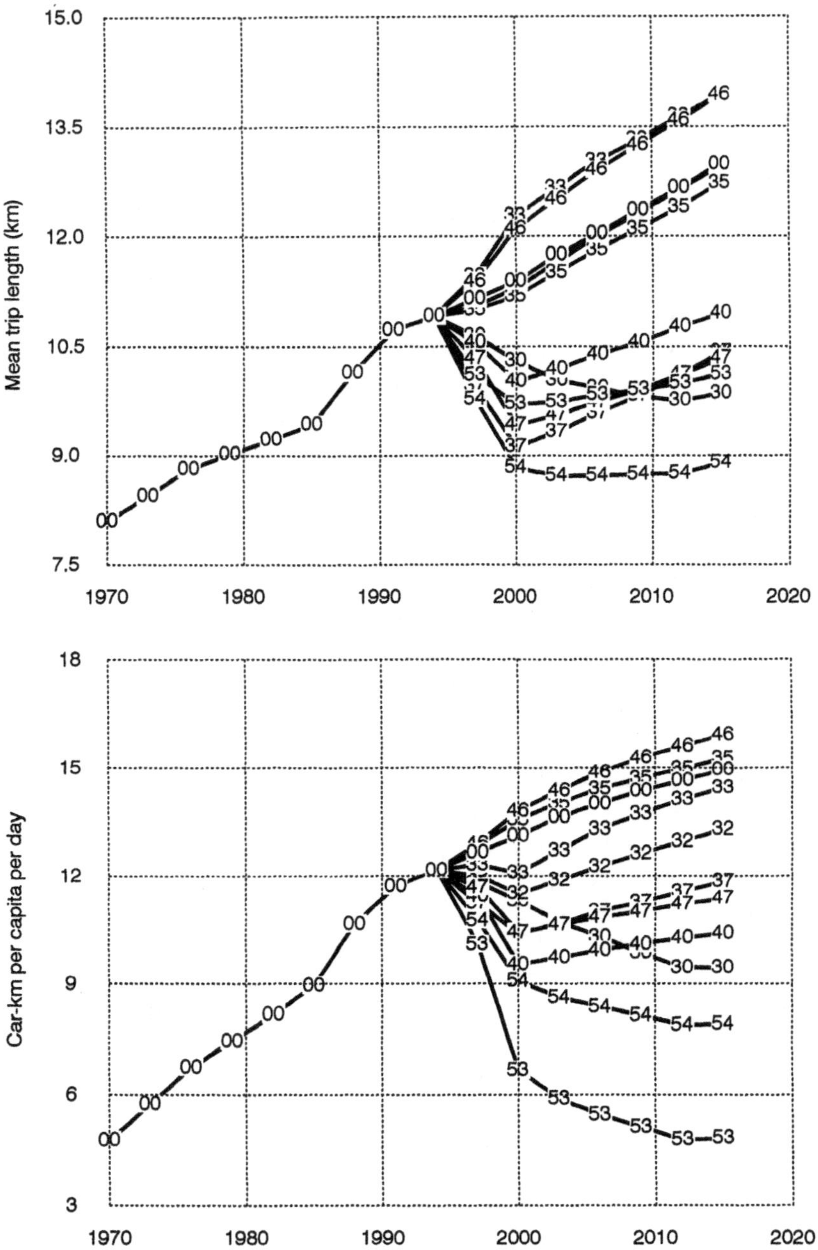

Fig. 6. Mean Trip Length (top) and Car-km Per Capita Per Day (bottom)

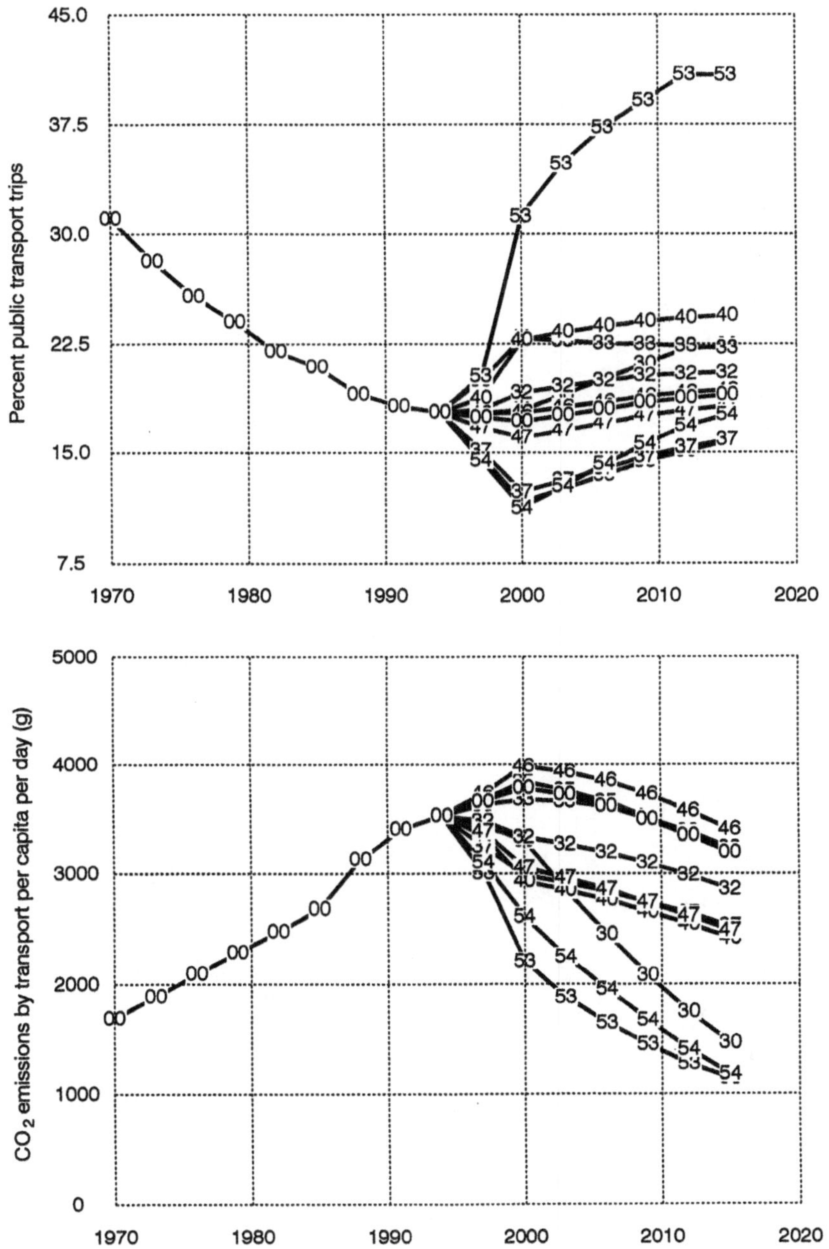

Fig. 7. Percent Public Transport (top) and CO_2 Emissions of Transport (bottom)

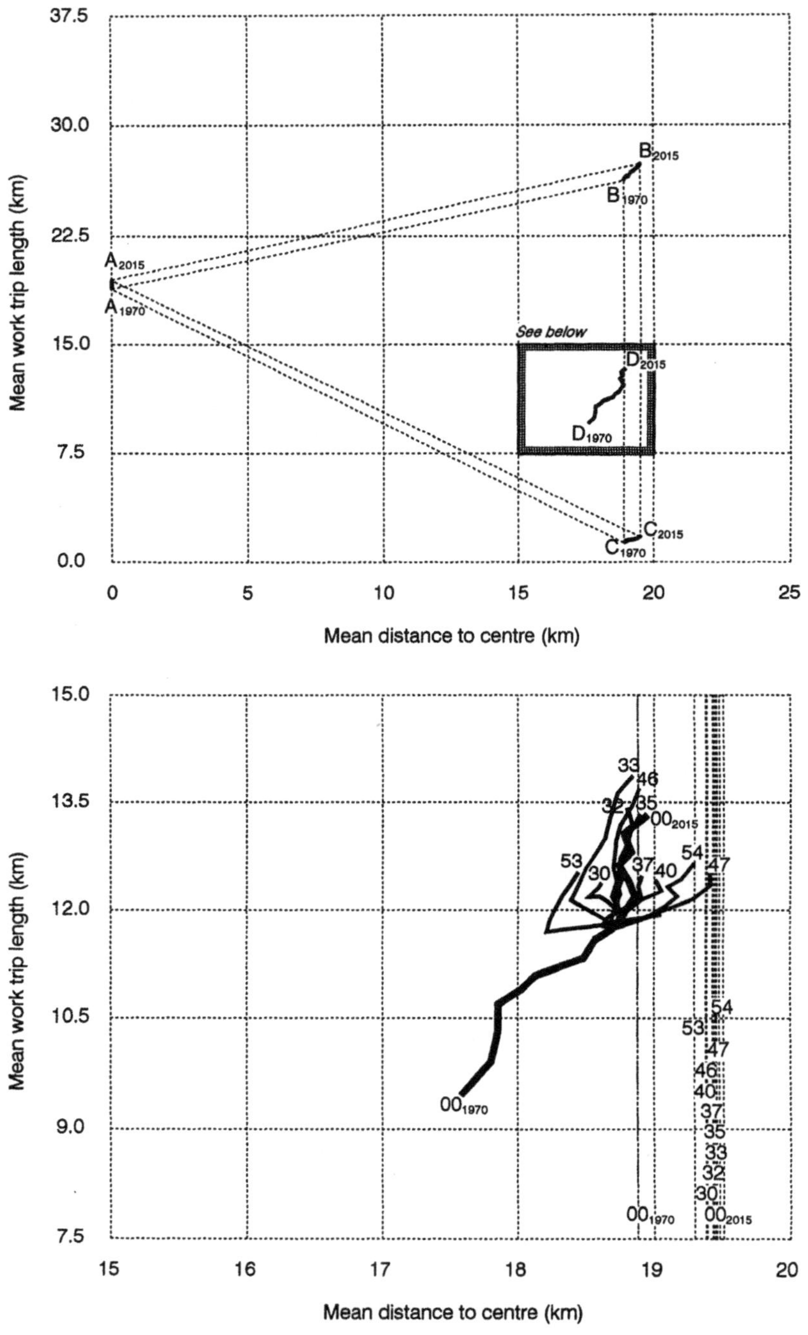

Fig. 8. The 'Brotchie Triangle': Base Scenario (top) and All Scenarios (bottom)

How much of this effect is due to changes in land use rather than travel behaviour? Figure 8 gives some idea. The top diagram shows the 'Brotchie Triangle' of the Dortmund region between 1970 and 2015. It is apparent how the region is drifting apart in the base scenario both in terms of workplaces and residences. The lower diagram is a blow-up of the highlighted area in the top diagram with the trajectories of the other ten scenarios added. The vertical lines at right indicate the line B-C of the Brotchie Triangle, i.e. dispersion of population.

In all scenarios the spatial structure of the region moves towards more dispersal and more travel. Changing the cost of travel (Scenarios 30, 32, 33, 35, 37) has only little effect on location, but substantial effect on distance travelled. If travel speeds are changed (Scenarios 40, 46, 47), the impact on location is stronger. In both cases the direction of change is related to the share of public transport trips (see Figure 7, top). Higher shares of public transport (Scenarios 30, 32, 33, 53) are associated with a more compact city, whereas car-dependent cities (Scenarios 37, 47, 54) tend to be more dispersed. Workplace location responds stronger to transport changes than housing. If travel speeds are reduced (Scenarios 40, 47, 54), workplaces move outward to be closer to residences. This explains why Scenario 40 (in which public transport becomes faster and car travel slower) combines a gain in public transport with more dispersed employment. If car travel costs rise in conjunction with higher public transport use (Scenarios 30, 53), both workplaces and residences centralise. Scenario 53 leads to the most compact cities of all scenarios both in terms of workplaces and residences.

The impacts on mean work trip length are as expected. Higher speeds (Scenario 46) lead to longer work trips, whereas slower speeds (Scenarios 40, 47, 53, 54) result in energy savings. Because of their limited adjustment potential, the savings of worktrips are small compared with those of all trips (see Figure 6, top); this suggests that the largest savings are made with respect to 'voluntary' trips to shopping and leisure. Remarkably, the moderate Scenario 53 has shorter work trips than the radical Scenario 54.

It is frequently argued that increasing the fuel tax as in Scenario 53 would be socially unfair as it would restrict automobility to the rich. Figure 9 (top) looks into that issue. It shows car-km per household per day for four household income groups for Scenario 53. As one might expect, poor households (1) drive less than the middle-class (2) or the more affluent (3-4), but all households increase their distance travelled by car during the 1970s and 1980s. The drop in all four trajectories illustrates that all households are affected by the policies of Scenario 53, but that the more affluent households give up more in absolute terms, with the effect that after 2000 the ratio of car travel between the four household groups is rather more balanced than in the 1970s and 1980s, though on a lower level.

Figure 6 (bottom) shows average travel speeds of all trips by household income group. The two top income groups, because of their level of car

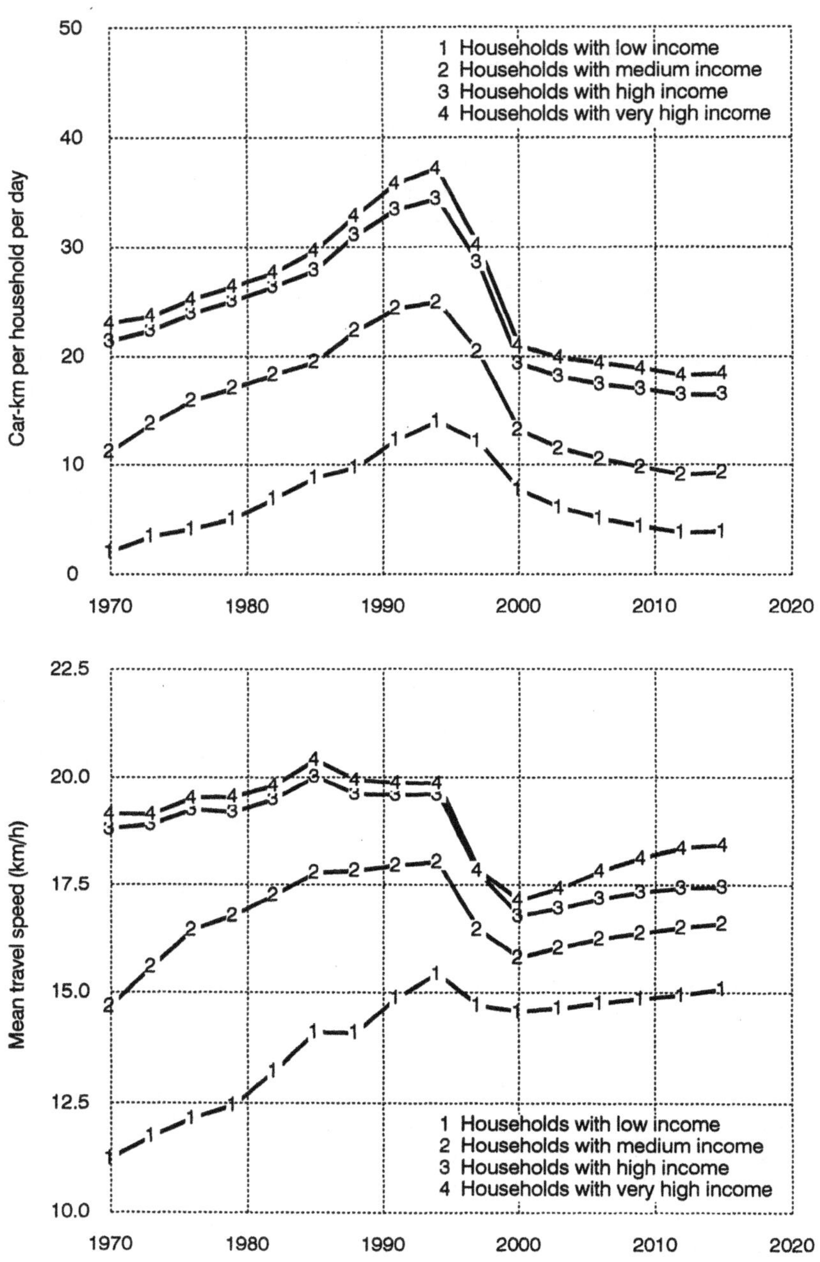

Fig. 9. Car-km Per Household Per Day (top) and Mean Travel Speed (bottom) in Scenario 53

ownership, already in the 1970s enjoyed high average travel speeds. From the mid-1980s even for them no further increases in travel speed have been possible because of increasing road congestion. Households with medium or low incomes, however, have been able to increase their average travel speeds by buying more cars. After Scenario 53 becomes effective, all income groups travel more by public transport and hence more slowly, but when more energy-efficient cars become available after 2000, gradually return to more car travel. As above, the ratios between the travel speeds of the four groups are more balanced after the introduction of the policies of Scenario 53 than before.

A final argument against Scenario 53 might be that the money required to improve public transport as substantially as assumed in the scenario would be unaffordable to local governments. A simple calculation as the one shown in Table 2 demonstrates that, despite the decline in car-km travelled, the additional revenue from the increased fuel tax would be sufficient to more than double the annual expenditure of the regional public transport authority. In combination with the expected increase in fare revenue, this would allow them not only to accommodate the additional passengers by running more trains and buses, but also to significantly improve the quality of service by more comfortable vehicles, more attractive rail stations and bus stops and better passenger information.

Table 2. Financial Impacts of Scenario 53

	Scenario 00 2015		*Scenario 53 2015*
Million car-km/year	7,748		2,578
Fuel price (DM/l)	2.7		12
Fuel tax revenue (million DM)	496		1,289
... Difference (million DM)		793	
... in DM of 1990 (million DM)		380	
For comparison: VRR[a]			
Total revenues (million DM)		148	
Total expenditures (million DM)		367	
Deficit (million DM)		219	

[a] *The Public Transport Authority Ruhr (VRR) is approximately five times as large as the study area. The numbers were scaled down accordingly.*

2.4 Relevance of the Results

The main conclusion from these results is that a combination of policies to increase the cost of car travel and to improve the quality of public transport would permit a significant reduction of energy use and CO_2 emissions of urban transport without unacceptable losses of mobility, without aggravating social disparities, and without additional costs for the public authorities. Other factors not considered in the analysis, such as car sharing (increase of car occupancy), chaining of trips (reduction of number of trips), information and marketing and a potential change of values in the direction of growing environmental awareness, all work in the same direction and would contribute further to energy conservation.

This result implies that the present settlement system of European cities contains a huge unused potential for reducing trip lengths and avoiding trips without fundamentally changing the physical layout of cities. It does not suggest that rebuilding cities does not make sense, but for other, such as ecological, social or aesthetic reasons, and not for energy conservation.

3. Future Work

To further test the hypothesis presented in this chapter, in a current project hypothetical *urban structures* are being compared in the light of new demographic developments, new lifestyles and new transport and information technologies, using criteria such as accessibility, total passenger-km, energy use, land requirement and other environmental indicators with respect to three objectives:

- *Equity*. The urban structure should be equitable. Social disparities should not be aggravated by the spatial organisation of the urban area. There should be no social or spatial discrimination in the distribution of accessibility, amenities or ecological disadvantages.
- *Sustainability*. The urban structure should be ecological in the sense of environmental sustainability. The consumption of non-renewable resources such as energy or land and the pollution of the environment such as noise intrusion, air pollution, water and soil contamination should be as low as possible.
- *Efficiency*. The urban structure should be efficient in that it satisfies the mobility needs of firms and households with as little effort and cost as feasible.

An urban structure is defined in the project as a combination of a *land use system* and a *transport system*:

- A *land use system* is a spatial configuration of dwellings, workplaces and public facilities within an urban area, i.e of land use categories such as high-density inner-city residential areas, large high-rise housing estates, low-density suburbs with detached houses, medium-density mixed-use areas, office parks, industrial estates or greenfield shopping centres. Land use scenarios considered for investigation are

- 'Compact City': intensified urban density in the inner city,
- 'Polycentric City': decentralised concentration in subcentres,
- 'Garden City': dispersed concentration around former village cores,
- 'Auto City': the abandonment of urban centres.

Within one land use system different patterns of activities and spatial interaction are possible. An infinite number of associations of workers and jobs via commuting is possible with the same spatial distribution of dwellings and workplaces. Also with a given distribution of shopping and service facilities an infinite number of spatial interaction patterns for shopping and service trips is possible. Which activity and interaction patterns emerge depends on the transport system connecting the locations of activities.

- A *transport system* is defined by the transport infrastructure, i.e. the road network and the public transport network, including the level of service of public transport, as well as cycling and walking. The transport system is also defined by policies that influence mobility in the form of technical standards, legal or institutional regulations, taxes and fees. Transport scenarios considered for investigation are

  - 'Star Network': radial public transport and highway network,
  - 'Grid Network': rectangular public transport and highway network,
  - 'Mixed Network': radial public transport and rectangular highway network,
  - 'Local Networks': loosely coupled local transport networks.

  Each transport scenario may be combined with different levels of service and fares of public transport and levels of car ownership and car travel costs and speed limits for the road network.

There have been only very few studies in which the social and environmental impacts of different configurations of urban form have been systematically compared. The ISGLUTI study (Webster et al, 1988) examined relatively small modifications of existing land use and transport systems and contained only a minimum of environmental indicators. Rickaby (1987; 1991; Rickaby et al, 1992) used the TRANUS land-use transport model (de la Barra et al, 1984; de la Barra, 1989) to compare spatial configurations of cities with respect to accessibility and energy efficiency, yet the results were inconclusive because of a too limited set of investigated alternatives. Roy (1992), using an analytical model of a circular city, confirmed the hypothesis that spatial reorganisation (moves) can contribute much more to reducing the need for travel than increasing density (rebuilding).

The model used in the analysis is a microsimulation model as a successor to the aggregate urban simulation model described in this chapter (Spiekermann and Wegener, 1992). In the model, decisions affecting the construction of buildings are exogenous, but decisions affecting the location of activities as well as travel decisions are endogenous subject to constraints such as job availability, housing supply, transport costs and traffic constraints. An important element of the new model is its combination with a geographical information system (Spiekermann and Wegener, 1993). The GIS is used to generate artificial urban structures from a spatial database of the Dortmund metropolitan area. Each land use system so created is characterised by empirical data describing the socio-economic composition of the population, the supply of jobs and the locations of public and private shopping, education, health and leisure facilities based on actual conditions in the Dortmund region.

4. Conclusions

This chapter reported on a project in which a land-use transport model was used to explore the impacts of strategies to reduce transport-related CO_2 emissions in the metropolitan area of Dortmund in Germany.

The simulations showed that a combination of policies to increase the costs of car travel and to improve the quality of public transport would result in a significant reduction of energy use and CO_2 emissions of urban transport without substantial land use changes and without causing unacceptable losses of mobility or increasing social disparities or involving additional costs for the public authorities. This result suggests a reassessment of the widely held opinion that the only way to reduce the need for automobile travel in urban regions is a return to mixed-use, compact land-use patterns.

The urgency of the need to reduce energy use and CO_2 emissions in urban areas may grant a new lease of life to comprehensive land-use transport models. This time the models will not be used, as previously, to forecast the direction of urban *growth*, but to guide the spatial *reorganisation* of metropolitan areas towards environmental sustainability. However, if the contribution of the models is to be useful, they must have the requisite variety to respond to the new issues relevant to today's cities. First of all they need to contain the necessary submodels to forecast the impacts of transport and/or land-use policies, not only in terms of travel cost or travel time or accessibility, but also in terms of environmental indicators such as energy use, air pollution, land consumption, noise intrusion and road accidents. Second, they must be able to forecast these indicators with sufficient spatial detail to assess their implications for various social or ethnic groups in the city. Lastly, they must be capable of reproducing the potential substitution between travel and location decisions regarding the spatial pattern of daily activities such as living, working, shopping or education or recreation. Today there are only few models satisfying all of these requirements (see Wegener, 1994). However, recent advances in data availability, modelling techniques and computer memory and speed make it likely that a new generation of urban models will emerge that could live up to the new challenge.

References

Banister, D. (1992) Energy use, transport and settlement patterns. In M.J. Breheny (ed.), *Sustainable Development and Urban Form*. European Research in Regional Science 2. Pion, London, 160-181

Breheny, M.J. (1992) Sustainable development and urban form: an introduction. In M.J. Breheny (ed.), *Sustainable Development and Urban Form*. European Research in Regional Science 2. Pion, London, 1-23

Breheny, M.J. (1995) Counterurbanisation and sustainable urban forms. In J. Brotchie, M. Batty, E. Blakely, P. Hall, P.W. Newton (eds.), *Cities in Competition: Productive and Sustainable Cities for the 21st Century*. Longman Australia, Melbourne, 402-429

Brotchie, J.F. (1984) Technological change and urban form. *Environment and Planning A 16*, 583-596

de la Barra, T. (1989) *Integrated Land Use and Transport Modelling*. Cambridge University Press, Cambridge

de la Barra, T., B. Pérez and N. Vera (1984) TRANUS-J: putting large models into small computers. *Environment and Planning B: Planning and Design 11*, 87-101

Newman, P., Kenworthy, J. (1989) *Cities and Automobile Dependence. An International Sourcebook*. Gower Technical, Aldershot

Rickaby, P.A. (1987) Six settlement patterns compared. *Environment and Planning B: Planning and Design 14*, 193-223

Rickaby, P.A. (1991) Energy and urban development in an archetypal English town. *Environment and Planning B: Planning and Design 18*, 153-175

Rickaby, P.A., Steadman, J.P., Barret, M. (1992) Patterns of land use in English towns: implications for energy use and carbon dioxide emissions. In M.J. Breheny (ed.), *Sustainable Development and Urban Form*. European Research in Regional Science 2. Pion, London, 182-196

Roy, J.R. (1992) Transport efficiency in cities with subcentres. Proceedings of the 6th World Conference on Transport Research, Lyon, 291-302

Schmitz, S. (1991) Minderung von Schadstoff- und CO_2-Emissionen im Straßenverkehr - eine Herausforderung für Raumordnung und Städtebau (Reduction of pollution and CO_2 emissions in road transport - a challenge for regional and urban planning). *Informationen zur Raumentwicklung* 1/2, 1-18

Spiekermann, K., Wegener, M. (1992) The ideal urban structure - efficient, equitable, ecological. Paper presented at the Advanced Studies Institute Transport and Environment, Free University of Amsterdam, 7-11 December 1992

Spiekermann, K., Wegener, M. (1993) Microsimulation and GIS: prospects and first experience. Paper presented at the Third International Conference on Computers in Urban Planning and Urban Management, Atlanta, 23-25 July 1993

Webster, F.V., Bly, P.H., Paulley, N.J. (eds) (1988) *Urban Land-Use and Transport Interaction. Policies and Models*. Avebury, Aldershot

Wegener, M. (1983) Description of the Dortmund Region Model. Working Paper 8. Institute of Spatial Planning, University of Dortmund

Wegener, M. (1985) The Dortmund Housing Market Model: a Monte Carlo simulation of a regional housing market. In K. Stahl (ed.), *Microeconomic Models of Housing Markets*. Lecture Notes in Economics and Mathematical Systems 239. Springer-Verlag, Berlin/Heidelberg/New York, 144-191

Wegener, M. (1994) Operational Urban Models: State of the Art. *Journal of the American Planning Association 60*, 17-29

Wegener, M. (1996) The End of the Industrial City. Berichte aus dem Institut für Raumplanung. Institute of Spatial Planning, University of Dortmund (forthcoming)

CHAPTER 7

SUSTAIN - A MODEL INVESTIGATING SUSTAINABLE URBAN STRUCTURE AND INTERACTION NETWORKS

John R. Roy, Leorey O. Marquez, Michael A.P. Taylor and Takayuki Ueda

1. Introduction

1.1 Recent Background

Immediately after the 1973 Energy Crisis there was a heightened interest in improved understanding and modelling of the interrelationships between urban development and transportation. Such an interest had been awakened already by the apparent neglect of longer term land use adjustments in demand assessments for urban freeways, which were stimulating rapid outward expansion of urban populations, re-introducing congestion as trip lengths increased. Although the resultant major boost in research and development funding did not last very long, progress continued in model formulation and application, especially in the comparative analysis by ISGLUTI (International Study Group for Land Use Transport Interaction) in Webster et. al. (1988). At the same time, politicians and auto-makers responded to increasing public awareness by steadily providing more fuel-efficient vehicles with improved emission standards.

Suddenly, in the 90's, knowledge of the potential for global warming cut through this comfortable scenario of apparent incremental improvements,

Y. Hayashi and J. Roy (eds.), Transport, Land-Use and the Environment, 125–145.

which were actually continuing to be swamped in aggregate by outward urban growth and increasing car ownership. The immediate challenge became to assess the long term sustainability of our urban and regional systems, particularly in the U.S., Canada and Australia. The two main indicators are CO_2 emissions and the relative reliance on and consumption of non-renewable vs. renewable energy resources, followed by maintenance of purity in our soil, water and air. As transport consumes about 30% of our non-renewable energy resources, and about 60% of this is expended in urban areas, the improvement of the performance of our urban land use and transport systems can yield substantial benefits. However, urban systems are inherently complex and idiosyncratic, and any mathematical model which attempts to forecast even the spatial behaviour of such a system must find a way of identifying the elements of the system and their interactions which are pivotal in determining its long term sustainability. For instance, social factors such as increased female participation in the labour force and concomitant reductions in household size can be just as important in influencing total energy consumption in commuting and housing as the more obvious influences, such as the gradual switch to more fuel-efficient vehicles and more energy-efficient domestic appliances.

Existing mathematical models in this field have arisen from two distinct directions, the engineering/planning tradition and the urban economics tradition. The former class is very well represented in Webster et. al. (1988) and the latter in the work of Mills and his school. In some of the models in Webster et. al. (1988), the supply side was being introduced into the submodels of the transport, housing and employment markets, with the corresponding introduction of price and density adjustments enabling them to reflect urban economics principles. At the same time, urban economics models were evolving from their earlier concentration on monocentric cities to a consideration of multi-centred arrangements (Helsey and Sullivan, 1991) and congested transport networks (Yinger, 1993). Nevertheless, although urban economics models can provide general explanations of observable phenomena in urban systems, their usual restriction to continuous systems with analytical solutions and deterministic choice criteria hinders the quantitative extrapolation of their results to actual cities. On the other hand, whilst the most advanced of the integrated land use transport models in Webster et. al. (1988) are becoming consistent with principles of economic equilibrium, their use in cities other than that in which they were conceived and first applied usually requires extensive data collection and computational effort. Also, whereas they perform well in providing guidance on expected land use patterns for a given transport network scenario, they are more cumbersome in providing guidance on actually generating desirable transport network scenarios. Finally, in the comprehensive assessment and reassessment of the multiple land use transport configurations and associated policies now required to come to grips with strategies for sustainability, the use of these models could become extremely time-consuming unless some expert guidance can be provided to reduce the set of spatial scenarios and policies to be tested.

1.2 The Urban Simulation Model SUSTAIN

The SUSTAIN model, which evolved from some preliminary analysis in Roy (1992), itself stimulated by work as early as that of Marksjö and Karlqvist (1970), and more recently by Brotchie (1984) and Anderson, Roy and Brotchie (1986), aims to fulfill the above purpose, and act as a compact interactive preliminary evaluation tool to provide insight and approximate quantitative comparisons of alternative land use and transport policies directed at urban sustainability. By rapidly 'creating' alternative cities, which, whilst having certain idealised geometry, still reflect average housing densities and their distributions, as well as associated average travel times, chosen from those observed in different categories of existing cities (Newman and Kenworthy, 1989), the model can start educating the user on the potential benefits attainable from alternative strategies. For instance, Figures 2 and 3 indicate that if outward 'sprawl' reactions can be reasonably contained, employment decentralisation policies can have a potentially greater effect on reducing per capita energy consumption in commuting than strong urban consolidation policies. Whilst the latter have identifiable benefits in the longer term, they may end up playing a more supportive role, rather than the major role, in the moves to make lower density cities more sustainable by 2005.

Other related models are being developed as educational tools for integrated planning of land use and transportation. These include Young's LAND model (see Chapter 8), as well as the PLUTO model of Bonsall (1994). Whilst LAND, PLUTO and SUSTAIN all have clear educational functions, SUSTAIN is distinguished by its further function as a preliminary design aid.

The model is structured such that, initially, the planner can work through an educative (calibration) phase, creating widely different cities from scratch, making various land use and transport changes and comparing results. He will gain insights on the sensitivity of the system's environmental performance to alternative land use patterns and transport networks, identifying those which appear the most spatially efficient and equitable. The next phase of the model starts with the planner choosing a base case configuration, usually denoting a representation of an existing city. He can then adopt a macro future growth scenario, within which, guided by the insights acquired earlier, he can experiment with various sets of land use and transport planning policies, pricing policies, technological changes and societal changes, to try to 'encourage' the growing city to move towards CO_2 and energy sustainability targets. Armed with this knowledge, the planner is then better placed to devise specific land use constraints and policy options, as reasoned input to a comprehensive integrated land use transport model using data for an actual city (Webster et. al., 1988).

The next section discusses desirable features of a simulation model capable of making a preliminary assessment of alternative options to move towards urban sustainability, which are being embodied in SUSTAIN. Then, the structure of the SUSTAIN model program is described, including its modularity and options for interactive use. Finally, intended improvements are outlined, achievable in our framework of international cooperation.

2. Elements of a Compact Urban Simulation Model

2.1 Urban Form and Housing

Clearly, in the search for an interactive model specification with modest data and computational demands, certain system idealisations must be made. Whilst some of these must be firmly set, others may evolve upon experience, especially if the model structure is modular. The main objective is to develop a model which, whilst not purporting to estimate system response in an absolute sense, at least yields an estimate of the relative order of magnitude of the changes induced by alternative policies. Important properties are now described.

2.2 Geometrical Shape

As many cities are centred around a port or at key crossing points of a river, it seems reasonable to consider a circular urban form. This is consistent with transit lines and freeways converging radially towards the CBD (Central Business District). If urban subcentres develop, these can be connected to each other by ring roads, which can reasonably be regarded as circumferential. Although local roads and some arterials are often arranged in a rectangular grid, the former are usually treated as ubiquitous in a model of this type, and a radial/circumferential approximation is not serious. However, there are some approximations in assuming that the arterials are either radial or circumferential, which could be handled by application of modest distance penalties. If a fully circular form is adopted with a symmetric radial/circumferential transport network and a radially symmetric distribution of housing and suburban employment, the entire trip pattern is radially symmetric. This has the unique advantage that the analysis can be confined to one sector only of the city enclosing its representative subcentres, yielding dramatic savings in computational effort and data input requirements, supporting convenient interactive use. The current version of SUSTAIN is based on this fully circular form. Note that, alternative radially symmetric forms are available, such as described in Rickaby (1987). As a future option, a non-radially symmetric capability is being introduced, allowing SUSTAIN to perform preliminary analyses of existing cities.

2.3 Housing Distribution

The most universal attribute of major cities is the increase of land price and concomitant housing density as the CBD is approached. Occasionally variations of this trend may occur because of desirable waterfront locations and hilly terrain. However, although unit land prices are often more expensive in such areas, the subdivisions are often designed to attract more affluent households, who can afford to retain the low density housing environment. Of course, if significant suburban subcentres develop, adjacent land usually rises in price, and densities may increase locally, but usually not to levels approaching those surrounding the CBD.

As a result, SUSTAIN has grouped housing into three rings, with decreasing density (user-specified) as one moves outwards: 1. the inner core, 2. the middle ring and 3. the outer ring. In addition, along ring roads where employment subcentres are located, a user-specified annulus of higher density housing can be defined, which may be coordinated with good access to public transport. Also, we intend adding a fourth annulus of housing at a specified distance beyond the urban periphery (i.e. potentially allowing for empty space in between) to simulate deurbanisation trends.

Because much of the lower density outer housing is occupied by families with children, a different average household size is allocated in each of the three rings. This variable can also be manipulated in the forecasting phase, to reflect factors such as deferral of the age of initial child bearing and a reduced number of children for upwardly mobile professional couples. The higher average household sizes in the outer ring imply that somewhat lower densities are feasible in these areas.

2.4 Spatial Distribution of Employment

The increasing dispersion of employment away from the CBD, which is an observable phenomenon in many large cities, seems to be occurring with or without active encouragement by planners. Thus, in a general purpose model, it is essential to examine alternative forms and levels of such dispersion. Also, as spatial mismatches between housing and jobs remain large in many cities, there is certainly potential to considerably reduce commuting trip lengths, thus lessening peak period congestion and pollution. Whilst it is true that commuting accounts for only 30-40% of urban travel, it is critical environmentally, because (i) its reduction is the most amenable to land use changes, (ii) it is usually non-discretionary and (iii) it occurs at (or defines) the peak, where the most serious congestion occurs and where the maximum strain arises on the capacity of the public transport system. Also, as more and more households are having both partners in employment, many previous home-based trips are becoming work-based or performed on the way home from work, and related to efficient employment locations. Thus, the SUSTAIN analysis will concentrate on work trips and the location of employment with respect to housing. Other trips, such as for shopping, may enhance sustainability more through transport and telecommunications innovations, such as local demand-responsive para-transit travel for convenience shopping, and increased teleshopping supported by home deliveries for comparison shopping. Para transit modes may also become important for the many non-radial trips to employment subcentres, enhancing the sustainability of urban forms with dispersed employment.

As a key constraint to improved spatial matching of housing and employment is the housing budget of lower income households, workers and their associated jobs should be subdivided into at least three classes. In SUSTAIN, these are classified as (i) knowledge-based workers, (ii) service workers and (iii) industrial/factory workers. This subdivision also relates to the more diverse (and scattered) job options of the knowledge-based workers, as well as their greater potential for tele-commuting. With this stratification, spatial housing/job mismatches can be more easily identified, as well as options for their potential amelioration. Also, policies which may be good overall, but potentially entrap poorer households in inaccessible areas, can be guarded against.

3. Transport Networks and Vehicles

3.1 Network Specifications

In the defined circular city, the transport network clearly is made up of radial and circumferential links. For the road network, it is reasonable to subdivide the roads into three classes (i) freeways, (ii) urban arterials and (iii) local streets. Generally, radial freeways will connect the major subcentres to the CBD, with radial arterials performing the same function for the smaller subcentres. Similarly, the major subcentres will be interconnected circumferentially with ring road freeways, with ring road arterials interconnecting the minor subcentres. For public transport, it is reasonable to specify major radial links as transit or buses, minor radial links as buses and circumferential links as buses or light rail. Connections to stations may be made with car, mini-bus, cycle or by walking. The assumed radially symmetric form of the transport network provides major economies in network definition. For the non-symmetric option, we have two possible forms of the transport network (i) a rectilinear form or (ii) a radial/circumferential form centred on the CBD, with partial ring road sections and freeways.

3.2 Vehicle Types, CO_2 and Energy Use

For cars, the main variables are average passenger occupancy and fuel consumption functions in terms of average speeds. These can be readily converted to CO_2 production and primary energy use, allowing for energy losses in refining. Whilst buses are handled in a similar way to cars, transit must consider the primary energy penalty for producing each KwH of electricity, which can be large for inefficient fuels, such as brown coal. In addition, all public transport values should be corrected for empty seats.

3.3 Job Choice and Mode Choice

Typically, job choice and mode choice behaviour is modelled using entropy/gravity models when just Census data is available, or logit models when individual surveys can be undertaken. In a preliminary analysis model such as SUSTAIN, the aggregate approach seems more plausible. As demonstrated by Evans (1976), the extremum version of the non-linear entropy model reduces to a linear transport cost minimising objective, which represents all commuters choosing an available job, consistent with the capacity constraints, in the centre reachable by the lowest generalised travel cost. The corresponding result for mode choice is use of the mode with the lowest generalised cost. These may be denoted as the deterministic cases. At the other extreme, we have random or travel cost-indifferent job or mode choice, where a worker's probability of choosing a job in any given job centre is equal to that centre's share of the total employment. In practice, job choice and mode choice are neither deterministic nor random, but stochastic. However, as the actual behaviour must be bounded by the deterministic and random results, the spread between the two for a given housing/job configuration represents the theoretical scope for adjustment which the particular urban form has, in the shorter term, to increased energy prices or decreased energy availability. Thus, SUSTAIN can handle any of the three options:—

Deterministic Stochastic Random

separately for job choice or for mode choice. If simulating an actual city, observed average travel times and costs can be used to calibrate the stochastic model, typically the doubly-constrained entropy model (Wilson 1970) for job choice. A further feature is to calibrate an impedance parameter β_i for each of the residential rings i, using for our 'experimental' cities, plausible intermediate cost values, eg. $C_{exp} = 0.6\ C_{min} + 0.4\ C_{max}$, obtained from the deterministic and random cases respectively. Radial symmetry produces dramatic computational savings for the deterministic case and significant savings for the random and stochastic cases.

As shown in the 'urban triangle' of Brotchie (1984) and later analytically in Anderson, Roy and Brotchie (1986), whilst increases of spatial employment dispersion may dramatically decrease average trip lengths when workers tend to find and choose jobs in the subcentre nearest home, it can somewhat increase average trip lengths when the workers are particularly discriminating in their choice of job, virtually irrespective of travel costs. Thus, in practice, employment subcentres need to be sufficiently large and diverse to contain suitable job options for most of the surrounding residents — unless smaller subcentres maintain this diversity, the apparent accessibility gains of their closer spacing may be illusory. Agglomeration economies for industry are a further reason for promoting larger centres (Helsey and Sullivan, 1991). Whilst experience in parts of the U.S. indicates that job dispersion is automatically decreasing average trip lengths (Gordon et. al., 1989), the identification in each zone of the job preferences of workers who commute the furthest may guide firms in their future location decisions, especially those in the more 'footloose' sectors. The SUSTAIN model can reflect these actions by increasing the calibrated gravity parameters β_i of the job choice model, independently for the two sets of workers.

Although SUSTAIN groups jobs in distinct employment centres, the residential rings are partitioned into discrete 'zones' of area ($r\ \Delta r_i\ \Delta\Theta_i$). Recognising the approximations implicit in the designation of three rings of distinct housing density, spurious accuracy should not be sought by making too fine a zonal subdivision. On the other hand, a reasonably fine representation does allow the zonal centroids to be readily used as nodes of our transport network for the assignment phase. Also, to simulate the variety of urban settlement patterns illustrated by Rickaby (1987) [eg. spoked forms, satellite towns etc.], we may in future denote alternative patterns of these zones to be empty of housing.

At this stage, SUSTAIN performs the mode choice step contingent on job choice. Many recent studies have shown that there is a complex interrelationship between job and mode choices within the household unit, where mode availability, car ownership, the job locations of family members and the spread between work starting and work finishing times all interact.

Such behaviour can be simulated using Monte Carlo techniques. These features could potentially be included in the SUSTAIN framework in scenarios where particular policies to influence mode choice were being investigated.

Finally, for each ring and each category of worker, the percentage captive to both public transport and private transport will be provided. A reduction in the proportion captive to public transport is the mechanism by which SUSTAIN handles expected increases in car ownership.

3.4 Route Assignment: The Road Network

Although congestion on urban road networks is often rationalised as a necessary deterrent to excessive travel, it is wasteful not only of time but also of fuel (Biggs and Akcelik, 1986). Whilst the application of narrow transport planning approaches has failed to solve this problem, the use of integrated land use transport planning techniques offers some promise. After all, a freeway with small fuel-efficient cars and mini-buses with high passenger occupancy moving smoothly at 70 to 80 kph is a highly efficient form of urban transport, so long as the average trip lengths are not excessive. Further efficiencies are provided by a high level of reverse commuting. Both of these benefits are a potential consequence of integrated land use transport planning. Of course, an automated guideway system would increase the carrying capacity dramatically, because of reduced headways.

Thus, although a model such as SUSTAIN must consider the modelling of congestion in the road network, it need not consider a network with thousands of links. Firstly, with the important assumption of radial symmetry, the analysis can be confined to one sector (typically 30°, 45° or 60°). Secondly, if the roads are classified into three classes:

Local Roads   Urban Arterials   Freeways

and if feed-back congestion effects are ignored for local roads, the congestion analysis can be confined to the urban arterials and freeways. This will yield efficiency in the route assignment algorithm, and enhance the interactive operation of the model. Currently, average travel speeds on the different road types in each of the three residential rings are provided exogenously to the model, based on observed speeds in comparable cities. At the moment , a congested equilibrium assignment algorithm is being tested prior to inclusion in SUSTAIN.

3.5 Public Transport

Currently, SUSTAIN uses exogenous speeds on the train and bus networks, as well as exogenous waiting times. A possible improvement is to reduce waiting times by increasing trip frequencies where demand is high, allowing iterative feedback on mode choice. Another improvement is the consideration of demand responsive para transit modes (eg. mini buses and shared taxis), especially in outer low density areas. They could transport commuters from their homes to combined suburban employment/transport nodes, where they would either exit or change to transit or circumferential light rail to reach their final destination.

3.6 Residential Location and Re-Location

Whilst it is useful in the calibration phase of SUSTAIN to adopt a scenario approach to residential location, allowing a very wide set of alternatives to be tested, the forecasting phase needs more sophisticated treatment. Of course, if one were examining the possible evolution of a city say from 1990 to 2005, it would be certainly instructive to postulate alternative internal migration moves, different levels of urban consolidation and different degrees of outward expansion at the urban periphery in relation to their expected energy consumption and CO_2 emissions. For those sets of changes which produced good environmental outcomes, it would then be necessary to work backwards, examining the combinations of physical planning policies and pricing instruments which may of themselves lead the city to evolve toward such configurations. Two approaches to such an analysis are being adopted (i) extensions of recent work of Mohring (1993) for the symmetrical city and (ii) adopting work in Japan on random bidding approaches from Ueda, Nakamura and Shimizu (1993) to the unsymmetrical city. The approach of Mohring (1993) is generalised by (a) allowing jobs in employment subcentres rather than just at the CBD and (b) replacing a continuous system by a discrete system of residential zones. The average accessibility of each zone is a log-sum average of its relative accessibility to each employment centre. Cobb-Douglas utility functions are used, and the threshold level of utility is adjusted such that all households are accommodated , and the outer boundary adjusts so that any rent discontinuity is removed between the outermost zone and the rent of developed agricultural land, with the maximum bidder always being allocated the land. For the more realistic unsymmetrical city option , the above deterministic bidding framework is replaced by a random bidding approach (Ueda, Nakamura and Shimizu, 1993) , the extremum solution of which is identical to that obtained in the deterministic procedure.

It is well known, and demonstrated elegantly in Parable 3 of Mohring (1993) , that if the access of workers to employment is improved, either by provision of more efficient transport or by bringing compatible jobs closer to workers, average housing densities will tend to decrease, potentially cancelling some or all of the apparent energy savings. This may be further accentuated, if, as intended, we replace trip cost by generalised cost, including the value of time. Of course, this may not necessarily be bad if appropriate jobs move out in unison. Alternatively, one may respond by imposing rigid physical planning constraints on further outward`movement or cost penalties for inward commuting across certain thresholds. However, as such measures may often produce dramatic falls in consumer surplus for many households, they should be used as a last resort if all else fails. Thus, an existing challenge is to experiment with a model such as SUSTAIN, to see what other transport investment and pricing policies, differential provision of public facilities and infrastructure and milder physical planning constraints, such as housing density zoning, can achieve in moving the city less painfully towards the desired targets. In many ways, evaluation of the slower adjustments of residential location behaviour to transport , job and service access changes is one of the most subtle and controversial areas in the current urban environmental debate, an area where pooled international experience is indispensable.

3.7 Response to Land Use and Transport Policy Instruments

The calibration phase of SUSTAIN gives the user/planner the opportunity to create alternative urban forms and their associated transport networks from scratch, potentially homing in on alternatives which appear to have good sustainability properties. However, in the forecasting phase, when he starts with a representation of an actual city, usually with a deficient environmental performance, he is faced with two tasks (i) identification from the calibration phase, which of the more desirable alternatives represent achievable targets (eg. by 2005) for the actual city under study and (ii) identification of the sets of policy instruments which may be applied to encourage the city to move towards such desired forms. The latter task is clearly a trial-and-error process, influenced not only by the apparent effectiveness of the policies, but by their social and political achievability. It is this final task which should help to build judgement on ranking the priorities of the diverse policy alternatives. Such judgement should be invaluable when the planner moves from SUSTAIN to apply one of the coordinated land use transport models to the city, where model specification and data demands are considerably higher, and where the preliminary assessments made through use of SUSTAIN may yield substantial dividends.

4. Structure of the SUSTAIN Program

4.1 An Urban Simulation Model with Object-Oriented Design

With the strong requirement for transparent interactive use, as well as flexibility in model development and application, a decision was made to introduce a Windows-based system using the object-oriented programming philosophy inherent in the C++ language. As SUSTAIN is probably one of the first land use transport models to adopt this new approach (which is also being applied in the development of the new TOPAZ model at CSIRO), its relevance is discussed in some depth, whereby certain parts (i.e. 'objects') are free to evolve relatively independently over time, leaving the rest of the system intact. The advantages and properties of this structure for SUSTAIN are now summarised.

SUSTAIN employs an object-oriented design (OOD) which offers several features that reduce the complexity of the model, expand its capability and generality, and simplify the operations needed to manipulate and maintain it. SUSTAIN starts with any given city as a top-level object. In OOD, the properties of an object are divided into two groups, the data members or attributes of the object, and the functions or services that are performed on these attributes. The data members represent the quantitative and qualitative properties of the class that are encoded in the variables, data structures and even other objects associated with the class. The member functions specify the manner in which the properties of the class change in response to conditions in the environment. With the OOD approach, any given city can be constructed from some generic class CITY, which possesses the attributes and services common to all cities. The property that makes the object-oriented approach different from other models that employ a top-down design is that the services or functions are defined as being internal to the object, not as part of the environment in which the object exists. With the object-oriented approach, SUSTAIN aims to encapsulate the structure and behaviour of any object of class CITY, to allow the modeller/planner using SUSTAIN to treat any CITY as a self-contained entity.

Listing 1 shows an abbreviated definition of the class City as implemented by SUSTAIN. The attributes of CITY include simple numeric values ("totHouseholds", "totWorkers", "nResRings") and character strings ("city_name").

More importantly, every instance of CITY is defined to possess a transportation network ("pTransNet"), an employment market ("pJobsMarket") and a set of residential rings ("resRings"). These attributes are themselves objects of the classes TRANSNET, JOBSMARKET and RESIDRING, respectively. Notice that TRANSNET, JOBSMARKET and

RESIDRING are themselves defined in terms of other classes. This results in a hierarchy of objects for the class CITY as illustrated by the diagram in Figure 1.

Listing 1. Definition of Class CITY

```
class City
{
   public:
      //data members
      char city_name[MAX_NAME_LENGTH];
      TransportNet* pTransNet;// pointer to the transport network object
      JobsMarket* pJobsMarket; // pointer to the jobs market object
      ResRing_Array resRings; //an array of residential ring objects
      // Constructors/destructors functions
      ~City() { }
      // Implementor functions
      double get_TotalHHolds();
      BOOL set_popn (ResRingInfoW*, BOOL, long);
      BOOL set_popn(ResRingInfoW*, BOOL, float, int);
      BOOL set_popn(ResRingInfoW*, BOOL, long, float, int);
      // trip functions
      void computeTrip(float, float, float, float, int, Trips*, int, int);
      void get_energy(float, float, TransPortMode, int, int, float*, float*);
      void RunTrips();
      // Accessors
      double get_popn() { return popn;}

   private:
      // Data members
      double popn; //city population
      int nResRings; //number of residential rings
      double totHouseholds; //total number of households
      double totKBWorkers; // total number of knowledge-based workers
      double totManWorkers; //total number of manual workers
      double totWorkers; //total number of workers
      // Implementors
      BOOL build_resrings(InputData& Indata);
      void update_sections(int resr);
      void clean_up();
}; // class City
```

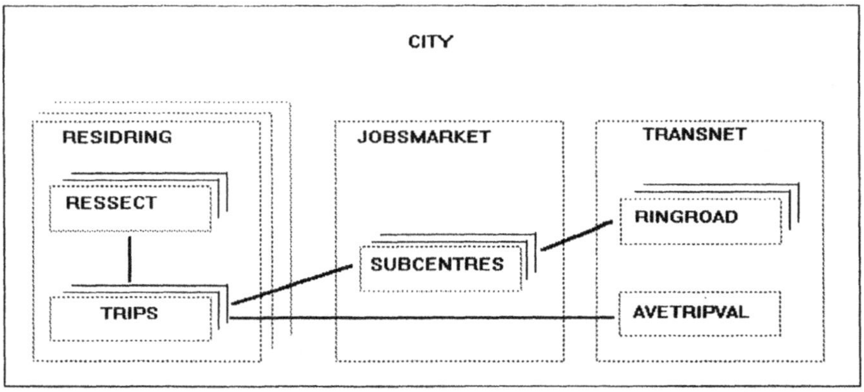

Fig. 1. Hierarchy of Objects in CITY

Apart from their attributes, CITY objects and sub-objects possess a rich set of functions that mirrors the complex interactions in an actual city. Encapsulation allows these functions to process entire groups of objects that may belong to different classes. For example, the trips made by workers in a residential section in a residential ring are computed based on the properties of the transportation network and the location of the subcentres where the jobs are located. The "computeTrip()" function of the CITY for a given RESIDRING will therefore call on the RESSECT object to provide the number of workers making the trip, on the TRANSNET object to provide the speeds used in the trip and on the SUBCENTRE object to provide the coordinates of the destination. The "computeTrip()" function then calculates the trip values, updates the average trip time in the TRANSNET object and finally increments the number of workers that went to the SUBCENTRE object.

Aside from designating the properties that a particular object possesses, encapsulation also specifies the level at which these properties are made visible or accessible to the outside world. This implied ability of objects to conceal their properties from the outside world, aptly called information hiding, prevents the modeller/planner from having to understand and unnecessarily deal with the complexities of the object. In return, the object is protected from unexpected intrusions into its domain. As a result, no matter how complicated are the structures underlying the transportation network, employment market and residential rings, the modeller/planner can choose (and may be allowed) to manipulate only those CITY members that are of direct relevance to him.

A class specifies the accessibility of a component by designating it as either public or private. Public class members are accessible to all functions and objects. The public designation is used for those class members that have global connections in the program. At the other extreme, private class members are only accessible to the object itself. The private designation is applied to those attributes and functions that are limited to the internal use of the object.

To illustrate, Listing 1 designates the three "set_popn()" functions of CITY as being public. This is because the population of the city can be set from several points; when a new city is created from an input file, when a new city is constructed through interaction with the modeller, or when the population of an existing city is changed. Since the calls to "set_popn()" can be requested by objects other than CITY, "set_popn()" is made accessible publicly. On the other hand, the "build_resring()" function has been assigned private accessibility because only the CITY object should be allowed to build its own residential rings. Actually, the "build_resrings()" function only defines the grouping of the residential rings. The details of how each residential ring and the residential sections within the rings are constructed are again hidden from the rest of the CITY by the RESIDRING and RESSECT classes, respectively.

The "set_popn()" function also illustrates another feature of object-oriented design called function overloading. In function overloading, functions and operators (such as <, >) that perform similar tasks can be defined to use the same name or symbol, as long as their parameters can be differentiated from each other. Thus, CITY has three "set_popn()" functions; one for each of the conditions that can produce the function call. Ambiguity is avoided since the "set_popn()" functions are distinguished by the arguments that are passed to them. More importantly, the modeller is able to economise on the number of functions used in the model. Function overloading also allows regular mathematical operations to be performed on objects. This feature makes it theoretically possible to add two cities, subtract a ringroad from a transportation network, or determine if two residential rings are equal.

Finally, the forecasting phase of SUSTAIN applies one of the most important characteristics of the OOD approach called class derivation . Class derivation refers to the ability to define new (child) classes from previously defined (parent) classes. The "child" class automatically inherits the properties of one or more "parent" classes, with perhaps modifications, in addition to new properties that are bestowed on it. For example, in SUSTAIN, a FUTURECITY object is constructed from a given CITY object to represent the status of the CITY object at some future point in time. The definition of the FUTURECITY class is then simply stated as:

```
class FutureCity: public City
{
   public:
      FutureCity () : City(HWND hwnd, InputData& Indata) {};
      FutureCity() {};
      compute_energy_change(City& base);
}
```

The above definition means that a FUTURECITY object can be constructed as completely identical , down to the last component, to a given base period city. Since the FUTURECITY object is itself a CITY object, the FUTURECITY object can also perform the same services that are performed by CITY objects. For example, "set_popn()" can be performed to give the FUTURECITY object a forecast higher population. Other changes can be administered to FUTURECITY while keeping its original state in the base. When all the changes have been applied , the FUTURECITY object can then compare itself with the base CITY using its new set of services.

In summary, SUSTAIN models a CITY using a hierarchy of objects and subobjects. The objects are defined to reflect the separate, self-contained but interacting components of a theoretical or real city. The complexity of the city representation is significantly reduced by encapsulating the properties of these components in classes and then constructing the needed objects from these classes. At any level in the hierarchy, information that is not of direct relevance to the modeller is hidden from him. Functions and operations that process different objects can be defined. New cities can be derived from existing cities to reflect future scenarios of development.

4.2 Principles of Model Operation

The initial phase of SUSTAIN is the 'calibration' or experimental phase, in which the user can create widely different urban forms with widely varying public and private transport networks and vehicles. The four key items of output are:

Trip Distances   Trip Times   Transport Energy CO_2 Produced

Graphical images depict the spatial variations of these values over the city sector being analysed. In the future, the emissions of other pollutants will also be included.

As the locations of ring roads (freeways or arterials) and their associated employment subcentres turn out to play a key role in transport energy use and CO_2 production, SUSTAIN provides an option to automatically move the radius of the ring road incrementally between two nominated extreme positions, identifying that which is the most transport efficient. Also, maximum traffic volumes are computed on the ring roads, as well as relative levels of radial and reverse radial commuting on the radial freeways, with the aim of recognising land use patterns yielding more balanced patterns of network loading. These loadings already indicate rough requirements for freeway capacities, prior to completion of the congested assignment algorithm. The planner has the option of interactively changing most of the parameters described in the previous chapter, gradually acquiring a knowledge of those which most strongly influence energy use. This quantitative/qualitative knowledge is one of the key benefits expected from manipulation of the model.

In practice, planners rarely have the 'luxury' of creating a new city from scratch, but are usually stuck with a seemingly recalcitrant city, which grew in times when the importance of environmental properties was unrecognised, and contains large fixed infrastructure investments which often promote waste, rather than conservation of resources. In the 'forecast' phase of the model, an attempt is made to simulate the current structure and activity pattern of an actual city, reflecting net average housing densities and their radial variation, the locations and magnitudes of CBD vs suburban employment, average speeds on the transport network, household attributes, vehicle properties and occupancy rates, the distinction between the two classes of workers and their jobs, the average trip lengths for commuting out of the inner, middle and outer zones, route capacities, fuel prices, public transport fares, modal split and % captive variations. Then, the planner is encouraged to look forward (say to 2000 or 2005) and to adopt a two level approach:

- Top Level
  Input total expected population growth to forecast period
- Lower Level
  Under the given 'global' scenario from the top level, adopt successively different sets of land use transport scenarios and policies, checking their sustainability properties and the relative improvements (or otherwise) with respect to the base period scenario.

Of course, the experience gained in the rapid creation, analysis and comparison of candidate cities in the calibration phase should assist the choice of well-coordinated sets of scenarios in the lower level forecast trials.

In the new version, residential location changes will not need to be specified exogenously, but will occur in an urban economics framework similar to that developed by Mohring (1993) and Ueda, Nakamura and Shimizu(1993), with the addition (optionally) of certain physical planning constraints, which may be varied in scope and intensity.

4.3 Some Preliminary Results

This section illustrates some early results of the model, restricted, for conciseness, to comparison of cities with the same population, that is, 2.5 million inhabitants.

In Figures 2 and 3, the results of Roy (1992) are enhanced by (i) including two ring roads and optimising their location simultaneously to minimise travel energy consumed (ii) defining two categories of workers and jobs, (iii) including an annulus of higher density housing around each ring road and (iv) adding modal split. This is performed for the 'Australian' city, with three rings of high, medium and low density housing respectively and for the 'Western US' city, with three rings of high, low and low density housing respectively. Both job choice and mode choice are taken as deterministic, with the number of jobs in each centre determined by the model such that all workers find jobs at the centre reachable by them at least generalised travel cost. Note that, each input data file of this calibration/testing phase represents a candidate city on which the planner can experiment with a wide range of transport and job location changes, creating modified 'what if' cities, without having to trace a path of transition from the input cities. The main results are classified as:

- Effect of Number of Subcentres.

  As the total number of subcentres increases from a moderately dispersed form (6 subcentres) to a more highly dispersed form (12 subcentres), we have the following per capita reductions in travel energy consumption.

  Australian City 16.4% US City 14.1%

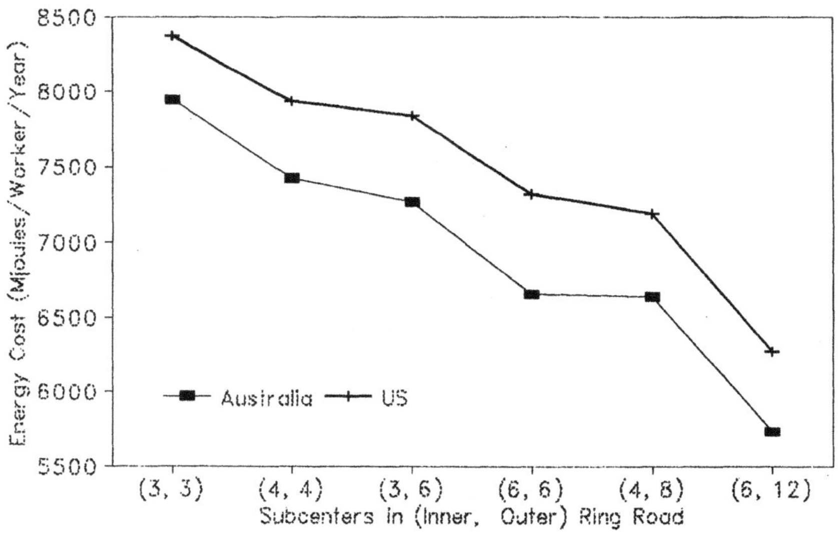

Fig. 2. Per Capita Energy Cost vs. Number of Subcentres

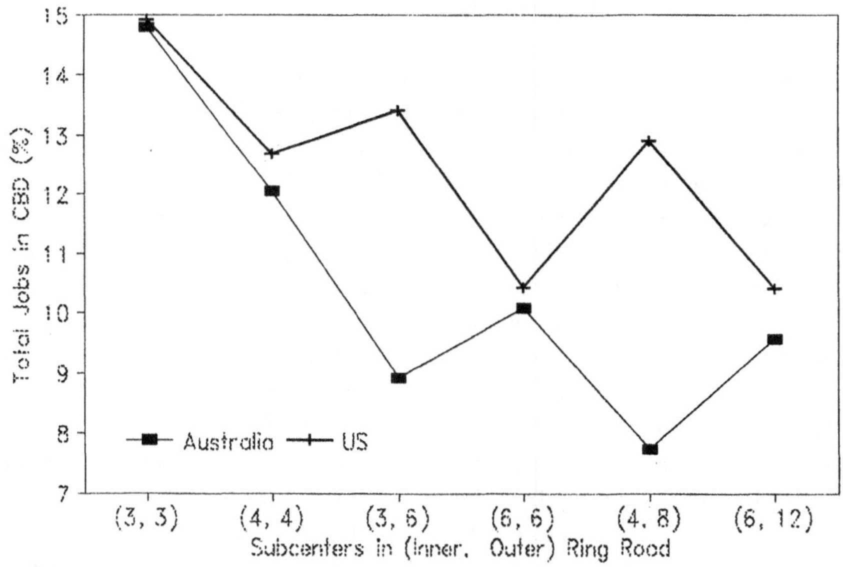

Fig. 3. Percentage of Total Jobs in CBD vs. Number of Subcentres

•Effect of Urban Consolidation.

If the middle ring of the US city were increased from low to medium density, yielding the Australian city, the effect would be

Average Travel Energy Savings 6.6%

•Relative Size of CBD.

Assuming that the two ring roads have been located independently for each case to minimise travel energy, their resulting outward shift for the US cities gives the following optimal % of CBD employment to total employment

US City 10.4 — 14.9% Australian City 7.7 — 14.8%

From the above, it is seen that the influences of moderate increases in employment dispersion may readily reduce per capita travel energy much more than quite considerable levels of urban consolidation. If the former policy produces more sprawl in the longer term, it may be contained by pricing policies and/or physical planning controls, or new jobs may drift out in sympathy. This phenomenon can be examined when the land price adjustment mechanism is added. Also, from the standpoint of transport efficiency, most CBD's are too large. Of course, if electricity generating losses could be strongly reduced, encouragement of transit would increase the optimal size of the CBD. In practice, the CBD has a special cultural/entertainment/financial role, and jobs in such sectors should be allocated there prior to the analysis.

5. Some Concluding Remarks

Certainly, the assumption in SUSTAIN of radial symmetry in networks, housing and employment, as well as the representation of the varying housing density as discrete values in three distinct rings represents a considerable simplification of many observed urban forms. However, we hold that many of the interesting questions on urban sustainability can be examined and their influence weighted relatively by analysis of such idealised cities, with the comparisons still being relevant for actual cities. With SUSTAIN often expected to precede the application of comprehensive land use transport models to an actual city, there remains the chance to use the future unsymmetrical version to account for special properties, including airports and significant natural features compromising radial symmetry. The potential educative benefits of using SUSTAIN can only help the planner approach the application of the comprehensive models more intelligently.

Two of the most interesting interactions in urban systems are route congestion and effects of changing accessibility on land prices, density and residential location. These subsystems are under preparation. In addition, when proceeding from a base case scenario (say 1990) to a future period (say 2005), it is necessary to evaluate the required public infrastructure investment corresponding to each chosen future scenario. Also, expected changes in consumer surplus should be monitored. With the rigorous determination of surplus benefits associated with entropy demand models performed by Miyagi and Morisugi (1993), consumer surplus reductions due to any restrictions on the scope of job choice can be evaluated from our trip distribution models. Rather than merely examining the alternatives with lowest energy use and CO_2 emissions, we may need to make some trade-offs between energy conservation goals and consumer satisfaction.

References

Anderson, M., Roy, J.R. and Brotchie, J.F. (1986). Some Aspects of Adaptability of Alternative Urban Configurations, *Environment and Planning B*, 305-318.

Biggs, D. and Akcelik, R. (1986). Car Fuel Consumption in Urban Traffic, *Proceedings of 13th Conference of the Australian Road Research Board*, 7, 123-132.

Bonsall, P.(1994). Computer-Aided Training for Planners - Experience with the PLUTO Package, *Proc. Conf. Computer - Aided Learning in Engineering*, Univ. of Sheffield.

Brotchie, J.F. (1984). Technological Change and Urban Form, *Environment and Planning A*, 16, 583-596.

Evans, S. (1976). Derivation and Analysis of Some Models for Combining Trip Distribution and Assignment, *Transportation Research*, 10, 37-57.

Gordon, P., Kumar, A. and Richardson, A. (1989). The Influence of Metropolitan Spatial Structure on Commuting Time, *Journal of Urban Economics*, 16, 138-151.
Helsey, R.W. and Sullivan, A.M. (1991). Urban subcentre formation, *Regional Science and Urban Economics*, 21, 255-275.
Marksjö, B.S. and Karlqvist, A. (1970). Mathematiska Modeller for Fysisk Planering. Meddelande 3, *Regionalplanekontoret*, Box 12257, S-10229, Stockholm.
Miyagi, T. and Morisugi, H. (1993). A Direct Measure of the Value of Choice Freedom, Presented at 13th Pacific Conference of the Regional Science Association International, Whistler, Canada.
Mohring, H. (1993). Land Rents and Transport Improvements: Some Urban Parables, *Transportation*, 20, 3, 267-283.
Newman, P.W. and Kenworthy, J.R. (1989). *Cities and Automobile Dependence: An International Sourcebook*. Gower, UK.
Rickaby, P.A. (1987). Six Settlement Patterns Compared, Environment and Planning B, 14, 193-223.
Roy, J.R. (1992). Transport Efficiency in Cities with Subcentres, *Selected Proceedings of 6th World Conference on Transport Research*, I, 291-302.
Ueda, T., Nakamura, H. and Shimizu, E.(1993). GIS Integrated System for Urban Transport and Development Planning, *Proceedings of Seminar Environmental Challenges in Land Use Transport Coordination*, Blackheath , Australia.
Webster, V., Bly, P.H. and Paulley, N.J. (eds) (1988). *Urban Land-Use and Transport Interaction*. Avebury, Aldershot.
Wilson, A.G. (1970). *Entropy in Urban and Regional Modelling*, Pion, London.
Yinger, J. (1993). Bumper to Bumper: A New Approach to Congestion in an Urban Model, *Journal of Urban Economics*, 34, 249-274.

CHAPTER 8

EDUCATING PLANNERS IN TESTING OF ALTERNATIVE STRATEGIES FOR CITIES OF DIFFERENT SIZES

William Young and Kevin Gu

1. Introduction

Transport, land use and the environment are interrelated. Changes in one will, in time, impact the other. The need to understand this relationship is increasing in importance as environmental problems increase and cities change. Further, the complexity of the interaction and the range of policy options that can be used to refine it require an understanding of the time scale and sensitivity of responses to policy changes. Understanding the interaction requires study and education. The study reported here aims to assist in the education process through the development of the land use, transport and environment interaction package: LAND (Location of Activities and Network Development).

The LAND package, is an educational tool. It aims to assist students in gaining an understanding of the relationship between transport, land use and the environment. This paper investigates the application of the LAND model to study changes in a hypothetical city similar in size and layout to Canberra over the period 1981 to 2001. The study reported here investigates the use of the LAND package by students at the Australian Defence Force Academy. The application of the package to a hypothetical city, similar in structure to Canberra, enabled the models appropriateness for general use and sensitivity to network description to be ascertained. Application of the model to Adelaide and Melbourne are also discussed.

Y. Hayashi and J. Roy (eds.), Transport, Land-Use and the Environment, 147–169.

2. The Land Computer Package

Planning students and the general community's understanding of the land use, transport and environment system, is often guided by limited experience. The complexity of the system is, therefore, often not fully understood. The LAND computer package is being developed to assist in the education of students and the community in the complexities of the system. This section of the paper discusses the main philosophies behind LAND and the overall structure of the package.

Education tools like LAND aim to increase the level of understanding. A major concern in developing an educational tool is to determine its main requirements. One approach that assists in determining the requirement of this approach are the developments in computer games. Computer games have a number of requirements. These include a high standard of user interface, a challenge, an interactive environment, an acceptable level of user input, and entertainment.

A major development in LAND is the user interface. Existing levels of expectation in the user environment require interactive editing, mouse facilities and pop up menus. The acceptance of the package therefore depends on creating such an interaction. To facilitate these developments and to provide access to the full memory capabilities of the new PC's, the LAND educational tool was developed using VISUAL BASIC 3.0 in the Microsoft WINDOWS environment. The LAND package is entered through a WINDOWS icon. A mouse click on the icon will bring up the front form. This form provides entry to the LAND main menu. The main menu shows the major components of LAND as the system control options, data input (Figure 1), modelling (Figure 1), output presentations (Figure 1), view, setting of key variables, and help facilities.

These components provide the user with the facility of developing a land use pattern and transport network, running the model and viewing the output. More details of the user interface can be found in Gu, Haines and Young (1992a, 1992b).

3. The Land Model

The modelling form (Figure 1) is accessed by clicking the mouse on the modelling option on the main menu. The remainder of this section discusses the activities that occur behind the modelling form when the simulation model is running.

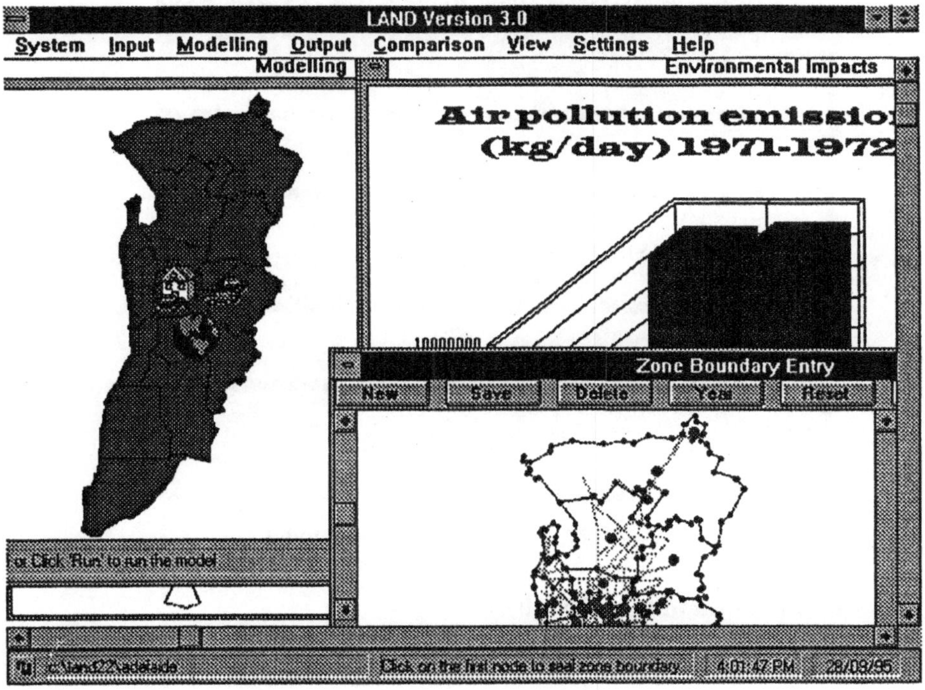

Fig. 1. LAND Data Input, Modelling and Output Forms

3.1 The LAND Model Structure

The structure of the LAND simulation model revolves around a dynamic updating of land use in a city using a one year cycle. The model requests that the base year and transport network for the study be set up. The city can start with no development or a city that is at a particular stage of development can be input. The input of this data is a relatively straight-forward task, provided the information is available (Gu, Young and Haines 1993).

The overall model structure is outlined in Figure 2. It follows a sequential path. LAND starts from the land use module which estimates the location of activities in the different zones within the study area. The location activities consists of population, houses, households, employment, unemployment and jobs. These in turn are inputs to the transport module in the form of trip matrices. The transport module determines the travel pattern, and the generalised travel costs which in turn impacts on further location decisions. After the travel pattern are estimated, the environmental impacts of the transport system are calculated. LAND updates location activities, transport patterns and environmental impacts every year.

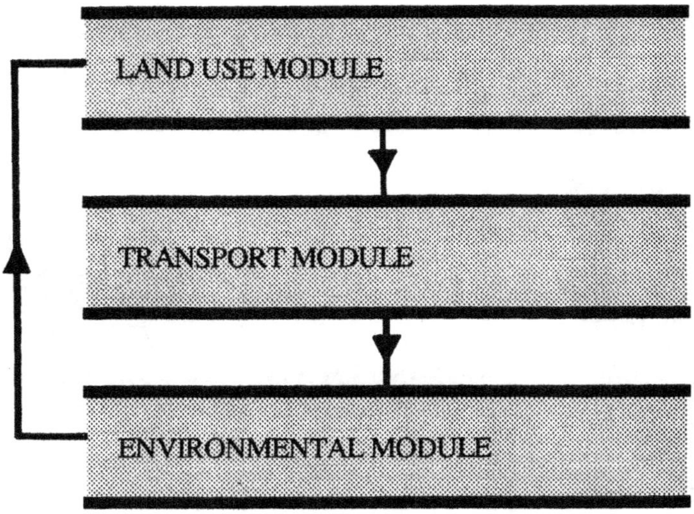

Fig. 2. General Structure of the LAND Simulation Model

3.2 The LAND Land Use Module

The LAND land use module takes as input in the base year information on the distribution of houses, jobs and households. It takes this information and develops home to work linkages. The general sequence of events is to age the population, determine the amount of unemployment and overcrowding, determine the movement of working households then determine the total population.

The age profile is updated simply by ageing the population and then considering the survival rate. Households are then varied depending on change in the age structure. The migration of households into or out of the city is considered when aging the population.

The total distribution of households is fed into procedures that determine the unemployment and proportion of overcrowding in the urban area. Unemployment results when the number of jobs are less than the number of workers. Overcrowding occurs when the number of houses are less than the number of households. Unemployment and overcrowding are distributed in proportion to population in each household type in the urban area. The proportion of overcrowding and employment is therefore assumed to be the same in each household group.

The module considers the people moving within the urban area, considering in sequence, those with no fixed job and home, a fixed home, and a fixed job. Households are divided into working households and retired households. Working households are further broken into three stages in 'life cycle' categories with household head of age 20-33, 34-47 and 48-60 years old. People age over 60 are considered as retired persons.

The number of movers and the household vacancies in the urban area are determined by LAND. Households are allocated to the vacancies in relation to the most favourable house and location combination. The quality of house/location combinations is a linear combination of the house quality, the distance to the households work place and the accessibility to employment. The accessibility to employment takes the form:

$$A_j^e = \sum_{\text{all } i} E_i \; e^{-\beta C_{ij}} \tag{1}$$

where A_j^e is the accessibility to employment e in zone j,
E_i is the employment in zone i,
C_{ij} is the average generalised travel cost for all modes from zone i to zone j, and
β is the travel impedance function coefficient.

Business is considered in a similar fashion to households. Businesses do not move, fail or move. New businesses consider the options available. The quality of job is a linear combination of the distance from the workers home and the best job location. The highest value of this combination is the most favoured job location. The accessibility to population (i.e. markets and work-force) takes the form:

$$A_j^p = \sum_{\text{all } i} P_i \; e^{-\beta C_{ij}} \tag{2}$$

where A_j^p is the accessibility to population, and
P_i is the population at zone i.

Retired people are distributed in proportion to population in each household type in the urban area. The zonal population distribution is a function of the resultant household pattern. It is the sum of number of people working, unemployed, young and retired.

3.3 The LAND Transport Module

The resultant land use pattern (households and employment) is used as a basis for determining travel patterns. The LAND transport module is a four step process which predicts the trips generated, the distribution of trips, the choice of mode and assigns the trips to a private and public transport network. The transport module uses as input the distribution of population and households determined in the land use module. It outputs information of the distribution of trips to the environmental module. Transport networks are represented using a node-link system. The accessibility functions, travel impedance function and model split utility function all involve a variable travel cost (C_{ij}). The generalised travel cost consists of three elements. The cost associated with travel time, the cost associated with distance travelled and the cost associated with waiting. It takes the form:

$$C_{ij}^{k} = a_1^k t_{ij}^k + (a_2^k + a_3^k) d_{ij}^k + a_4^k w_{ij}^k \qquad (3)$$

where C_{ij}^{k} is the generalised link cost of travel from zone i to zone j by mode k (cents),
d_{ij}^{k} is the link distance in kilometres from i to j,
t_{ij}^{k} is the link travel time in minutes from i to j by mode k which is calculated from the Davidson's congestion function introduced below,
w_{ij}^{k} is the walking, waiting and other access times in minutes associated with the journey i to j for mode k,
a_1^{k} is the average value of time in monetary units (cents/min) in mode k,
a_2^{k} is the average running cost of travelling unit distance (cents/km) in mode k
a_3^{k} is the travel fare per unit distance (cents/km) for mode k, and
a_4^{k} is the value of time (cents/min) for walking and waiting associated with the journey for mode k.

3.4 The LAND Environment Module

The LAND environmental module takes as input information on the distribution and speed of travel output from the transport module. It transforms this information into measures of emissions and fuel consumption.

Within an urban area, transport pollutant emission rates and energy consumption depend on the varying levels of congestion. The higher the levels of congestion (and thus the lower the travel speed), the higher are the levels of energy use and pollutant emissions (Sharpe, 1979). Therefore the traffic condition and location of the individual links in a network, and the interaction between links strongly influence transport energy consumption and pollution levels within a city. Other factors such as the type of vehicles travelling on a network and the fuel they use also need to be considered.

Road vehicles in LAND are divided into two major categories cars and equivalents, and heavy vehicles. Cars and equivalents are further broken up into four classes according to the fuel type they use leaded petrol, diesel, unleaded petrol, and liquid petroleum gas (LPG). Heavy vehicles are broken into petrol-engine and diesel-engine vehicles. The total rate of emission or fuel consumption E(x) can be defined as:

$$E(x) = \sum_{m} E_m(X) p_m q \tag{4}$$

where $E_m(x)$ is the mean rate of pollutant emission or fuel consumption (Taylor, 1992).
p_m is the proportion of type m vehicles, and
q is the link traffic flow of vehicle.

This link based model was established by Taylor and Anderson (1982) and Taylor (1992). As long as we can predict $E_m(x)$ for a range of traffic condition, the total pollution levels and fuel consumption can be estimated.

Since LAND presently concentrates on the urban level, the pollutant emission and fuel consumption are estimated using simple relationships based on vehicle type and travel conditions on network links. The generation rates $E_m(x)$ of pollutant emission or fuel consumption are assumed to be directly related to mean travel time per unit distance on a link, link flow and road type.

Simple linear relationships between $E_m(x)$ and unit distance travel time ω have been shown to be appropriate in estimating pollution generation and fuel consumption for urban transport planning models (Evans and Herman, 1976; Taylor and Anderson, 1982; Taylor and Gipps, 1982).

The linear function suggested by Evans and Herman (1976) takes the form:

$$E_m(X) = k_1 + k_2 \omega \tag{5}$$

where $E_m(x)$ is the generation rate (g/km) of pollutant x or fuel consumption rate (litres/100 km) by a type m vehicle,
k_1, k_2 are coefficients, and
ω is the mean travel time per unit distance on a link (min/km).

The relationships have been used in the URPOL model (Taylor, 1982) and have been calibrated using the data collected by Kent (1980). These data cover the range 0.8-5.0 min/km for unit distance travel times (mean travel speeds 12-75 km/h) and thus adequately represent traffic conditions in most Australian urban areas.

The air pollutants considered in LAND include carbon dioxide (CO_2), carbon monoxide (CO), nitrous oxides (NO_x), hydrocarbons (HC) and lead (Pb). Four types of fuel consumption for motor vehicles are considered: leaded petrol, diesel, unleaded-petrol and LPG.

For petrol-powered cars and equivalents, the coefficients of k_1 and k_2 used in the study are outlined in Gu, Young and Haines (1993). The emissions and fuel consumption for other types of vehicles were assumed to relate to those for petrol-powered vehicles (Taylor, 1992).

For diesel-powered cars and equivalents, the fuel consumption model was derived by a comparative study of fuel consumption on petrol and diesel vehicles

(Weeks, 1981) and heavy vehicles, Glazebrook's (1980) fuel consumption relation was adopted. The ratios of fuel consumption for unleaded-engine and LPG-engine vehicles to fuel consumption of petrol engine cars and equivalents are 0.9 and 1.25 respectively.

4. Application of Land to a Hypothetical City: The Canberra Study

The application of LAND to Canberra was carried out by obtaining data, from the Australian Bureau of Census and Statistics and the Australian Capital Territory Planning Authorities, and its predicted growth over the period 1981 to 2001. A further dimension to this exercise was introduced since the study was not carried out by the developers of LAND, but was undertaken by a number of students from the Australian Defence Force Academy (ADFA). This enabled novice users access to the software and provided an indication of their reaction to the software. The study reported in this section was carried out by three student groups. The groups were:

- Major Michael G. Healy, Captain Dechlan R. Ellis and Mr Anton R. Kandiah (Group 1),
- Anthony Farrel, Stephen Fester, George Yacoub and Graham Tippets (Group 2), and
- Lcdr John Jacobi, Captain Kerry Marshall and Captain Ian Stoppard (Group 3).

The reporting of the results consists of taking one of the studies (Group 1) and reporting it in the following sections. These are compared with the results obtained from the other two groups.

4.1 Development of Canberra Network and Strategies

The data, obtained from the Australian Bureau of Census, provided information on 82 suburbs. This data was provided to the students. However, running the model at this level of detail would have increased the workload on the students to an inappropriate level and defeated the general educational philosophy built into LAND. Future applications of LAND at this finer detail may, however, be a useful exercise.

Group 1 aggregated the data on the 82 suburbs into 10 zones (see Figure 3). Ten zones provided an appropriate level of detail for the student application of LAND. The basic zone information is presented in Table 1.

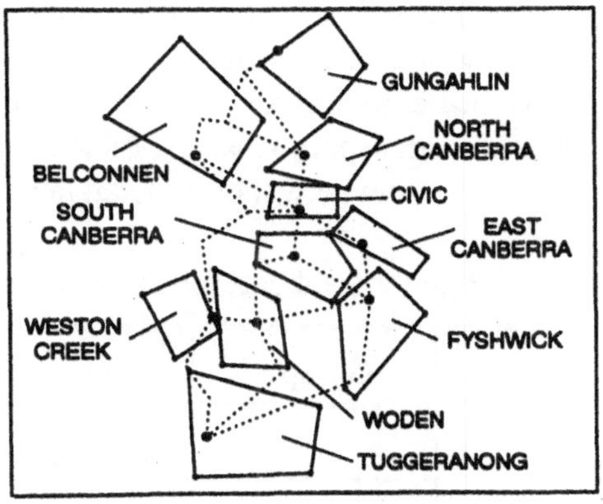

Fig. 3. Canberra Zonal Pattern Used by Student Group 1

Table 1. Basic Zonal Information for Canberra Used by Student Group 1

Zone number	Zone name	Area (km^2)	Population (1981)	Jobs available (1981)
1	Belconnen	48.45	80,127	14,448
2	Nth Canberra	16.73	34,128	4,719
3	Civic	9.24	9,283	22,4547
4	Sth Canberra	18.66	25,968	18,230
5	Woden	21.13	39,471	12,936
6	Fyshwick	25.80	832	8,713
7	Tuggeranong	47.01	29,100	1,454
8	East Canberra	10.75	5,096	9,713
9	Weston Creek	16.34	29,112	2,541
10	Gungahlin	21.23	291	798
	Total	235,34	253,408	96,099

Although the zones were not equal in size they were encompassed by natural boundaries. Importantly, the study did not take into account the population of Queanbeyan and its effect on the Canberra employment market and therefore could not be considered a completely accurate study of Canberra.

The road network (Table 2) was represented by a total of 20 nodes, 18 roads and 27 links. The capacity of each road is modelled as the road capacities in 1993. No attempt was made to improve the road network throughout the modelling period. The public transport network (Table 2) was represented by 14 nodes, 14 roads and 14 links. Both networks are relatively simplistic.

Table 2. Network Description for Canberra for Three Student Groups

Network information	Group 1	Group 2	Group 3
Road			
Zone centroids	10	12	12
Centroid connectors	0	0	0
Nodes	20	50	28
Road names	18	36	27
Links	27	71	54
Public transport			
Zone centroids	10	12	10
Centroid connectors	0	0	0
Nodes	14	28	10
Road names	14	14	12
Links	14	36	12

Overlaid on the zone and transport network is demand information. The land use information for the year 1981 is presented in Table 3. The housing ranged from a low of 79 in Gungahlin to a high of 22,437 in Belconnen. The distribution of households was similar to that of houses. There was a high of 24,171 households in Belconnen and a low of 87 households in Gungahlin. Jobs were located primarily in Civic (22,547) with a few jobs in Gungahlin (798).

Changes in the land use pattern, due to house building, job increases and household migration, were assumed to take place after the 1981, base year. To investigate the predictions made by LAND several strategies were chosen. The strategies chosen by student Group 1 were:

Table 3. Base Year Land Use Information for Canberra (1981)

Zone name	Housing	Workforce	Households
Belconnen	22,437	14,918	24,171
Nth Canberra	9,817	4,919	10,295
Civic	2,742	22,601	2,800
Sth Canberra	7,342	18,382	7,833
Woden	11,007	13,167	11,907
Fyshwick	265	8,717	251
Tuggeranong	8,126	1,625	8,778
East Canberra	1,482	9,742	1,537
Weston Creek	8,152	2,712	8,782
Gungahlin	79	798	87
Total	71,449	97,581	76,441

Do Nothing Strategy. The do-nothing option involved the running of the model for 20 years without inputting any growth in housing or jobs, and without any migration into or out of Canberra. The option provided a base for the comparison of the other strategies.

Centralise option. In the centralised option, all growth in housing and jobs was put in one zone, namely Civic, for each of the 20 non-base years. No limit was set on the number of houses that could be located in this zone. Although this is unrealistic it served to demonstrate the transport and environmental trends that would be present under this type of strategy.

Decentralised strategy. In the decentralised strategy, all growth in housing and jobs was divided between three outer zones, namely Belconnen, Gungahlin, and Tuggeranong. It was assumed that the percentage of total growth in housing and jobs is spread over each of these zones. The growth was varied over time in five year increments over the 20 year period. The distribution used is shown in Table 4. This strategy is more realistic than the centralised option although it is still an extreme. It serves to illustrate the opposite extreme to the centralised option.

Development option. Each student group was allowed to develop a proposed strategy for Canberra and compare this to the other three strategies. The Healy, Ellis and Kandiah group (Group 1) chose the growth splits shown in Table 5.

Table 4. Percentage of Growth Splits in Canberra Zones for the Decentralised Option by Student Group 1

	1982-86	1987-91	1992-96	1997-2001
Belconnen	30	30	20	20
Gungahlin	0	0	40	60
Tuggeranong	70	70	40	20
Total	100	100	100	100

Table 5. Growth Split in Canberra Chosen by Group 1 for Development Strategy

Zone	Growth (% of total growth)
Civic	30
Belconnen	20
Woden	20
North Canberra	10
South Canberra	10
South Canberra	10
Total	100

Group 1 chose this strategy since it aimed to encourage strong growth in housing and jobs in the Civic zone, recognise that there will be continued growth in the regional centres in Belconnen and Woden, actively limit growth in Tuggeranong and Gungahlin to stop urban sprawl, and encourage moderate growth in the inner zones adjacent to Civic (ie North, South and East Canberra).

4.2 Comparison of Canberra Strategies

The comparison of the strategies by group 1 was carried out by investigating the land use, transport and environmental dimensions separately. The finding of group 1 are presented in the following subsections.

LAND produces six outputs that can broadly be described as land use information. These outputs are the distribution of population, households, houses, jobs available, employment and unemployment. Only population and employment will be discussed in detail here.

Population distribution. Canberra in 1981 had a population of approximately 250,000 people spread over an area of 235 km^2. The average density was therefore slightly more than 100 persons/km^2. Vast areas were unoccupied and therefore the average density tended to be misleading, but even so the maximum density of a zone was 1868 persons/km^2. The results of modelling change over the period 1981 to 2001 for each strategy is shown in Table 6.

Table 6. Population Characteristics in Canberra in 2001 for Student Group 1 for each Strategy

	Do-Nothing	Centralised	Decentralised	Develop-ment
Belconnen	98,080	73,092	132,226	108,940
Nth Canberra	40,765	26,750	31,725	50,053
Civic	13,021	225,350	9,311	68,800
Sth Canberra	33,990	19,361	24,040	42,332
Woden	53,257	29,986	38,011	74,339
Fyshwick	1,178	657	866	812
Tuggeranong	40,537	9,512	106,680	20,503
Est Canberra	7,091	4,215	5,105	24,993
Weston Creek	40,148	26,665	27,632	24,687
Gungahlin	457	181	40,172	246
Total	328,524	415,769	415,768	415,765

The do-nothing option had little effect on population densities. The population grew slowly and the density grew proportionally in all zones, increasing by approximately 25%. Is it notable that Gungahlin did not attract any increase in development with the population density remaining at about 10% of other dormitory areas such as Belconnen and Tuggeranong.

The centralised option increased the population in Civic from 9,200 in 1981 to over 225,000 in 2001. Although the 1981 density is low the 2001 density is unrealistically high. The model showed that as the Civic area grew the surrounding areas slowly declines.

The decentralised option resulted in strong growth in Belconnen and Tuggeranong. Belconnen increased from 80,000 to 132,000 and Tuggeranong from 29,000 to 106,000 over the 20 year period. The density of the two zones increased proportionally, and was approaching 2,000 persons/km^2 in Belconnen in 2001. Given that the overall zone density is similar to that of Central Melbourne it is unlikely that Belconnen would have the vast open spaces which it had in 1981. The impact on Civic and the surrounding North, South and East Canberra areas was minimal. The population of these areas fluctuated by up to 25% with an initial reduction in population but then returned to close to the 1981 figure after 20 years.

The development option resulted in a significant growth in Civic. The population increased from 9,200 to 68,800 with a commensurate increase in density from 1,005 to 7,445 persons/km^2. Unlike the centralised option the zones surrounding Civic also increased their population slightly, except for East Canberra. East Canberra's population increased by a factors of four to bring its density to a level similar to that of the other zones. The strategy also allowed the Belconnen area to increase its population slightly. Tuggeranong decreased progressively from 29,000 to 20,503.

Employment distribution. Canberra, in 1981, had 96,000 jobs available and this increased to 160,000 by 2001 in all but the do-nothing case. The four development strategies considered allocated these jobs to workers in different proportions (Table 7).

Table 7. Employment Characteristics in Canberra in 2001 for Student Group 1 for each Strategy

	Do-Nothing	Centralised	Decentralised	Development
Belconnen	14,448	14,448	34,268	28,175
Nth Canberra	4,719	4,006	4,719	11,775
Civic	22,547	102,947	16,626	43,070
Sth Canberra	18,230	13,443	18,230	24,245
Woden	12,936	12,936	12,936	26,779
Fyshwick	8,713	6,425	6,425	5,942
Tuggeranong	1,454	305	40,534	1,341
Est Canberra	9,713	3,086	7,162	16,384
Weston Creek	2,541	2,541	2,541	2,345
Gungahlin	664	181	16,396	57
Total	95,965	160,318	159,837	160,113

The number of jobs allocated in the do-nothing case remained static throughout the study period. In 1981, 1400 people were unemployed. With natural growth in population this increased to 16,300 by 2001.

The centralised option located new jobs in Civic, increasing the jobs available form 22,500 to 103,000 progressively over the 20 year period. The effect of this was to increase the density of jobs in Civic by more than 10 times.

The decentralised option concentrated on the Belconnen, Tuggeranong and Mitchell/Gungahlin zones. Initially this had little effect on the traditional employment zones of Civic and East Canberra, but in the last quarter of the modelling period people were drawn from their jobs in these areas to take up jobs closer to where they live. As there was an oversupply of jobs from 1987, the jobs left vacant were those in the inner area. The model therefore clearly showed that people prefer to live and work in the same location.

The development option doubled the number of jobs in Civic, East Canberra, Belconnen and Woden. The jobs in the dormitory areas were all taken up.

This section compares the three strategies with the do-nothing strategy based on transport implication. The daily travel characteristics, at the end of the 20 year period, for each strategy are shown in Table 8. There was a linear change between 1981 and 2001.

Road network comparisons. As expected the centralised option has the lowest number of vehicles kilometres travelled, with a reduction of around 20% from the do nothing option. The real benefit of the centralised option in reducing total vehicle kilometres would be greater than this figure, as it includes growth, whereas the do nothing option did not. This also applies to all further comparisons. The decentralised option almost doubles the total vehicles kilometres travelled compared to the do-nothing option. The development option reduced the total vehicle kilometres travelled by approximately 5%. The centralised option located all growth in housing and jobs into Civic, which is already a large source of jobs. Hence fewer and fewer people had to drive outside the zone to get to work. The decentralised strategy located all housing and jobs on the fringe, but there was already many people living closer to Civic, hence more and more people had to drive outside the zone to get to work. The development solution was somewhere in between the centralised and decentralised options, and reduced the total number of vehicle kilometres travelled compared with the decentralised case. This trend is also reflected in the figures for total number of trips, mean trip distance and mean trip time, for the same reasons.

The mean trip speed shows a different trend. The decentralised option had the lowest mean speed, followed by the development option, the centralised option, and as expected the do-nothing option had the highest speed. It would intuitively be expected that the centralised option would have the lowest speed, as all growth was in the built up area. Similarly it would be expected that the centralised option would have the highest speed, as the network links were 80 km/h and, in some parts 100 km/h dual carriageways for most part. However, it could be that the increased traffic caused by the decentralised option is causing increased congestion and thus decreasing mean speeds.

Table 8. Daily Travel Characteristics in Canberra for the Year 2001 for Student Group 1

Transport statistics	Do-Nothing	Centralised	Decentralised	Development
Road network				
Vehicle kilometres	6,429,757	5,234,035	12,528,969	7,976,338
Vehicle hours	103,817	87,049	249,923	143,467
Trips	594,452	566,184	826,386	762,778
Mean travel speed (km/h)	61.9	60.1	50.1	55.6
Mean trip distance (km)	10.8	9.2	15.2	10.5
Mean trip time (minutes)	10.5	9.2	18.1	11.3
Public transport network				
Trip kilometres	883,757	1,006,143	938,216	944,483
Trip hours	15,448	18,281	16,264	16,887
Trips	115,310	156,574	97,144	129,00
Mean travel speed (km/h)	57.2	55.0	57.7	55.9
Mean trip distance (km)	7.7	6.4	9.7	7.3
Mean trip time (minutes)	8.0	7.0	20.0	7.9

Public transport comparison. Comparing to the do-nothing option, the centralised strategy increased the total trip kilometres of travel on public transport by approximately 14%, the decentralised strategy by 6% and the development strategy by 7%. The reason for these increases are different in each case. The centralised strategy also increased the total trip hours of travel and the total number of trips, but decreased the mean trip distance and time. This indicates the more people were using public transport, due to the centralised growth they had shorter trips on average. This trend is probably a function of the fact that the Civic zone is the best served by public transport as it is the central node of the five major routes. The decentralised strategy on the other hand decreased the total number of trips, and increased the total trip hours, mean trip hours and average travel time. This indicates that fewer people were using public transport and due to the decentralised growth they had longer trips on average. The development solution showed the same trend as the centralised option, but not as drastically, due to the distribution of growth selected.

LAND provides information about fuel consumption and air pollution. Group 1 discussed each of these. Their discussion is presented below.

Fuel consumption figures are summarised in Table 9.

Table 9. Fuel Consumption Comparisons in Canberra for 2001 for Student Group 1 (kl/yr)

Fuel type	Do-Nothing	Central	Decentralised	Development
Petrol	122,000	99,000	240,000	151,828
Unleaded Petrol	16,800	13,700	33,200	20,998
Diesel	62,500	50,700	122,000	77,320
Liquid petroleum gas	9,350	7,600	18,400	11,643

As expected, the centralised strategy had the lowest fuel consumption, with a reduction of approximately 18-19% in the total fuel consumed compared to the do-nothing strategy. This is because the centralised option encouraged an increased in housing density and jobs in the Civic zone, which decreased the number of people having to travel long distances to work. However, in the decentralised option people had to travel long distances to their job locations, which increased fuel consumption by approximately 96-98% compared to the do-nothing strategy. The development option showed an increase in fuel consumption of approximately 24% compared to the do-nothing strategy, which is a significant improvement on the decentralised option.

Air pollution figures for the year 2001 are summarised in Table 10.

There is a strong correlation between fuel consumption and air pollution for all strategies. The percentage changes in air pollution are very similar to those for fuel consumption for all three strategies when compared to the do-nothing option.

Table 10. Emissions Comparisons for Canberra for 2001 for Student Group 1 (ton/yr)

Air pollution	Do-Nothing	Central	Decentralised	Development
Carbon monoxide	12,000	9,908	24,619	15,307
Hydrocarbons	2,800	2,319	5,632	3.556
Nitrogen Oxide	5,450	4,400	10,519	6.711
Lead	47	39	94	59

4.3 Comparison of Student Projects on Canberra

The previous subsection briefly described the study carried out by student group 1. A useful extension of this study is to compare the results of a number of groups to obtain an indication of the sensitivity of the model to the differing assumptions made. In particular, does the number of zones and the network developed influence the model output?

Table 2 presents a comparison of the networks developed by groups 1 to 3. It can be seen that group 1 divided Canberra up into 10 zones while the other two groups used 12. Group 3 chose a similar road network to group 1 but a slightly simpler public transport network. The most complex road and public transport networks was developed by group 2. It contained almost twice the number of roads and links as the other two network.

Comparison of the outputs of the models for the do nothing option are shown in Table 11. This was the only option studied by each group which had the same distribution of land use and any difference could be attributed to differences in the networks chosen. It can be seen that the land use predictions for the three models are relatively stable. The transport, fuel consumption and air pollution predictions for groups 1 and 2 were also relatively similar although group 1 did predict higher public transport use. Group 3, however, predicted considerably more travel on public transport and less travel on the road system than the other two groups. This resulted in lower fuel consumption and air pollution estimates. The mode usage obtained by group 1 given car occupancies level used in the model of 1.25, was of the order to 25 % compared with 12 % for group 2. This variation was primarily due to the different specification of the transport networks and indicates that care must be taken in developing an appropriate level of detail in the network description.

Table 11. Daily Travel Characteristics in Canberra for Do Nothing Option for the Year 2001 for Three Student Groups

	Group 1	Group 2	Group 3
Land use			
Population	328,524	329,468	334,348
Available jobs	96,099	96,103	96,290
Employment	95,965	95,845	95,723
Road Network			
Vehicle kilometres	6,429,757	6,266,269	4,842,140
Vehicle hours	103,817	114,003	76,528
Trips	594,452	583,912	472,280
Public transport			
Trip kilometres	883,757	609,424	1,576,939
Trip hours	15,448	10,493	27,462
Trips	115,310	85,164	193,420
Petrol (kl/yr)	138,800	140,032	103,270
Diesel (kl/yr)	62,500	62,962	46,793
LPG (kl/yr)	9,350	9,458	6,978
Carbon Monoxide (tons/yr)	12,000	12,477	8,935
Hydrocarbons (tons/yr)	2,800	2,886	2,121
Nitrogen Oxides (tons/yr)	5,450	5,436	4,089
Lead (tons/yr)	47	48	35

Another comparison that can be carried out relates to the outputs of the four strategies. These strategies were not specified in exactly the same manner by each groups but were similar in general trend. Table 12 presents the results.

A ranking was given to each strategy depending on the level of pollution, vehicle travel and fuel consumption. The best ranking (1) was given to the strategy with the lowest vehicle travel, pollution and fuel consumption. Higher rankings were given as the vehicle travel, pollution and fuel consumption increased. It can be seen that the do-nothing and centralised options tended to get the best ranking, whilst the decentralise and development strategies received the poorest ranking. However, the rankings provided by each group were not consistent. This could be due to differences in their specification of each strategy, but is also the result of the differing choice of public and private transport network.

Table 12. Daily Travel Characteristics in Canberra for the Year 2001 for Three Student Groups

	Group 1	Group 2	Group 3
Do nothing strategy			
Road Vehicle kms	6,429,757	6,266,269	4,842,140
Petrol Consumption (kl/yr)	122,000	123,000	90,708
Carbon Monoxide (tons/yr)	12,000	12,477	8,935
Ranking	2	2	1
Centralised strategy			
Road Vehicle Kms	5,234,035	5,613,730	5,196,826
Petrol Consumption (kl/yr)	99,000	111,000	98,627
Carbon Monoxide (tons/yr)	9,908	11,823	9,937
Ranking	1	1	2
Decentralised strategy			
Road Veh Kms	12,528,969	11,502,485	8,290,806
Petrol consumption (kl/yr)	240,000	233,000	157,000
Carbon Monoxide (tons/yr)	24,619	25,980	15,759
Ranking	4	3	4
Development strategy			
Road Veh Kms	7,976,338	12,865,036	5,498,325
Petrol consumption (kl/yr)	151,828	252,000	103,000
Carbon Monoxide (tons/yr)	15,307	26,730	10,155
Ranking	3	4	3

5. Overview of Melbourne, Adelaide and Canberra Studies

The previous discussion has concentrated on a study of Canberra. LAND has also been applied to larger cites. Table 13 summarises the finding of applications to Adelaide and Melbourne. Information on Canberra is summarised for comparison. It can be seen that there are considerable differences between the three cites Melbourne population varies between a population of 2,452,646 (1971) and 3,106,722 (1991). Adelaide has a population range of 842,698 (1971) and 1,046,929 (1991). Canberra population varies between 253,408 (1981) and 326,374 (1991). The relative size in population is also indicated in the road vehicle kilometres of travel, public transport usage, petrol consumption and carbon monoxide emissions.

6. Concluding Remarks

This study has outline the LAND model and its application to a city similar in structure to Canberra by university students. The application of LAND, by Australian Defence Force Academy students, to the city of similar character to Canberra showed that the model could be applied to investigate different growth strategies and their impacts. Results of the application of the model to Melbourne and Adelaide were also presented. The application of LAND to a number of cities has shown that the package can be used to investigate the interaction between land use, transport and the environment. LAND can also be used as a planing tool to assist engineers and planners to develop a variety of planning scenarios.

Acknowledgements

The author acknowledges the Bureau of Transport and Communications Economics (BTCE) for providing the support for the verification of the LAND model. Within the Bureau Dr Maurice Haddad, Dr Leo Dobes, Dr Franzi Poldy, David Gargett and Kay Loong provided considerable support and advice during the project. Their assistance is gratefully acknowledged.

Table 13. Melbourne Land Use Information from the Census 1971-91

Year	1971	1981	1991
Melbourne Population	2,452,646	2,748,908	3,106,722
Employment	1,074,747	1,203,601	1,295,988
Road Vehicle kms	$28.60 * 10^6$	$40.52 * 10^6$	$77.33 * 10^6$
Public transport trip kms	$7.6 * 10^6$	$6.15 * 10^6$	$8.71 * 10^6$
Petrol consumption (l/day)	$2.83 * 10^6$	$3.75 * 10^6$	$6.04 * 10^6$
Carbon Monoxide (kg/day)	$0.12 * 10^6$	$0.15 * 10^6$	$0.30 * 10^6$
Adelaide Population	842,698	952,858	1,046,926
Employment	345,576	396,917	461,330
Road vehicle kms	$13.16 * 10^6$	$18.81 * 10^6$	$26.15 * 10^6$
Public transport trip kms	$1.20 * 10^6$	$1.69 * 10^6$	$2.39 * 10^6$
Petrol consumption (l/day)	$1.63 * 10^6$	$2.22 * 10^6$	$2.55 * 10^6$
Carbon monoxide (kg/day)	$0.047 * 10^6$	$0.061 * 10^6$	$0.066 * 10^6$
Canberra Population		253,408	326,324
Employment		96,099	126,768
Road vehicle kms		$2.68 * 10^6$	$2.93 * 10^6$
Public transport trip kms		$0.20 * 10^6$	$0.21 * 10^6$
Petrol consumption (l/day)		$0.64 * 10^6$	$0.72 * 10^6$
Carbon monoxide (kg/day)		$0.029 * 10^6$	$0.045 * 10^6$

References

Evans, L. and Herman, R. (1976). A simplified approach to calculations of fuel consumption in urban traffic systems. *Traffic Engineering and Control 17* (8/9), 352-354.

Glazebrook, G.J. (1981). Fuel consumption of motor vehicles in Sydney. Joint SAE/ARRB Seminar Can traffic management reduce vehicle fuel consumption and emissions and affect vehicle design requirements", Melbourne.

Gu, K., Haines, A. and Young, W. (1992a). The Development of a Land Use and Transport Interaction Model". Report 2. Department of Civil Engineering, Monash University. June 1992.

Gu, K., Haines, A. and Young, W. (1992b). GAMES: Their role in transport planning. Proc. Australasian Transport Research Forum 17(2), 377-93.

Gu. K., Young, W. and Haines, A. (1993). LAND version 2.1 user manual". Monash University, Department of Civil Engineering.

Kent, J.H. (1980). "Relationships between driving cycles and urban driving patterns. Program and Papers". Joint SAE/ARRB Seminar. Melbourne, July 1980.

Sharpe, R. (1979). "Improving energy efficiency in community land use-transportation systems". Environment and Planning 12A, 203-216.

Taylor, M.A.P. (1982). "Pollution Generation and Fuel Consumption in Urban Transport Networks. Computer User Manual for Program URPOL". Division of Building Research, CSIRO, Melbourne, 1982.

Taylor, M.A.P. (1992). "Modelling Environmental Impacts of Road Traffic for Transport Network Analysis". Proc. the Australasian Transport Research Forum, 17(2), 529-549.

Taylor M.A.P. and Anderson, M. (1982). "Modelling pollution and energy use in urban road networks", Proc. 11th ARRB Conference 11(6), 1-17.

Taylor, M.P.A. and Gipps, P.G. (1982). "On the modelling of flows in transport networks". Program and papers. Joint SAE/ARRB 2nd Conference on Traffic, Energy and Emission, Melbourne, May, 1-23.

Weeks, R. (1981). "Fuel consumption of a diesel and a petrol car". Transport and Road Research Laboratory Report LR 964, Crowthorne, Berkshire.

CHAPTER 9

NOx REDUCTION EFFECTS OF THE POLICY TO REDUCE DIESEL AUTOMOBILES AND ITS INFLUENCE ON PRICE CHANGE

Hisa Morisugi and Eiji Ohno

1. Introduction

Although motorisation has contributed significantly towards Japan's present stable economic growth, the increase in automobiles has also cause social problems such as traffic jam, traffic noise, air pollution due to automobile exhaust gas and so on. NOx emission has especially become a more serious problem in the urban areas. This is because it brings about some diseases such as lung cancer and asthma. The nitrogen oxidant pollutant (NOx) volume is increasing and has exceeded the environmental quality standard in the urban areas. Although most of the air pollutant problems have improved in line with the appropriate regulation policies, the NOx problem has still remained. The main cause of this NOx pollutant problem is due to the use of the diesel automobiles. As the statistical data shows, NOx emission from automobiles accounts for 70% of the overall emission caused by human activities, where, out of this 70%, about 50% is attributed to diesel automobiles [EA 1992]. This is despite the fact that diesel automobiles only accounts for 20% of the overall total number of automobiles. The technical improvement of the diesel engine to reduce the environmental problems has some trade-off relations between the NOx reduction and the suspended particulate matter. Moreover, the number of diesel automobiles has steadily

Y. Hayashi and J. Roy (eds.), Transport, Land-Use and the Environment, 171–190.

increased and it has become more widely used because of its economical advantages. If no measure is taken towards this form of motorisation, the NOx problem will become more and more serious for human life, especially in the metropolitan areas.

One of issues in the public opinion for natural environment is the countermeasure for the NOx reduction especially by emission control of the diesel automobile. Although, as mentioned above, the emission volume from the diesel automobile is more than that from the gasoline automobile, the light-oil is cheaper than the gasoline due to the tax difference policy in Japan. This policy was made in the 1960's in order to give priority to automobiles for industry use over private use. Considering the fact that the NOx problem has become so serious, it seems natural to increase the light-oil tax in order to attain the welfare efficiency and equity of the polluters pay principle (PPP) [for example, IEA 1993; Iwata 1990; Nagai 1987]. In the United States, the gasoline automobile takes an advantage over the diesel automobile in the fuel pricing policy against the NOx problem, as shown in Table 1 [MOC 1993]. This paper aims, therefore, to construct a simulation model for both, predicting the number of diesel automobiles and the NOx volume, and analysing the impacts of the policy to reduce diesel automobiles, where the policy on the fuel price/tax control may be included.

Table 1. Fuel Price and Tax in Developed Countries

	Gasoline				Light-oil			
	A yen/*l*	B yen/*l*	A-B yen/*l*	B/A %	A yen/*l*	B yen/*l*	A-B yen/*l*	B/A %
US	30.1	8.8	21.3	29.2	28.5	11.0	17.5	36.8
UK	86.2	63.1	23.1	73.2	82.7	53.6	29.1	64.8
France	99.8	76.5	23.3	76.7	72.5	45.3	27.2	62.5
German	87.8	64.8	23.0	73.8	72.8	45.3	27.5	62.2
Japan	128.3	57.5	70.8	44.9	79.1	30.0	49.1	37.9

Notes 1) A: price, B: tax, A-B: price without tax, B/A: tax ratio.
2) exchange rate: 1 dollar = 132.73 yen, 1 pound = 229.20 yen, 1 franc = 23.57 yen, and 1 mark = 79.95 yen (Jun. 1993).
3) US: by DOE/EIA (Dec. 1992), UK, France, and German: by OPAL (Mar. 1993), Japan: by MOC (Apr. 1993).

The policy to reduce diesel automobiles such as the fuel price/tax control, however, may cause the raising in commodity prices through the increase of transport cost. If the policy makes the commodity prices increase so high, then even if it has positive effects on NOx reduction, it may not be socially acceptable. So it is necessary to predict price change by the policy in its evaluation. In this paper we construct a model to predict the price change by the policy based on the price equilibrium analysis utilising the input-output analysis.

In order to achieve the goal mentioned above, first, we construct a cohort type of simulation model, considering the ownership transfer between diesel automobile and gasoline automobile in the market of sedan and small-truck. Secondly, we predict the number of diesel automobiles and the NOx volume by using this simulation model, and analyse the NOx reduction effect of the policy to reduce diesel automobiles by carrying out this simulation model. Thirdly, we construct a model for predicting price change by the policy, based on the price equilibrium analysis utilising the input-output analysis. Finally, we predict the price change by the policy, and propose some recommendations for the policy against the NOx problem.

2. Trend of Diesel Automobiles' Share and Fuel Price

The trend of number of registered automobiles has been increasing as shown in Figure 1 [AIRA 1973-1992]. The number of sedan increased from 11 million vehicles in 1974 to 29 million vehicles in 1991. Remarkably, the share of diesel automobiles in the sedan market was increasing rapidly from only 0.02% (2.5 thousand vehicles) in 1974 to about 9% (2.76 million vehicles) in 1991. The number of small-truck increased from 5.3 million vehicles in 1974 to 6.3 million vehicles in 1991, after reaching a peak stage of over 6.8 million vehicles in 1979-80. The diesel automobiles' share in the small-truck market also increased rapidly from 7% (380 thousand vehicles) in 1974 to 56% (3.5 million vehicles) in 1991. During the same period, the fuel price underwent changes as shown in Figure 2 [MOC 1993]. The gasoline price fluctuated from 97 yen/*l* in 1974 to 128 yen/*l* in 1991 through 168 yen/*l* in 1982, and the light-oil price from 51 yen/*l* in 1974 to 74 yen/*l* in 1991 through 122 yen/*l* in 1982.

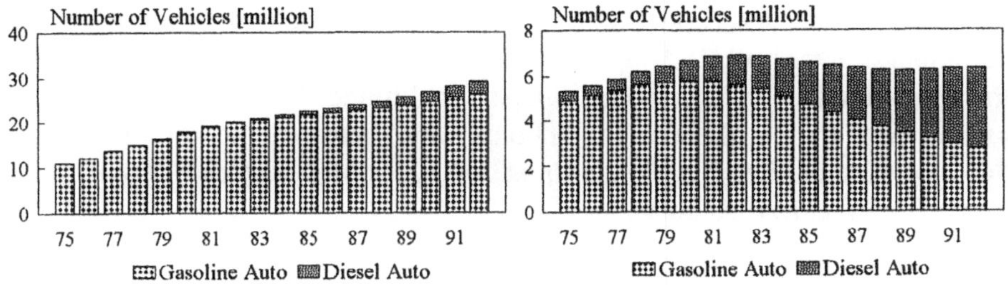

(a) sedan (b) small-truck

Fig. 1. Change of Diesel Automobiles' Share

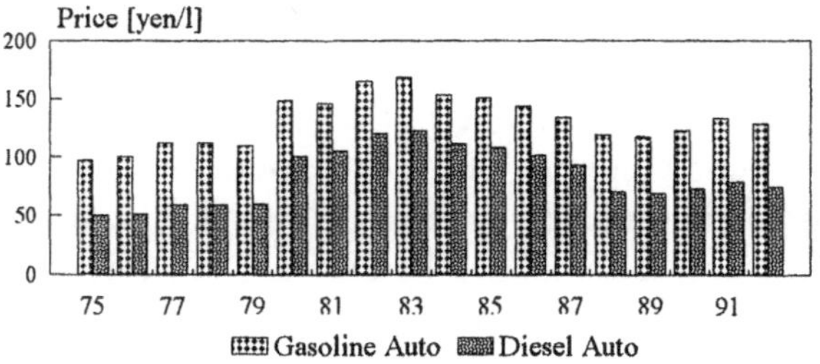

Fig. 2. Change of Fuel Price

3. Cohort Model for Predicting Number of Diesel Automobiles

3.1 Scope of Model

Focusing on the market for sedan and small-truck, and considering the ownership transfer between the diesel and the gasoline automobile, we construct a cohort type of simulation model both, for predicting the constitution of automobile ownership in the society, and analysing the impacts of the policy to reduce diesel automobiles [Morisugi et al. 1990]. The reason why we do not treat the whole range of automobiles, except for sedan and small-truck, is because of the very low elasticity of constitution, and because of the fact that there are very few gasoline automobiles in the market of large-truck and bus nowadays in Japan.

To date, some studies on the forecasting of automobile ownership have been carried [for example, Allanson 1982; Nakamura et al. 1984-1985; SRI 1981]. Most of them, however, focus mainly on the relation between the economic growth and automobile ownership. Therefore these studies could not analyse the constitution of automobile ownership on the fuel price elasticity. Oyama and Kawashima [1983] constructed a model to estimate the automobile ownership and suggested the application of the cohort model for the estimation. However, their model is limited by the constant ratio assumption, that is the constant choice ratio between the diesel and the gasoline automobile, and the constant survival ratio, whereas in reality the ratio for each cohort changes.

3.2 Model Structure

A population born in the same year is called a cohort in the field of population study, and the cohort model is utilised to estimate the total population by monitoring the birth rate, the migration and the mortality rate of each cohort in the time series. This method, in the context of this paper, will be used to predict the constitution of automobile ownership. Although the Oyama-Kawashima model adopted the same method, if only forecasts the total number of vehicles. By comparison, our model analyses the fuel price elasticity of diesel automobiles' share. Figure 3 shows the flowchart of a cohort model to predict the number of diesel automobiles. First, we classify vehicles in the term <t-1> into the diesel and the gasoline automobile category. Their cohort pyramids are derived from the age of the automobiles. When the term changes, the new cohort pyramid of the term <t> is derived by multiplying the old cohort pyramid in the term <t-1> by the survival ratio of each cohort. These are formulated as:

$$D_{t,M+1} = DR_{t-1,M}\, D_{t-1,M}\,, \tag{1.a}$$

$$G_{t,M+1} = GR_{t-1,M}\, G_{t-1,M}\,, \tag{1.b}$$

where D is the number of diesel automobiles; G is the number of gasoline automobiles; DR is the survival ratio of diesel automobiles; GR is the survival ratio of gasoline automobiles; subscript <t> means the term; subscript <M> means the age of automobile.

Secondly, we obtain the number of newly registered vehicles. Although this number seems to depend upon the number of people who replace their automobiles and the number of first time buyers, it is difficult to observe the number in time series. In this study, the number of newly registered vehicles is defined by subtracting the number of the over-two-years-old automobiles from the total number of automobiles in each term. This is formulated as:

$$N_t = T_t - \Sigma_M\, DR_{t-1,M}\, D_{t-1,M} - \Sigma_M\, GR_{t-1,M}\, G_{t-1,M}\,, \tag{2}$$

where N is the number of newly registered vehicles; T is the number of total automobiles.

Thirdly, by multiplying the number of newly registered vehicles by the choice ratio between the diesel and the gasoline automobile, we derive the number of one-year-old vehicles and complete their cohort pyramids. These are formulated as:

$$D_{t,1} = Y_t\, N_t\,, \tag{3.a}$$

$$G_{t,1} = (1-Y_t)\, N_t\,, \tag{3.b}$$

where Y is the choice ratio of the diesel automobile between the diesel and the gasoline automobile.

Considering equations (1.a), (1.b), (2), (3.a) and (3.b), we can construct the cohort model for predicting diesel automobiles' share. Note that the survival model which gives the values DR in equation (1.a) and GR in equation (1.b), the trend model which gives the values T in equation (2), and the choice model which gives the values Y in equations (3.a) and (3.b) in the following sections.

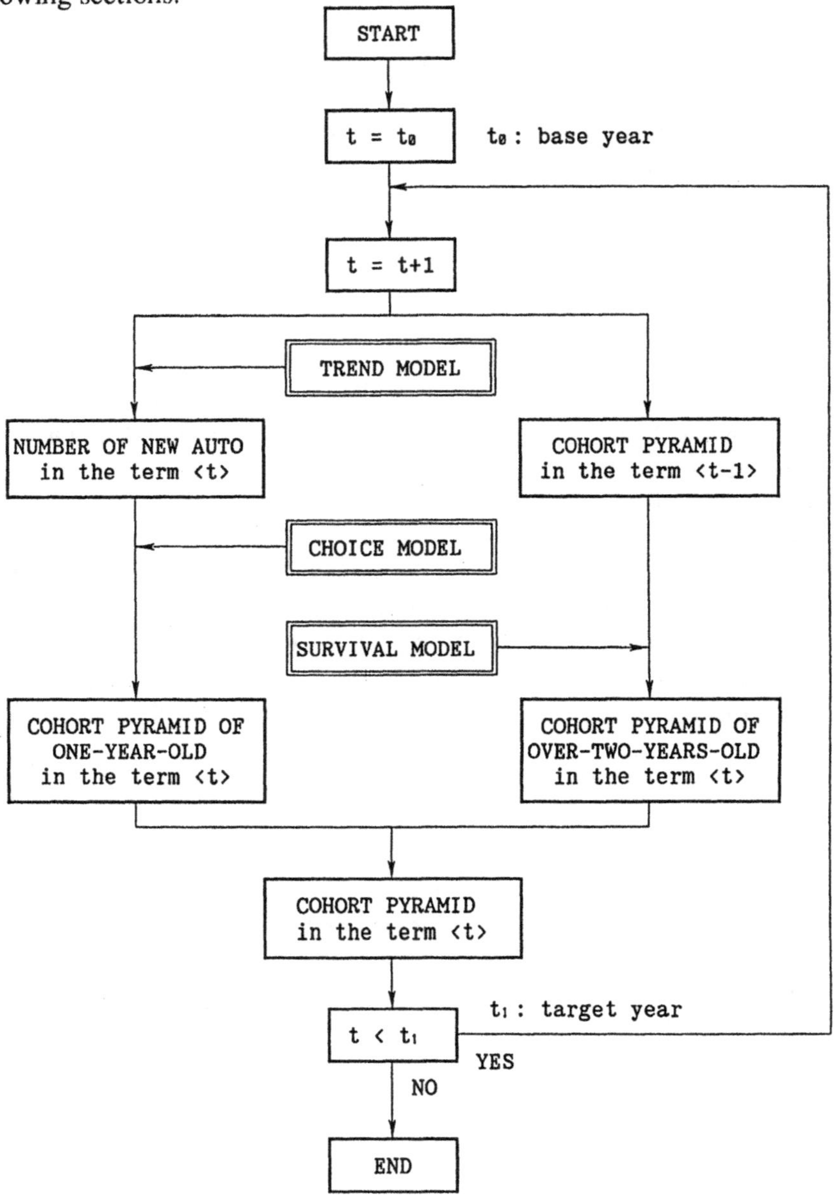

Fig. 3. Flowchart of Cohort Model

3.3 Survival Model

An automobile user might decide to replace his current automobile if it no longer functions satisfactorily, or, when he loses interest in it. By modelling their behaviour within the framework of the random utility theory, using the same procedure as in the case of the choice model, we construct the survival model in the binary logit form, which gives the survival ratio DR in equation (1.a) and GR in equation (1.b). There seems to be three factors which influence the replacement of automobiles: first, is the durability of automobile, which has been improved by technological innovation; second, is the decline of durability of an automobile; and third, is the price of light-oil and gasoline. By using the year as the first index of durability and the automobile's age as the second index, we define the utility function of automobile users on replacement, which has the random error additionally. Assuming that the error distributes as the Gumbel distribution in the probability, the survival model is derived as the binary logit form for each type of automobile and engine:

$$DR_{t,M} = \frac{\exp[DL_t]}{\exp[DL_t] + \exp[DD_t]} = \frac{1}{1 + \exp[DD_t\text{-}DL_t]}, \qquad (4.a)$$

$$DD_t\text{-}DL_t = \alpha_0 + \alpha_1(t\text{-}t_0) + \alpha_2(MD_{t-1}\text{-}M) + \alpha_3 \ln[DP_{t-1}/GP_{t-1}], \quad \text{for sedan,} \qquad (4.b)$$

$$DD_t\text{-}DL_t = \alpha_0 + \alpha_1 \ln[t\text{-}t_0] + \alpha_2(MD_{t-1}\text{-}M), \quad \text{for small-truck,} \qquad (4.c)$$

where DD is the utility on renewing of the diesel automobile; DL is the utility on continuing of the diesel automobile; DP is the price of the light-oil [yen/l]; GP is the price of the gasoline [yen/l]; MD is the average age of diesel automobiles; M is the age of automobile; t is the year; t_0 is the base year; α_0, α_1, α_2, and α_3 are unknown parameters.

$$GR_{t,M} = \frac{\exp[GL_t]}{\exp[GL_t] + \exp[GD_t]} = \frac{1}{1 + \exp[GD_t\text{-}GL_t]}, \qquad (5.a)$$

$$GD_t\text{-}GL_t = \beta_0 + \beta_1(t\text{-}t_0) + \beta_2(MG_{t-1}\text{-}M), \quad \text{for sedan,} \qquad (5.b)$$

$$GD_t\text{-}GL_t = \beta_0 + \beta_1 \ln[t\text{-}t_0] + \beta_2(MG_{t-1}\text{-}M), \quad \text{for small-truck,} \qquad (5.c)$$

where GD is the utility on renewing of the gasoline automobile; GL the utility on continuing of the gasoline automobile; MG is the average age of gasoline automobiles; β_0, β_1, and β_2 are unknown parameters.

For estimating equations (4.a)-(4.c) and (5.a)-(5.c), we used the time series data from 1974 to 1991 [AIRA 1973-1992], and got enough values of the correlation coefficient between the observed and the estimated values. According to the t-values, it should be appropriate to take those factors into the survival model. The base year t_0 in these equations is assumed to be 1974, which gives the better t-values and the better correlation coefficient to the model.

Table 2. Estimated Parameters of DR Model

(a) sedan		(b) small-truck	
parameter	estimated value (t-value)	parameter	estimated value (t-value)
α_0	-5.763 (19.32)	α_0	-3.706 (35.64)
α_1	4.917×10^{-2} (4.321)	α_1	2.311×10^{-1} (5.088)
α_2	-4.077×10^{-1} (25.62)	α_2	-3.134×10^{-1} (31.08)
α_3	-2.885 (6.157)		
correlation coefficient: 0.876		correlation coefficient: 0.905	

Table 3. Estimated Parameters of GR Model

(a) sedan		(b) small-truck	
parameter	estimated value (t-value)	parameter	estimated value (t-value)
β_0	-3.652 (30.33)	β_0	-2.943 (27.44)
β_1	4.450×10^{-2} (4.324)	β_1	2.989×10^{-1} (6.238)
β_2	-4.596×10^{-1} (29.46)	β_2	-3.236×10^{-1} (30.68)
correlation coefficient: 0.901		correlation coefficient: 0.903	

3.4 Trend Model

There may be some relationship between the total number of sedan and that of small-truck because of the ownership transfer between them. However, since it is difficult to observe these relationship in the market, the trend model to predict the total number of automobiles is constructed in each market, the sedan and the small-truck market. This model is constructed by assuming a linear function as:

$$T_t = \gamma_0 + \gamma_1(GP_{t-1}+DP_{t-1})/2 + \gamma_2 T_{t-1}\,, \tag{6}$$

where γ_0, γ_1, and γ_2 are unknown parameters.

For estimating equation (6), we also used the time series data from 1974 to 1991 [AIRA 1973-1992], and got enough high values of the correlation coefficient and the t-values.

Table 4. Estimated Parameters of T Model

(a) sedan		(b) small-truck	
parameter	estimated value (t-value)	parameter	estimated value (t-value)
γ_0	2.393×10^6 (8.513)	γ_0	1.082×10^6 (1.640)
γ_1	-6.072×10^3 (2.572)	γ_1	-3.026×10^3 (1.145)
γ_2	9.652×10^{-1} (106.0)	γ_2	8.877×10^{-1} (6.354)
correlation coefficient: 0.999		correlation coefficient: 0.941	

3.5 Choice Model

When automobile users buy and register their new vehicles, they choose between the diesel and the gasoline automobiles. By modelling their behaviour within the framework of the random utility theory, we construct the choice model in the binary logit form, which gives the choice ratio Y in equations (3.a) and (3.b). There seems to be three kinds of factors for choosing the type of new vehicle: the motive power factor which includes its power, weight, combustion efficiency and durability; the environmental factor such as its emission volume, noise and vibration; and, the economic factor such as its fuel saving and fuel price, and so on. Considering the fact that most of the automobiles may be used by the users as the means for their travel, thus, the third factor must be the most important consideration. The number and the share of the vehicle in the automobile market may also be taken into consideration as its social reliability on choosing.

By using these kinds of factors, we define the utility functions of the automobile users for choosing between the diesel automobile and the gasoline one, which has the random error additionally. Assuming that the error distributes as the Gumbel distribution in the probability, the choice model is derived as the binary logit form:

$$Y_t = \frac{\delta \exp[GL_t]}{\exp[DU_t] + \exp[GU_t]} = \frac{\delta}{1 + \exp[GU_t\text{-}DU_t]}, \tag{7.a}$$

$$GU_t\text{-}DU_t = \varepsilon_0 + \varepsilon_1(GP_{t-1}\text{-}DP_{t-1}) + \varepsilon_2 \ln[\Sigma_M G_{t-1,M}/\Sigma_M D_{t-1,M}], \tag{7.b}$$

where DU is the utility on choosing diesel automobile; GU is the utility on choosing gasoline automobile; ε_0, ε_1, ε_2, and δ are unknown parameters.

For estimating equation (7.a)-(7.b), we also used the time series data from 1974 to 1991 [AIRA 1973-1992], and obtained enough high values of the correlation coefficient between the observed and the estimated values. The parameter δ is assumed as a certain value which gives the better t-values and the better correlation coefficient to the model.

Table 5. Estimated Parameters of Y Model

(a) sedan		(b) small-truck	
parameter	estimated value (t-value)	parameter	estimated value (t-value)
ε_0	6.088×10^{-1} (0.323)	ε_0	-1.037×10^{-1} (1.000)
ε_1	-6.052×10^{-2} (1.414)	ε_1	-1.376 (5.662)
ε_2	9.362×10^{-1} (12.19)	ε_2	2.717×10^{-1} (34.87)
δ	0.3 :	δ	1.0 :
correlation coefficient: 0.954		correlation coefficient: 0.995	

3.6 Total Test

By using this cohort model, we carried out the simulation of change in the diesel automobiles' share. Figure 4 shows the simulation result by indicating the observed values on the estimated curve. For the small-truck, the result exhibits high fitness. The total number of sedan, on the other hand, tends to be overestimated.

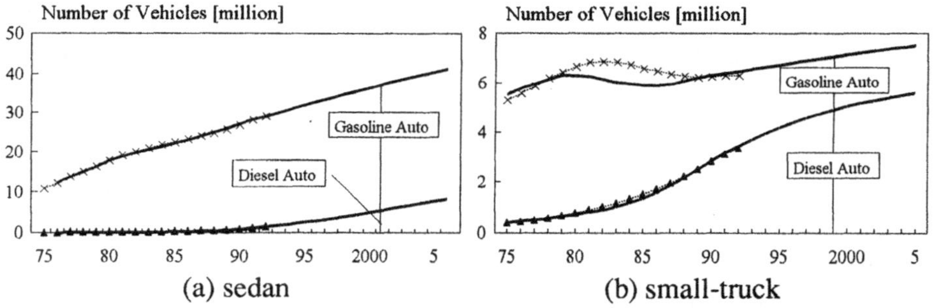

Fig. 4. Estimation of Diesel Automobiles' Share

4. Simulation Analysis on NOx Reduction by the Policy

This study assumes that the policy to reduce diesel automobiles is carried out in 1992, and analyses its impacts not only on the share of number vehicles but also on the NOx volume by using this cohort model. The NOx regulation level in each year is used as the NOx volume for each automobile of each cohort, where its unit is [ton/h] (ton per hour). The policy alternatives are indicated as follows.

POLICY 1: to stop producing the diesel automobile

POLICY 2: to increase the light-oil price/tax
<gasoline> 128 [yen/*l*]
<light-oil>
- Case 1 ; 102 [yen/*l*] (=128×0.8)
- Case 2 ; 128 (=128×1.0)
- Case 3 ; 154 (=128×1.2)

POLICY 3: to set the upper limit of age on usage of the diesel automobile
<upper limit of age>
- Case 1, 2, 3, 4, and 5; 14, 12, 10, 8, and 6 [years-old], respectively.

The impact of POLICY 1 is shown in Figure 5. If the present situation continues (without-policy situation), first of all, the diesel automobiles' share in the sedan market will increase from 4% in 1991 to 20% in 2005, and the share in the small-truck market will also increase from 50% to 75%. Although the NOx volume reduction has been achieved through the numerous NOx emission regulation, a without-policy situation makes it increase in both markets. Figure 5 indicates that POLICY 1 makes both the diesel automobiles' share and the NOx volume decrease drastically, albeit over a long time period.

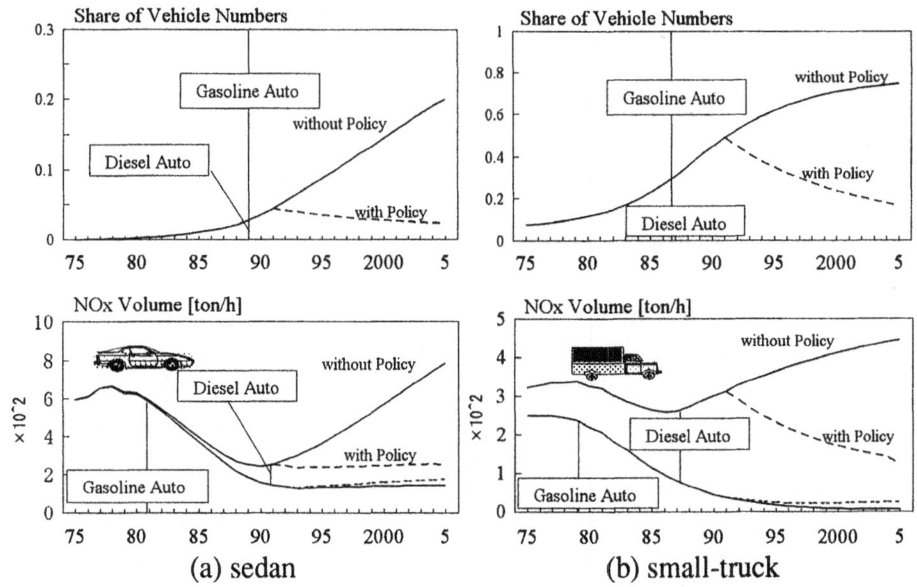

Fig. 5. Impact of POLICY 1

POLICY 2 is the fuel price control policy and its impacts is shown in Figure 6. This result indicates that the light-oil price should be higher than the gasoline price at least in order to reduce the NOx volume from automobiles. From the viewpoint of the market mechanism, this policy should induce the high price of commodities because it increases the transport cost.

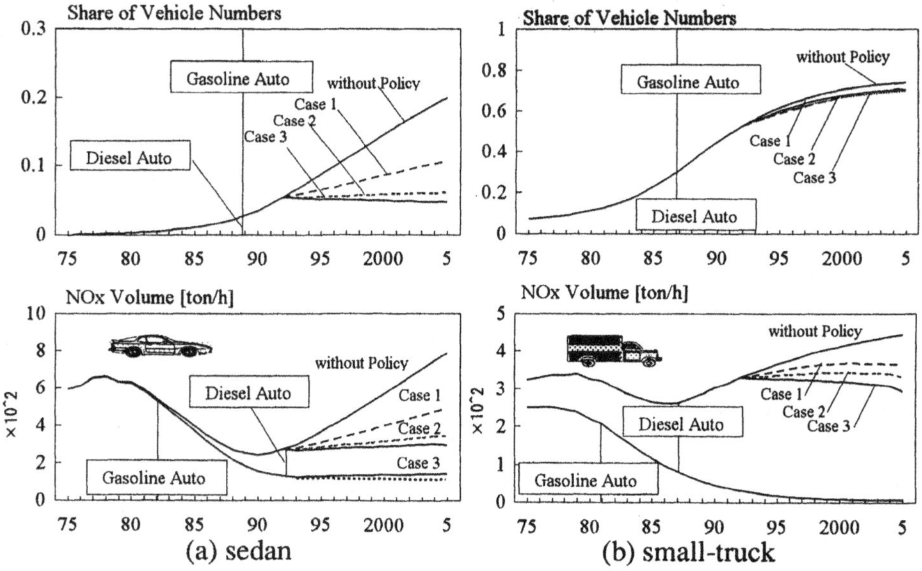

Fig. 6. Impact of POLICY 2

POLICY 3 requires the discontinuation of use of the old automobiles because older cars tend to emit much more NOx than the new ones. Its impacts is shown in Figure 7. In order to achieve a much higher reduction of the NOx volume, there is a need to change old automobiles to new ones much earlier than the present practice. However, this may cost the user more than the accepted level. This policy would also induce an increase in the commodity prices.

From these results, it can be seen that the policy to stop producing the diesel automobile and the light-oil price increase may lead to the achievement of high reduction of the NOx volume. There may be some other policies, for example the policy that the gasoline automobile should be required to sell a certain ratio of the number of diesel automobiles sold in the year before. Although we have tried those kinds of policy, the results show that the differences are not that significant. We tried, therefore, some combined policies as shown in Figure 8. From this case which combines POLICY 2 (Case 1) with POLICY 3 (Case 1-5), we may understand that a certain combination of policies can achieve enough results even if each policy is not so heavy.

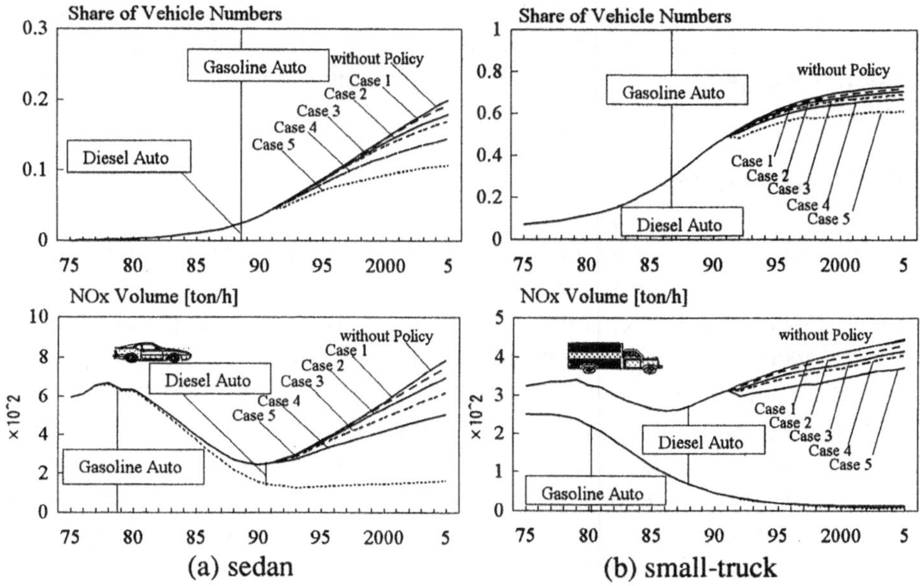

Fig. 7. Impact of POLICY 3

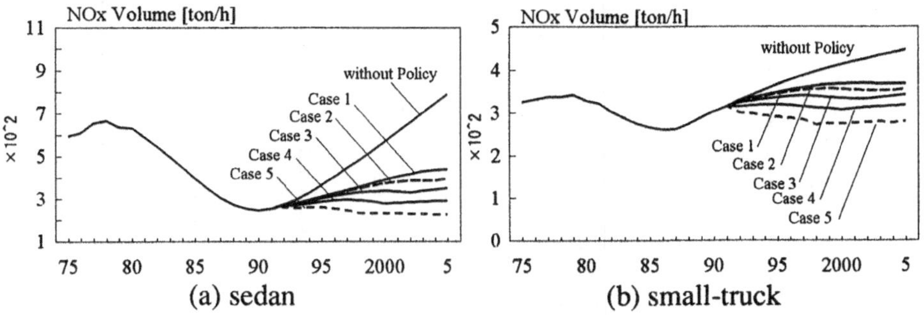

Fig. 8. Impact of POLICY 2 + POLICY 3

5. Model for Predicting Price Change

5.1 Assumption

Since the policy to reduce diesel automobiles, as mentioned above, may cause a rise in commodity prices, we constructed a model for predicting price change by the policy and forecast price increasing ratio. Now, the policy to be analysed here is restricted to the economic policy that increases the light-oil price/tax since this is the only policy that might be feasible and acceptable in our society, and has a good enough effect on NOx reduction as shown in the previous discussion. The subject of analysis is also restricted to the freight transport by truck because only the truck transport has enough data to carry out the price equilibrium analysis by its price increasing, where the analysis is to predict price change by using the input-output analysis [Kaneko 1967].

5.2 Model Structure

First of all, we prepare I-O table evaluated by producers price and the freight input table as shown in Figure 9 [MCA 1989]. In Figure 9, x_{ij} is the input of sector j from sector i; X_j is the total output, Y_i is the final demand, V_j is the value added, superscript <s> means the transport input, subscript <1, ..., m> means the intermediate sector, subscript <n> means the transport sector.

Using x_{nj} and X_j in Figure 9, θ (x_{nj}/X_j) means a direct rise in product price of industry j when θ represents increasing ratio of freight rates. A rise in this price causes the increase in cost of production of each industrial sector because each sector uses freight transport as input of production, and which in turn may influence to increase the price of products.

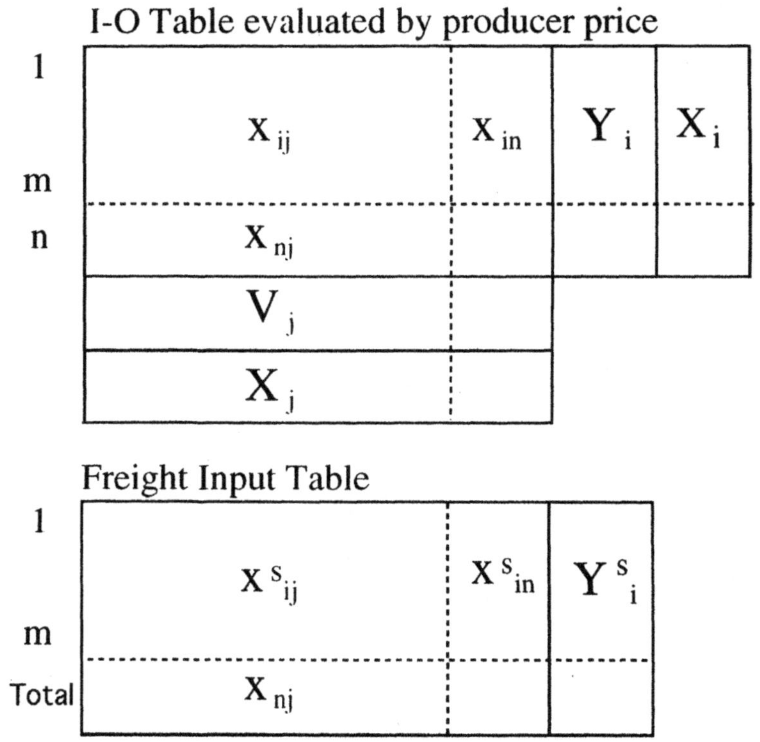

Fig. 9. I-O and Freight Input Tables

Increasing ratio of product price from the viewpoint of producer, ΔpP_j, is expressed as:

$$[\Delta pP_j] = \theta\, [x_{nj}/X_j][I\text{-}A]^{-1}\,, \qquad (8)$$

where I is the unit matrix (m×m matrix); A is the input coefficient matrix (m×m matrix); [ ] is expression of matrix or low vector. From the viewpoint of consumer, on the other hand, increasing ratio of product price $\Delta pP_i'$ is based on the cost of insurance and freight (CIF) price base as:

$$\Delta pP_i' = \frac{X_i \Delta pP_i}{X_i + (\Sigma_j x^s_{ij} + Y^s_i)}. \qquad (9)$$

As consumers buy goods through the transport sector, increasing ratio of price which is paid by consumers, Δp^t_i, is:

$$\Delta p^t_i = \frac{\theta\, (\Sigma_j x^s_{ij} + Y^s_i)}{X_i + (\Sigma_j x^s_{ij} + Y^s_i)}. \qquad (10)$$

Then, the increasing ratio of consumer price from the viewpoint of consumers, Δp^c_i, is given by $(\Delta pP_i' + \Delta p^t_i)$. By using the price equilibrium analysis as mentioned here and the input-output table, we forecasted the price change by the policy to reduce diesel automobiles in order to reduce the NOx volume.

5.3 How to Derive θ (Increasing Ratio of Freight Rates)

The freight rates can be changed by permission of the Minister of Transportation [MOT 1990]. However, only in the case within ±10%, can it be changed freely by each transport company. In this study, we derived the increasing ratio of the freight rates by assuming that the increase of total operating expenses of the freight transport works is covered by freight rates revenue. This assumption means that the increase of operating cost causes a direct increase in the freight rates. Now, the studied sectors are set limited to the road freight transport and the self-freight transport by private motor cars.

The increasing ratio of the total operating expenses of the freight transport can be derived by using the increasing ratio of the light-oil price and the cost composition ratio of the light-oil. The cost composition ratio of the former sector can be used without any transformation, but that of the latter sector needs some transformation because the cost composition of the sector does not include value added, for example the wages. Then we set two cases in the self-freight transport by private motor cars; in one case (NVA) value added is equal to 0, and in the other case (VA) value added is set in the way that value added composition ratio is equal to a half of that in the road freight transport. As a result, the cost composition ratio is given in Table 6. Increasing ratio of freight rates θ, therefore, can be given by multiplying the increasing ratio of the light-oil price by the cost composition ratio of the light-oil.

Table 6. Cost Composition Ratio of Freight Transport Works

	Road Freight Transport %	Self-freight Transport (NVA) %	Self-freight Transport (VA) %
Intermediate Sectors			
Gasoline	0.25	21.90	10.55
Light-oil	8.59	13.08	6.30
Repair of Motor Vehicles	5.59	27.11	13.06
:	:	:	:
Total	31.72	100.	48.16
Value Added Sectors			
Wages and Salaries	47.89		36.35
:	:		:
Total	68.28		51.84
Total Domestic Products	100.	100.	100.

6. Forecast of Increasing Ratio of Freight Rates and Prices

The increasing ratio of the light-oil price and the freight rates as a result of the policy to increase the light-oil price/tax are shown in Table 7. The input-output table and the fuel prices which are used to calculate values in Table 7 are for 1985, where the light-oil price is 100.4 yen/*l* and the gasoline price is 143.0 yen/*l* in national average values.

Table 7. Increasing Ratio of Light-oil Price and Freight Rates

	Light-oil Price Increase Rate %	Road Freight Transport Price Increase Rate %	Self-freight Transport Price Increase Rate (NVA) %	Self-freight Transport Price Increase Rate (VA) %
Case 1	13.9	1.198	1.824	0.879
Case 2	42.4	3.645	5.551	2.673
Case 3	70.9	6.093	9.277	4.468

Notes 1) Case 1 : (light-oil price)/(gasoline price) = 0.8
2) Case 2 : = 1.0
3) Case 3 : = 1.2

The increasing ratio of prices of the first 20 sectors, by the policy that increases the light-oil price to 120% of the gasoline price (Case 3), is shown in Figure 10. Only in 8 sectors in the case of NVA, or 7 sectors in the case of VA, the increasing ratio of price is over 1%. In the commerce and mining sectors that are influenced by transport rates, the increase at the consumption stage is greater than at the production stage. In the processing and construction sectors, on the other hand, the increase at the production stage is greater.

The increasing ratio of average price by the policy is shown in Table 8. Even in the case of the policy that increases the light-oil price to 120% of the gasoline price (Case 3), the increasing ratio of average price is only 0.306% in the case of NVA, or 0.235% in the case of VA. Compared with the policy to introduce consumer tax (3%), these results on price change are almost negligible.

Table 8. Increasing Ratio of Average Price

(a) Case of NVA				(b) Case of VA			
	A %	B %	A+B %		A %	B %	A+B %
Case 1	0.040	0.021	0.060	Case 1	0.030	0.016	0.046
Case 2	0.120	0.063	0.183	Case 2	0.092	0.049	0.141
Case 3	0.201	0.105	0.306	Case 3	0.153	0.082	0.235

Notes 1) A : increasing ratio of price at the production stage,
2) B : increasing ratio of price at the consumption stage,
3) A+B : increasing ratio of price.

7. Conclusions

This paper presents a cohort type of simulation model both, to predict the number of diesel automobiles and the NOx volume, and to analyse the impact of the policy to reduce diesel automobiles. This model is estimated by using the time series data from 1974 to 1991, and the estimated model can explain the observed reality well. By using this cohort model, this study came out with a model for predicting the number of diesel automobiles and the NOx volume under certain policy conditions. According to this simulation,

1) without any policy in 1992, the NOx volume will increase, where the diesel automobiles' share in the sedan market will change from 4% in 1991 to 20% in 2005, and the share in the small-truck market will also increase from 50% to 75%,
2) the policy to stop producing the diesel automobile makes its share and the NOx volume decrease drastically, however it seems difficult to have the consensus because it makes producer loss and user inconvenience,

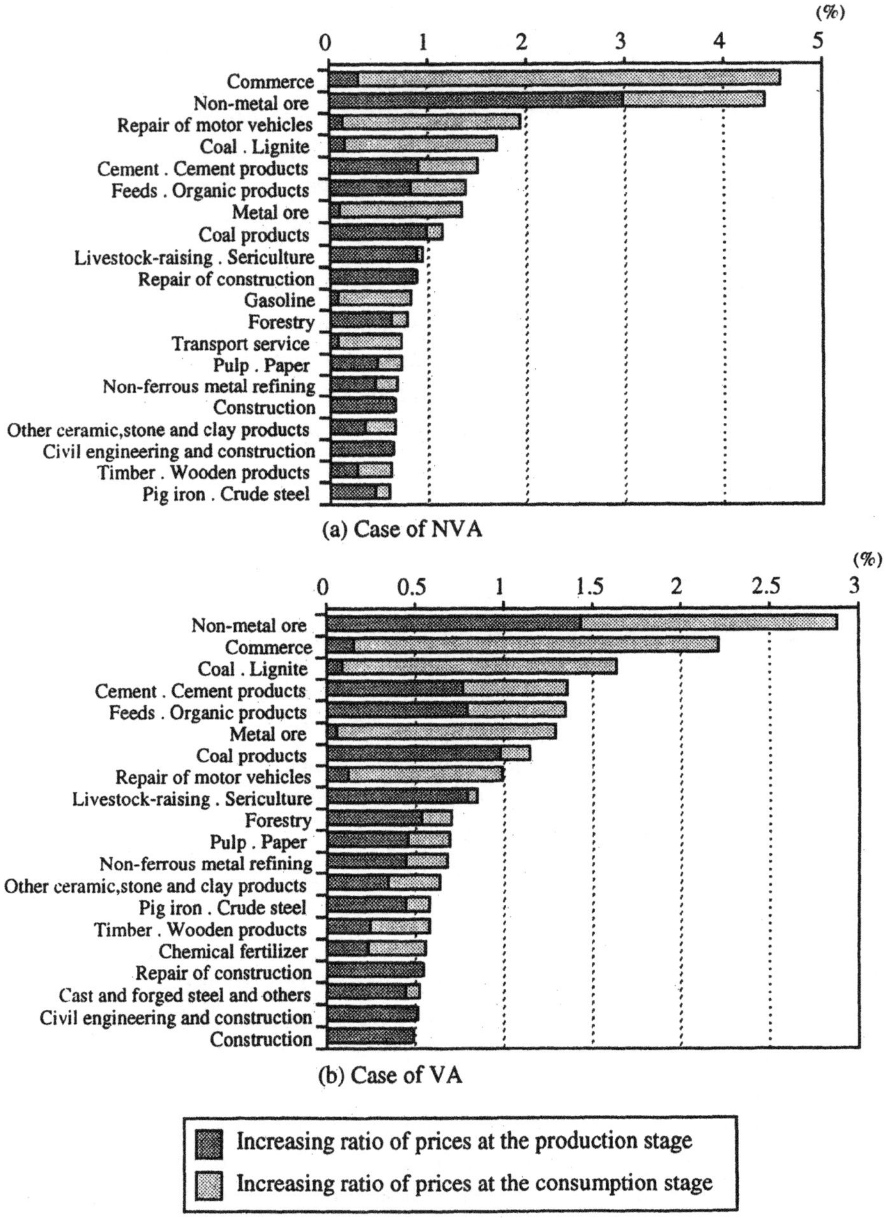

Fig. 10. Increasing Ratio of Prices

3) the light-oil price/tax increase policy might be the most appropriate because it has an efficient effect on reduction of the NOx volume and might obtain the consensus if it does not result in a big commodity price increase,
4) the policy to set the upper limit of age on the diesel automobile are not so efficient, and may induce an increase in commodity prices, and

5) a certain combination of policies can achieve enough results even if each policy is not so heavy.

The economic policies such as the fuel price/tax control, however, may cause an increase in commodity prices through the increasing of transport prices. From the viewpoint of necessity to predict price change by the policy in its evaluation, this paper presents a model based on the price equilibrium analysis by using the input-output analysis, and predicts price change by those policies. According to the prediction,

1) the policy to set the light-oil price to 120% of the gasoline price makes prices of 7 or 8 sectors increase over 1%, where
2) influence of transport rates increasing at the consumption stage is greater than at the production stage in the commerce and the mining sectors, and that at the production stage is greater in the processing and the construction sectors, and where
3) increasing ratio of average price is only 0.306% or 0.235%, where this result of price change is almost negligible, compared with the policy to introduce the consumer tax (3%), then
4) the light-oil price/tax increase of 120% does not have that much effect on the general price index, so this paper recommends this policy as the best among the alternatives.

References

Allanson, E. W. (1982). Car Ownership Forecasting, *Transportation Studies*, 1, 17-27 and 143-145, Gordon and Breach Science Publishers.

Automobile Inspection and Registration Association [AIRA] (1973-1992). *Number of Registered Automobiles*.

Environment Agency [EA] (1992). *Quality of the Environment in Japan.*

International Energy Agency [IEA] (1993). *Energy and the Environment Series: Cars and Climate Change*, OECD, Paris.

Iwata, K. (1990). Policy to Environmental Pollution by Automobile Traffic, *Environmental Pollution Study*, 19, 3, 38-43.

Kaneko, Y. (1967). Economic Change and Input-Output, *The New Criticism*, 143-184.

Management and Coordination Agency [MCA] (1989). *Input-Output Table in 1985.*

Ministry of Construction [MOC] (1993). *Road Pocket Book.*

Ministry of Transportation [MOT] (1990). *Freight Rates System for Transportation Works.*

Morisugi, H., Ohno, E. and Kawamata, T. (1990). A Cohort Model for Predicting Diesel Automobile Share and Dynamic Analysis on Fuel Price Elasticity, *Infrastructure Planning Review*, 8, 41-48.

Nagai, S. (1987). Urban Environment and Traffic Pollution, *Education Seminar*, 49-53.

Nakamura, H., Kashima, S. and Hidano, N. (1984-1985). Present Situation of Automobile Ownership and its Forecast, *Expressways and Automobiles*, 27, 11, 22-32; 27, 12, 25-36; 28, 1, 38-47.

Oyama, T. and Kawashima, H. (1983). Analysis and Forecast of Automobile Demand Structure, *Operations Research.*

System Research Institute [SRI] (1981). *Change of Automobile and Its Fuel Demand.*

CHAPTER 10

SHORT-TERM IMPACT ANALYSIS OF PRICING STRATEGIES ON VMT REDUCTION

Tschangho John Kim and Paul F. Hanley

1. Introduction

In the United States, the basic approach to regional air quality management was established by the federal Clean Air Act of 1970. Under the Act, the US Environmental Protection Agency (US EPA) sets National Ambient Air Quality Standards (NAAQS) for air pollutants. Areas which fail to meet NAAQS are called non-attainment areas.

The Clean Air Act Amendments of 1990 (CAAA 1990) imposed new and more stringent requirements for ozone non-attainment areas, and in particular, for the reduction of volatile organic compound (VOCs) emissions from mobile sources. Ozone non-attainment areas are classified on the basis of how badly the NAAQS for ozone are exceeded. Los Angeles is classified as the worst area, and seven other non-attainment areas fall into the severe classification. The Chicago region, composed of the city of Chicago and its suburbs, is one of the seven severe regions.

According to the mandates imposed by the 1990 CAAA, Chicago must be brought into compliance with the ozone standards by the year 2007. By November 15, 1992, the State of Illinois must submit to US EPA a Chicago area emission inventory for VOC, oxide of nitrogen (NO_X), and carbon monoxide (CO) emissions in 1990 (US EPA 1991), which is the base year used in setting targets for demonstrating reasonable further progress (RFP).

Y. Hayashi and J. Roy (eds.), Transport, Land-Use and the Environment, 191–212.

Also by that date, the State of Illinois must submit its plan for tightening up on vehicle inspection and maintenance, and for implementing the employer trip reduction (ETR) requirements, and other transportation control measures sufficient to offset any growth in emissions resulting from growth in VMT or the number of vehicle trips (Kane, John and Kim 1992).

In addition, the CAAA of 1990 require each state to prepare and submit a State Implementation Plan (SIP) by November 1992 that outlines how the state proposes to accomplish reductions in vehicle miles travelled (VMT) and emissions of VOC from mobile non-point sources. By 1993, each state's SIP must show reasonable progress toward achieving a 15% reduction in VOC by 1996 and a 3% further reduction per year thereafter until the non-attainment areas meet the standard. By 1994, major employers in each state must submit for state certification Employer Trip Reduction (ETR) plans that give convincing evidence that strategies for compliance will be achieved by the statutory date of 1996. Penalties will be levied for failure to comply.

In order to comply with these mandates, many strategies or TCMs have been suggested for reducing emissions and VMT. Some of those suggestions are shown in Appendix A. Some are voluntary in nature such as "Sib Bu Je" in Seoul, Korea. This is a campaign by a public-private coalition in Korea to encourage drivers to not drive their automobiles on days when the last digit of their license plate number matches the last digit of the day's date. This could effectively reduce the use of automobiles up to 10% per day. Some TCMs shown in Appendix A are pricing strategies and some are regulatory (US EPA 1989, 1990).

With these new mandates of the 1990 CAAA, the purpose of the paper is to assess the impact of some of these transportation control measures, particularly pricing strategies, on the reduction of mobile source emissions and VMT.

2. Compliance Issues in the Chicago Area

A brief assessment of the implications of these amendments indicates that the amount VMT reduction required in the Chicago non-attainment area would be enormous. Consider the following data.

The 1990 Baseline Profiles of the Chicago Non-attainment Area are as follows:

a. VMT:	128 million miles/day
b. Emissions:	1,270 tons/day
c. 15% of b:	190 tons/day

Even if we assume that as much as two thirds of the required 190 tons/day reduction would be achieved by technological enhancements, a reduction of 63 tons/day would still need to be realised. Assuming that 1 million miles/day VMT produce about 4.85 tons/day of emissions, this 63 tons/day reduction could be realised by eliminating about 13 million miles/day of VMT. We can deduce the following information:

d. Target VMT (128 mm-13 mm):	115 million miles/day
e. VMT growth at 2%/year:	144 million miles/day by 1996
f. VMT growth at 3.5%/year:	157 million miles/day by 1996
g. Needed reduction from e:	25% if VMT grow at 2%/year
h. Needed reduction from f:	37% if VMT grow at 3.5%/year

Explanations of the above profiles are as follows. According to Illinois Environmental Protection Agency (IEPA) estimates for 1990 (Illinois EPA,1992), the VMT was 128 million miles/day and the baseline emissions were 1,270 tons/day in the Chicago non-attainment area. Applying the 15% reduction guideline, the 1996 emission level must be below 1,080 tons/day for the area, requiring a reduction of 190 tons/day by 1996. Assuming that two thirds of this 190 tons/day reduction requirement or 127 tons/day can be achieved by implementing such enhanced technologies as a tougher inspection and maintenance (I/M) method, vapour recovery technology, reformulated gasoline, and intelligent vehicle highway systems that may reduce congestion and thus emissions, a reduction of 63 tons/day will still have to be achieved by other means. Further, assuming that 63 tons/day can be reduced by VMT reduction of 13 million miles/day, the 1996 VMT will have to be below 115 million miles/day. Let's mark 115 million miles/day as the target.

Assuming that the forecast by the Chicago Area Transportation Study (CATS) for VMT increase at 2% per year (CATS 1980) is accurate, the forecasted 1996 VMT/day would be 144 million miles/day, a 25% increase from the target of 115 million miles/day. If we use 3.5% per year as the rate of growth, which has been the rate during the period of 1984-1990 (IDOT 1980-1990), the 1996 VMT would be 157 million miles/day, a 37% increase from the target line of 115 million miles/day. The above analysis shows that VMT will have to be reduced by 25% to 37% in the next four years, a goal that seems extremely difficult to achieve, if not impossible.

3. Data

3.1 Delineating Analysis Zones

As required for any transportation study, this study area has been divided into several analysis zones for the impact analysis of transportation control measures (TCM's) on mode choice and vehicle miles travelled (VMT). Among the virtually infinite number of ways to delineate the area into analysis zones, we chose to divide the study area into 21 zones, plus one external zone, after our consultation with officials at the Chicago Area Transportation Studies (CATS).

We felt that a total of 22 zones are detailed enough to analyse impacts of TCM's on travel choice and, at the same time, a manageable size for operating spreadsheet type calculations. Travel choice analyses on a spreadsheet provide us with many insights on computational processes and their results, when compared to the conventional programming-based model implementation.

Figure 1 shows the boundaries and names of the analysis zones for this project's study area: the Chicago Region's Ozone Non-attainment Area.

3.2 Trip Tables for 1990 by CATS

The 1990 base year trip tables have been created by CATS according to the 22 zone system described above. The titles of the trip tables are:

Person Trips by Automobiles
- [IA]: Home to Work
- [IB]: Work to Home
- [IC]: Home to Non-Work
- [ID]: Non-Work to Home
- [IE]: Non-Home to Non-Home
- [IF]: Total Person Trips by Automobiles

Person Trips by Transit
- [IG]: Home to Work
- [IH]: Work to Home
- [II]: Home to Non-Work
- [IJ]: Non-Work to Home
- [IK]: Non-Home to Non-Home
- [IL]: Total Person Trips by Transit.

Total Person Trips
- [IM]: Total Person Trips

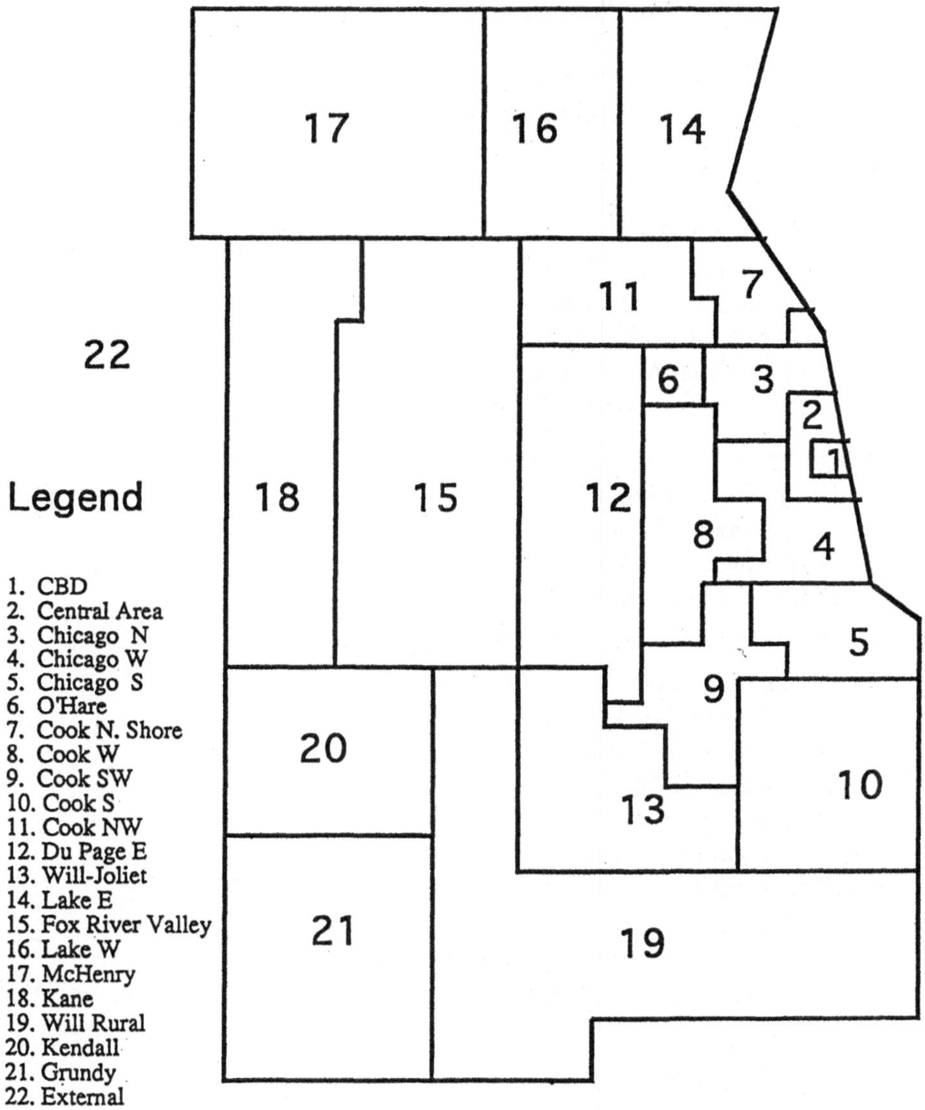

Fig. 1. Analysis Zones

Mode Share by Transit
- [IN]: Home to Work
- [IO]: Work to Home
- [IP]: Home to Non-Work
- [IQ]: Non-Work to Home
- [IR]: Non-Home to Non-Home
- [IS]: Total Person Trip Mode Share by Transit

Daily Automobile Trips
- [IIA]: Home to Work
- [IIB]: Work to Home
- [IIC]: Home to Non-Work
- [IID]: Non-Work to Home
- [IIE]: Non-Home to Non-Home
- [IIF]: External Automobile Trips
- [IIG]: Total Automobile Trips

Average Vehicle Occupancy Rates
- [IIH]: Home to Work
- [III]: Work to Home
- [IIJ]: Home to Non-Work
- [IIK]: Non-Work to Home
- [IIL]: Non-Home to Non-Home

Daily Truck Trip Tables
- [IIIA]: B Plate Truck Trips
- [IIIB]: Light Truck Trips
- [IIIC]: Medium Truck Trips
- [IIID]: Heavy Truck Trips
- [IIIE]: Total Truck Trips
- [IIIF]: Total Truck Trips: As Percent of Vehicle Traffic
- [IIIG]: External Truck Trips
- [IIIH]: Total Truck Trips including External Trips
- [IIII]: Total Truck Trips including External Trips: As Percent of Vehicle Traffic

Speed and Travel Time Tables
- [IVA]: Automobile Vehicle-Minutes
- [IVB]: Automobile Vehicle Miles
- [IVC]: Truck Vehicle Equivalent Minutes
- [IVD]: Truck Vehicle Equivalent Miles
- [IVE]: Average Interchange Speeds (Miles per Hour)
- [IVF]: Average Minutes between Districts
- [IVG]: Average Miles Between Districts

Highlights of the base year trip tables for the study area in 1990 are as follows:

A total of 20.4 million person-trips are made per day in 1990, of which about 18 million person-trips or 88 percent are using automobiles and 2.4 million person-trips or 12 percent are using transit.

Among the 18 million person-trips using automobiles, about 4.5 million person-trips or 25 percent are using them for work related trips.

A total of about 14 million automobiles are used by the 18 million persons using automobiles per day, which indicates that the average vehicle occupancy ratio is 1.29 for all purposes. A total of 3.9 million automobiles are involved in work related trips made by 4.5 million persons, which means that the average vehicle occupancy ratio is 1.15 for work related trips.

In terms of vehicle miles travelled (VMT), a total of 103.8 million miles are being travelled by 14 million automobiles per day, which gives an average distance travelled of 7.4 miles per car, per day. Of the total VMT, about 40 percent or 42.3 million miles are from work related trips made by 3.9 automobiles. The average distance travelled by an automobile for work related travel is about 10.8 miles.

Among the 2.4 million persons using transit per day, 1.5 million persons or 61.5 percent are using transit for work related trips.

3.3 Selected Average Distance between Zones and Person Trip Mode Share by Automobile

In order to estimate the distance between two zones, the question of appropriate measurement must be addressed. Is it the distance between the centres of zones? If so, what constitutes the centre of a zone? One might define zone centre as the physical centre of a zone or the activity centre of a zone. In either case, however, the estimated distance between centres of the 22 zones for this study may not adequately represent the actual distance between the original CATS' traffic zones, especially since our study zones are defined by the arbitrary delineation of boundaries combining a number of the CATS' traffic zones.

An alternative approach has been taken here. Average distances between study zones have been estimated by dividing each zonal value in the matrix for the Automobile Vehicle Miles [IVB] by the corresponding number of automobiles in each zonal pair in the matrix of the Total Automobile Trips [IIG]. Using this method, a zero is produced for any zone pair without trips between them. Thus, zeros exist not because the physical distances between zones are zero, but because of the method of estimating the composite distances explained above.

The Person Trip Mode Share by Automobile has been calculated by subtracting the Person Trip Mode Share by Transit [IS] from the unity.

4. Transportation Control Measures For Emission/VMT Reductions

Among many TCM's that may affect mode choice which, in turn, affect emission levels and VMT, we have chosen the strategies that affect the costs of automobile driving. We simulated mode choice changes using combinations of various levels of parking fee changes and several different price increases for gasoline as basic scenarios. We assume three levels of parking fee increases, i.e. $1.00, $2.00, and $3.00. In addition, four levels of gasoline price increases are assumed, i.e. $0.05, $0.25, $0.50, and $1.00. The combinations of these changes used for the simulations are:

Scenario 1: $1.00 parking fee plus $0.05/gallon gas tax
Scenario 2: $2.00 parking fee plus $0.25/gallon gas tax
Scenario 3: $3.00 parking fee plus $0.50/gallon gas tax
Scenario 4: $1.00 parking fee only
Scenario 5: $2.00 parking fee only
Scenario 6: $3.00 parking fee only
Scenario 7: $0.05/gallon gas tax only
Scenario 8: $0.25/gallon gas tax only
Scenario 9: $0.50/gallon gas tax only
Scenario 10: $1.00/gallon gas tax only
Scenario 11: $1.00 parking fee plus $0.25/gallon gas tax
Scenario 12: $1.00 parking fee plus $0.50/gallon gas tax

In estimating the impact of these scenarios, parking fee increases are assumed to affect only the work related trips, i.e. Home to Work and Work to Home trips. Thus, in estimating the impacts of the scenarios 1 to 3 and 11 to 12, we first obtained the costs of travelling between each pair of zones by various gas tax increases using the average distance between zones as calculated above, assuming the 16 mile/gallon fuel efficiency. We then added parking fee increases for the estimation of the work related.

For simulating scenarios 4 to 6, we applied cost increases due to the parking fee changes for work related trips only. We assume that non-work trips are not affected by the increases in parking fees. For simulating scenarios 7 to 10, we applied cost increases due to higher gas taxes to all-purpose trips.

5. Pivot Point Analysis and Analysis Model

Each of the scenarios listed above assume that the interzonal trip distribution is the same as the distribution given by the 1990 base year trip tables. Therefore, it is not necessary to relate all steps of the transportation planning modelling procedures in order to analyse impacts of price increases on mode choice. The pivot-point model determines the changes in mode choice given set of trips based on changes in cost of trips. The trip tables produced by the model show the shift of person trips from the automobile to transit induced by the increase in driving costs imposed by each scenario.

5.1 The Model

The pivot-point model used for the impact analysis of TCM's on mode choice is as follows:

$$\Delta p^{sk}(A) = -\beta[1 - p^{sk}(A)][p^{sk}(A)][\Delta c^{sA}] \qquad (1)$$

where,

$p^{sk}(A)$: Changes in probability of choosing automobile for the trip purpose k under scenario s

k: trip purposes; = 1, Home to Work trip
= 2, Work to Home trip
= 3, Home to Non-Work trip
= 4, Non-Work to Home trip
= 5, Non-Home to Non-Home trip

s: scenarios as described above, s = 1 to 12

β: mode choice coefficients related to automobile driving estimated by CATS
0.0072 for Work related trip and for non-CBD destination
0.0085 for Work related trip and for CBD destination
0.0329 for Non Work related trips

Δc^{sA} : changes in costs for driving an automobile under scenario s

See Appendix B for the derivation of the pivot-point model introduced above.

5.2 Estimating Probability of Choosing Automobiles

As briefly described above, we assume that parking fees affect only the work related automobile trips, while the gasoline tax increases affect trips for all purposes. For the estimation of changes in probability of choosing automobiles under scenarios 1 to 3 and 11 to 12, we first calculated cost increases due to gasoline taxes, based on 16 miles per gallon fuel efficiency for all trips.

For example, if a particular pair of zones are 8 miles apart and if gasoline tax is assumed to be increased by \$ 0.25/gallon, the cost increase due to gasoline tax increase to travel this particular pair is \$ 0.25 since it consumes 1 gallon for a round trip between this pair, assuming 16 mile per gallon fuel efficiency. Next, we added \$ 1.00, \$ 2.00 or \$ 3.00 parking fee increases for Δc^{sA} to estimate probability changes for work related trips. Using the above example, if parking fees increase by \$ 2.00, the total cost increase for driving an automobile between the pair of zones (Δc^{sA}) is \$ 2.25 for work related trips and \$ 0.25 for non-work related trips.

In estimating new probabilities (Δp^{sk}) due to these cost increases (Δc^{sA}) we simulated the model by imputing a \$ 0.01 increment due to the non-linear nature of the model. That means that if the total cost increase is \$ 2.25, we simulated the probability changes between the pair by 225 interactions (\$ 0.01 times 225), in order to estimate the final change in probabilities as accurately as possible.

Figures 2, 3, and 4 show the changes in probability for each \$0.01 increment assuming \$1.00 cost increases in travelling from zone 8 to zone 9, from 3 to 1, and from 10 to 11, respectively. These three figures indicate how the cost increase in automobile driving affects the probability of choosing automobiles in these three different starting conditions. The initial probability of choosing the automobile mode in Figure 2, 0.94, has been reduced to about 0.89, about a 5% reduction, by imposing a \$1.00 cost increase. Figure 3 shows the simulation results for a cost increase of \$1.00 between the origin zone 3 and the destination zone 1. The initial probability of 0.53 for the automobile choice has been decreased to 0.33, about 38% reduction. In Figure 4, the reduction is about 56% for the same amount of cost increase. In general, when the initial probability for choosing automobiles is high, the marginal decrease in the probability at the unit cost increase is relatively smaller than in other conditions where the initial probabilities is low. This is, in fact, in conformity with our intuition that the increase in automobile driving costs would encourage more drivers to shift to transit if the initial conditions attract more people to transit.

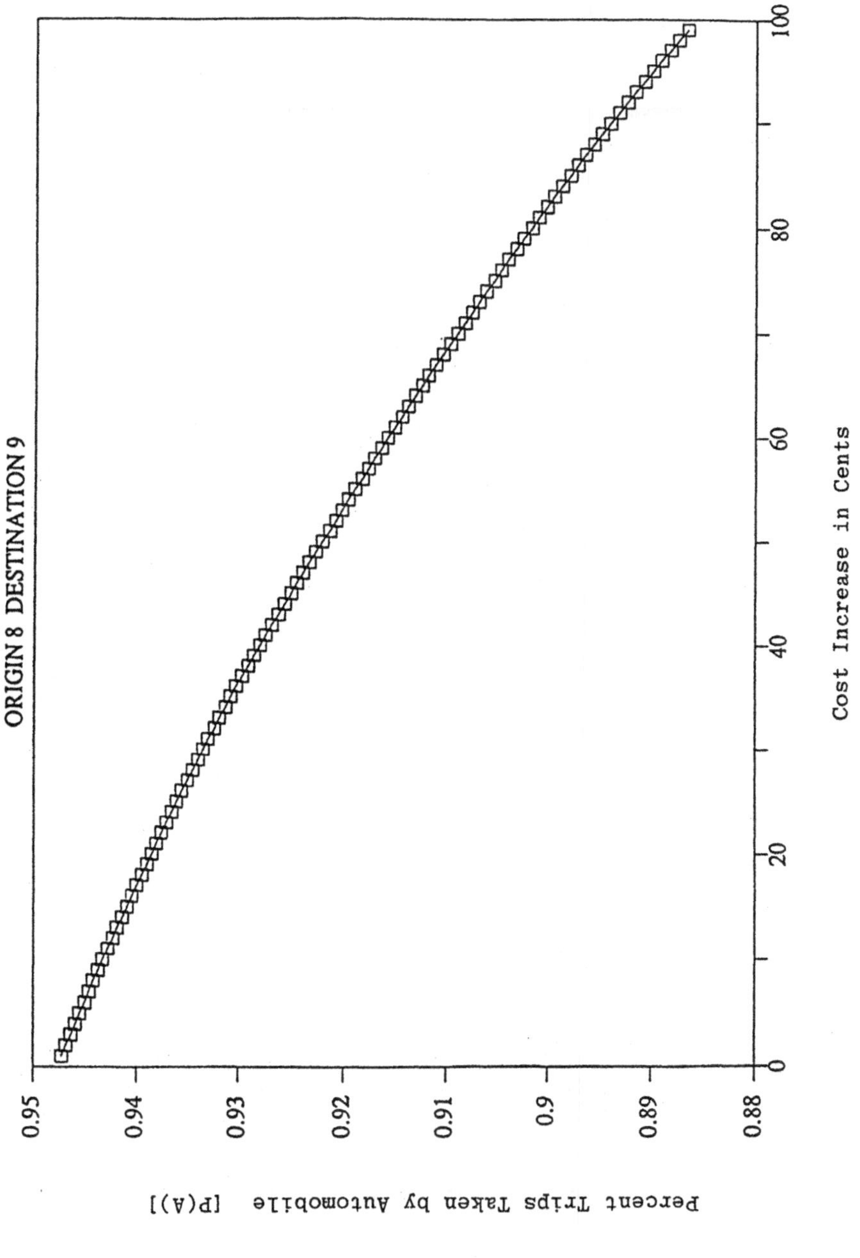

Fig. 2. Percent Trips Taken by Automobile as Function of Costs (1)

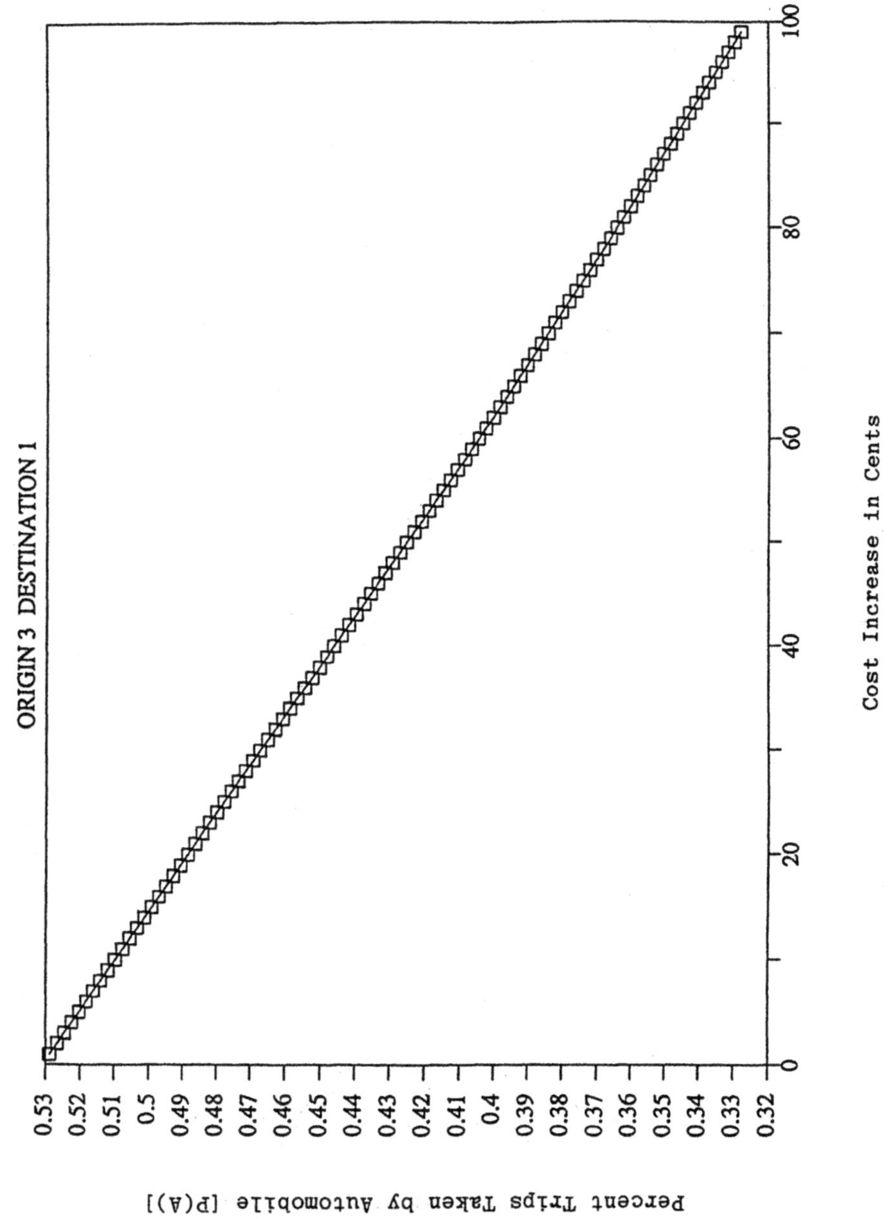

Fig. 3. Percent Trips Taken by Automobile as Function of Costs (2)

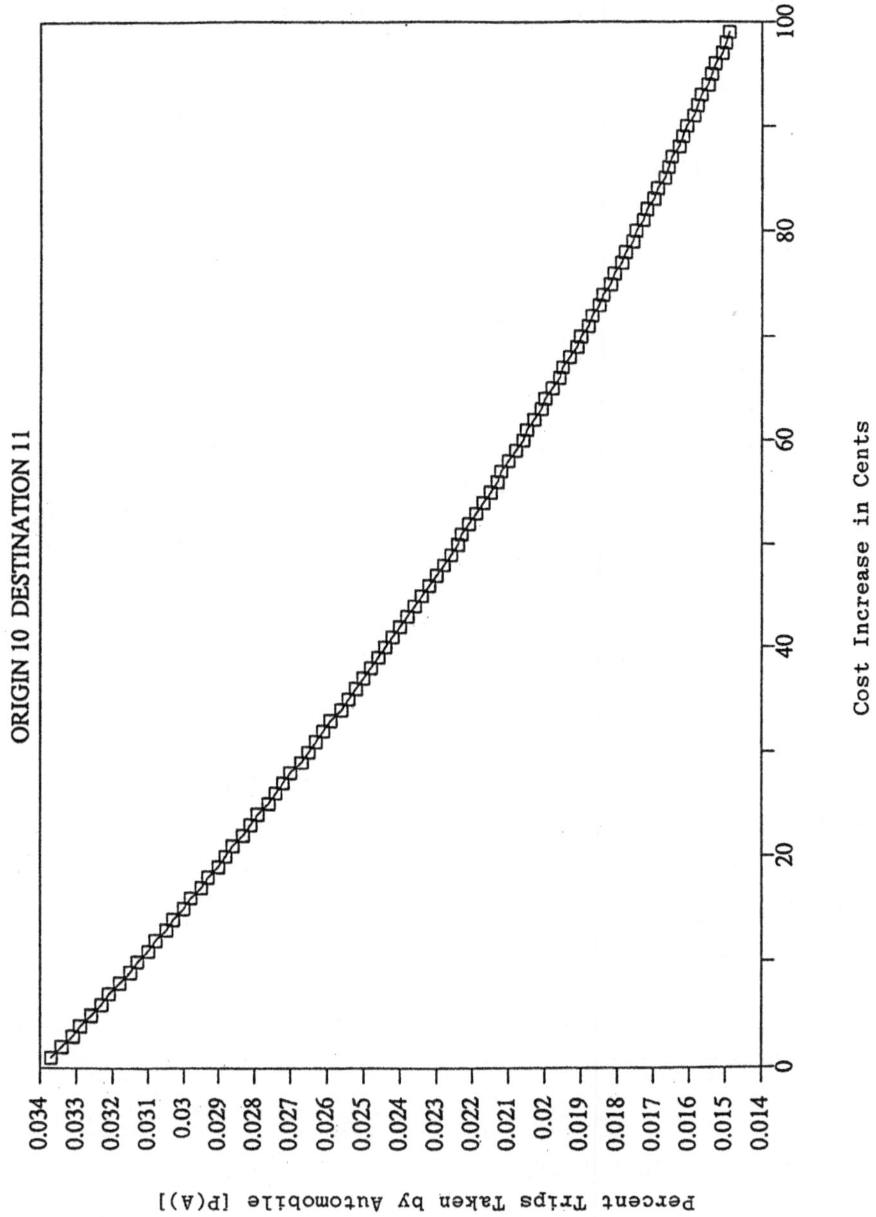

Fig. 4. Percent Trips Taken by Automobile as Function of Costs (3)

5.3 Estimation of Changes in Person-Trips and VMT

The new changes in the probabilities (Δp^{sk}) due to the cost increases (Δc^{sA}) are multiplied by the corresponding trip tables for each of the 5 purposes (p^k), in order to calculate the new total person trips by purpose. That is:

$$[\Delta p]^{sk} \times [P]^k = [PT]^{sk} \tag{2}$$

Next, the new person trip tables for each scenario are converted to automobile trip tables by dividing each zonal value $[PT(i,j)]$ in the tables by each corresponding vehicle occupancy rate for all zones $[VOR(i,j)]$ in tables for all-purpose trips. The vehicle occupancy rate tables are provided by CATS, as described earlier (see Tables IIH, III, IIJ, IIK, and IIL). The resulting value for each zone in the automobile trip tables are multiplied by the corresponding distance between zones to obtain the new VMT tables as shown in the following

$$\{[PT(i,j)]^{sk} / [VOR(i,j)]^k\} \times [D(i,j)] = [VMT]^{sk} \tag{3}$$

where,

$[PT(i,j)]^{sk}$: person trip tables by scenario s and trip purpose k

$[VOR(i,j)]^k$: vehicle occupancy rate tables by trip purpose k

$[D(i,j)]$: the distance table between zones

$[VTM]^{sk}$: matrices for the vehicle miles travelled by scenario s and trip purpose k

The resulting matrices by k purposes are combined into two: work related trips and non-work related trips.

6. Simulation Results: Interpretations

The simulation results showing the impacts on VMT reductions of implementing 12 scenarios are reported in Kane, John and Kim (1992). The summary of the results is shown as follows:

Scenario	VMT (million miles)	% change	Rank	Remarks
0	103.8			1990 base table
1	97.5	-6.0	10	$1.00 parking + $0.05/gal
2	87.0	-16.2	4	parking+$0.25/gal
3	76.4	-26.4	1	$3.00 parking + $0.50/gal
4	98.9	-4.7	11	$1.00 parking
5	94.2	-9.2	8	$2.00 parking
6	90.1	-13.2	6	$3.00 parking
7	102.4	-1.3	12	$0.05/gal
8	96.2	-7.3	9	$0.25/gal
9	88.9	-14.4	5	$0.50/gal
10	78.1	-24.8	2	$1.00/gal
11	91.4	-11.9	7	$1.00 parking + $0.25/gal
12	84.0	-19.12	3	$1.00 parking + $0.50/gal

Ranks indicate the order of scenarios most effective in reducing VMT. Figures 5 and 6 summarise the above findings. Some highlights of the simulation results are as follows:

1. The most effective TCM for reducing VMT is either the $3.00 additional parking fee plus $0.50/gallon gasoline tax (26.4% VMT reduction) or the $1.00/gallon (1990 price) gasoline tax increase (24.5% VMT reduction). It is interesting to note that Ruth and Kim (1992) conducted a time series regression analysis for the estimation of gasoline price elasticity of demand for VMT for the same study area. They estimated that about a 21% VMT could be reduced for a $1.00/gallon gasoline price increase at the 1988 constant price. While both numbers are different, the impact of a $1.00/gallon gasoline price (in 1990 price) increase resulted in 21% to 25% reductions in VMT for the study area.

2. The next effective TCM is scenario 12, which would impose a $1.00 parking fee for work related trips and $0.50/gallon gasoline tax for all purpose trips. The implementation of this scenario would reduce VMT by 19.1%. The next effective TCM is scenario 2, which is the imposition of a $2.00 parking fee plus $0.25/gallon gasoline tax. Scenario 2 would result in a 16.2% reduction of VMT. The reason why scenario 12 would reduce more VMT than scenario 2 is that the parking fee affects only the work related trips, which is 40% of the total daily VMT in the study area, while gasoline tax would affect all-purpose trips.

3. The next effective TCM is scenario 5, which would impose a $0.50/gallon tax on all-purpose trips. The first $0.05/gallon increase in gasoline tax

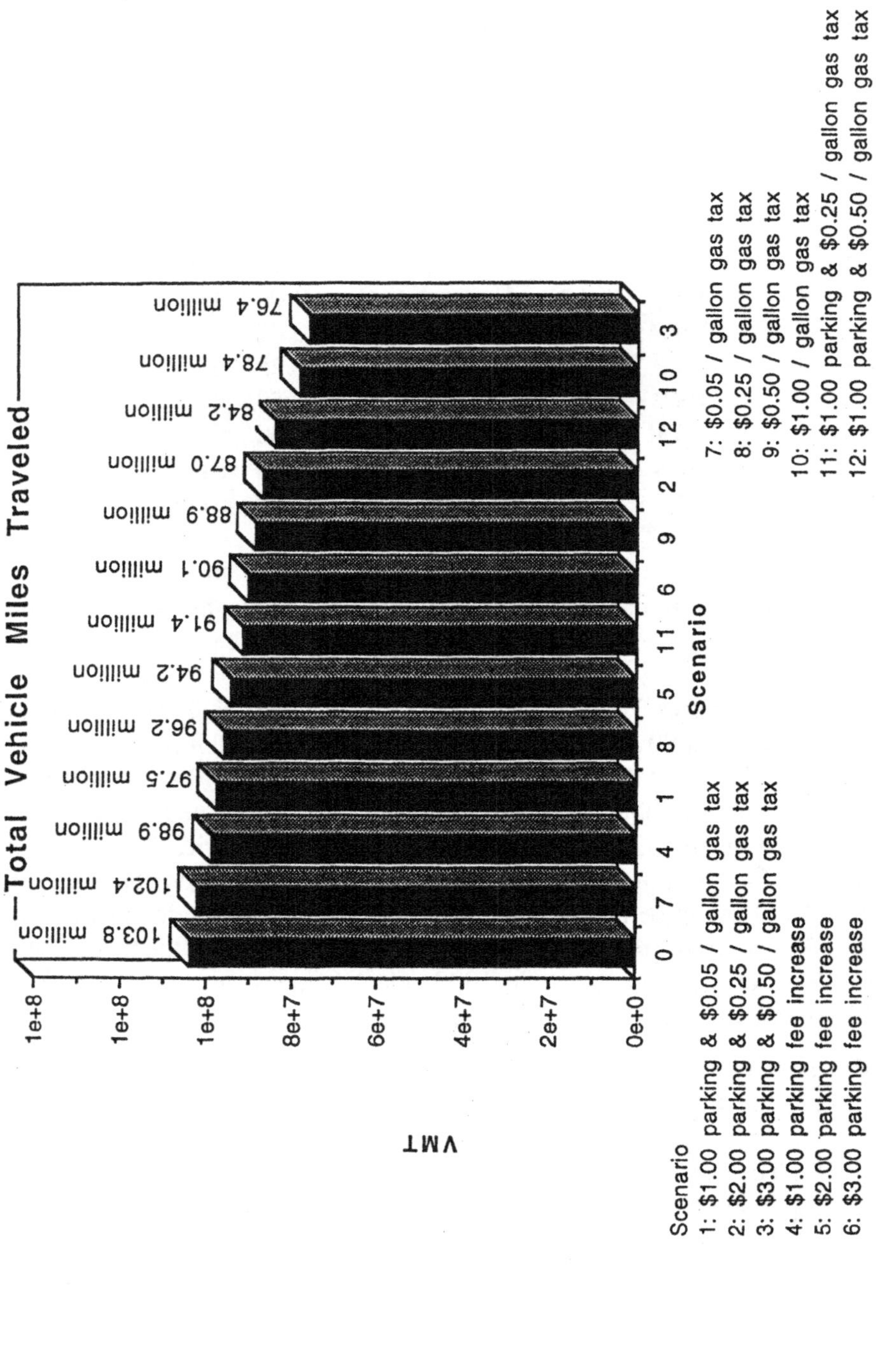

Fig. 5. VMT Reduction by Alternatives Scenarios

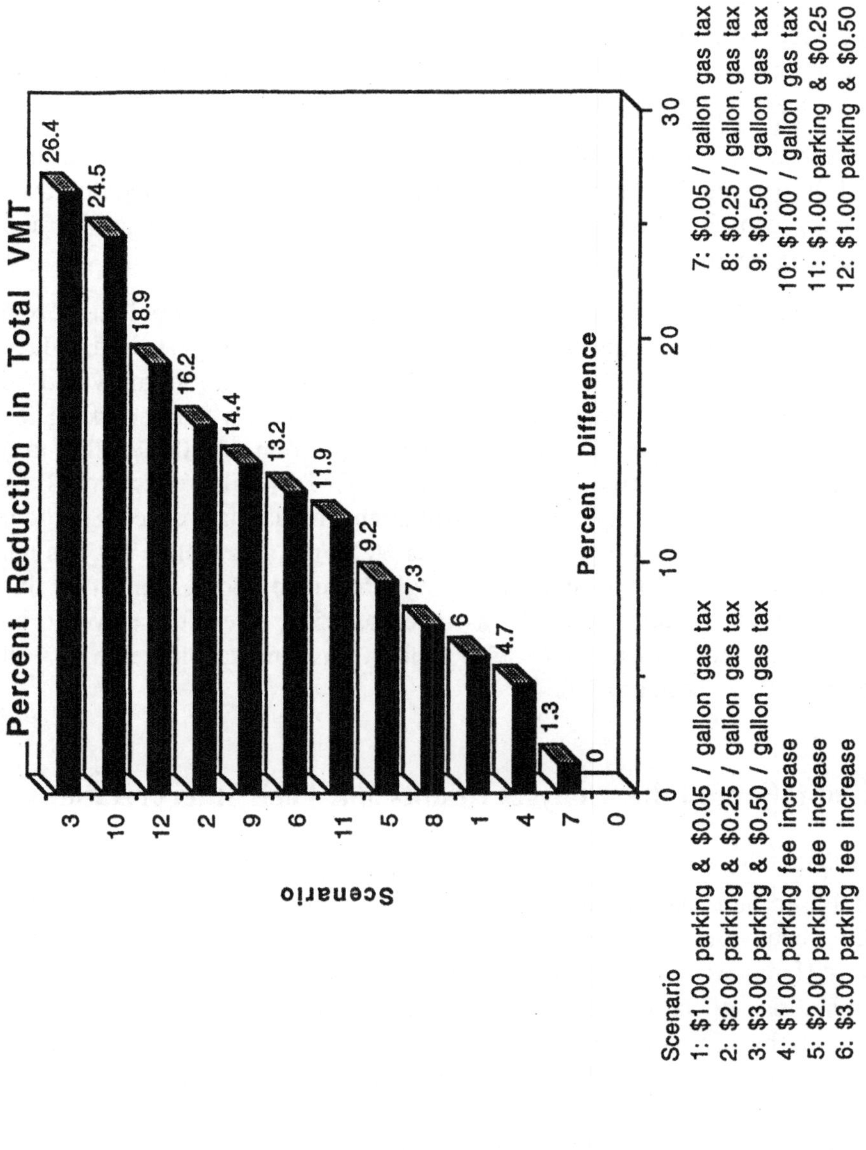

Fig. 6. Percent VMT Reduction by Alternatives Scenarios

(scenario 7) would reduce the total VMT only by 1.3%. The $0.25/gallon increase (scenario 8) would reduce VMT by 7.3%, which is more than 5 times the reduction from scenario 7, due to the non-linear structure of the mode choice model. This non-linear relationship can also be seen in the results of the $0.50/gallon increase (scenario 5), which resulted in a VMT reduction by 14.4%. However, due to the S-shaped, sigmoid curvature of the model results, the $1.00/gallon increase would result in a VMT reduction of 24.8%, less than one would expect by multiplying the results from scenario 7 by 20.

4. The least effective TCM's are scenario 7 ($0.05/gallon tax), scenario 4 ($1.00 parking fee), and scenario 1 ($1.00 parking plus $0.05/gallon tax). These scenarios would result in VMT reductions by 1.3%, 4.7%, and 6.0%, respectively.
5. The most plausible TCM option may be scenario 2, which would impose a $2.00 parking fee plus $0.25 gasoline tax. This strategy would reduce VMT as much as 16.2%. If the imposition of $0.50/gallon tax would be politically feasible, either scenario 12 (imposition of a $1.00 parking fee plus $0.50/gallon gas tax) or scenario 5 (imposition of $.0.50/gallon tax) could be very effective strategies. These TCM's would reduce VMT by 19.1% and 14.4%, respectively. Although an alternative scenario that would impose a $2.00 parking fee plus $0.50/gallon gasoline tax has not been simulated in this study, the results obtained above imply that this strategy could reduce VMT by as much as 25%, a result that would be similar to the imposition of $1.00/gallon gasoline tax increase for all purpose trips (scenario 10).

7. Limitations of the Analyses Results and Their Interpretations

As implied in several of the statements presented above, the pivot-point model used here assumes all other variables affecting mode choice decision making behaviours to be equal, except the costs of automobile driving. The cost differences in automobile driving in 12 different scenarios are combinations of various parking fees and gasoline taxes. By assuming all other variables to be equal, we have to assume that travel choice behaviours by trip makers in the study region are currently the same as they were in 1990.

In addition, even if we assume that travel behaviours now are the same as 1990, the analyses results are not the same as if we were to ask trip makers what they would do under various cost increase conditions. The results obtained here are the outputs from mechanical calculations for the mode choice, assuming that these changes are the reflections of trip behaviours when the cost of driving an automobile have been increased.

Despite these limitations, we are very excited about the simulation results because they can shed light on strategies that would reduce VMT, which in turn would reduce total vehicle emissions.

References

Chicago Area Transportation Study (CATS) (1980). Year 2000 Transportation System Development Plan, September.

Illinois Environmental Protection Agency (IEPA) (1992). Compendium of Employee Trip Reduction Task Force: Draft Legislation, January.

Kane, Douglas N., Stephen F. John, and Tschangho John Kim (1992). Clearing the Air. Choosing the Future: Reducing Highway Vehicle Emissions in the Chicago Ozone Nonattainment Area, Draft Report prepared for the State of Illinois Department of Energy and Natural Resources, Springfield, IL.

O'Melveny and Myers (1991). *The New Clean Air Act: Strategic Opportunities for Transit*, prepared for the American Public Transit Association, Washington, D.C.

Ruth, Matthias and Tschangho John Kim (1992). "Price Elasticities of Demand for Vehicle Miles Travelled", in Kane, John, and Kim (1992). Clearing the Air. Choosing the Future: Reducing Highway Vehicle Emissions in the Chicago Ozone Non-attainment Area, Appendix E, Draft Report prepared for the State of Illinois Department of Energy and Natural Resources, Springfield, IL.

United States Environmental Protection Agency (US EPA) (1989). Procedures for Emission Inventory Preparation, Volume IV: Mobile Sources, Technical Support Division, Office of Air Quality Planning and Standards and Emission Control Technology Division, Office of Mobile Sources.

United States Environmental Protection Agency (US EPA) (1990). Guidance for Initiating Ozone/CO SIP Emission Inventories Pursuant to the 1990 Clean Air Act Amendments, Office of Air Quality Planning and Standards, February 1990.

United States Environmental Protection Agency (US EPA) (1991). Emission Inventory Requirements for Ozone State Implementation Plan. Office of Air Quality Planning and Standards, EPA-450/4-91-010, March 1991.

Appendix A

A.1 Incomplete List for Emission Reduction Strategies

1.1 Alternative fuels
1.2 Vehicle emissions standards
  1. Federal standards
  2. California standards
1.3 Reformulated gasoline
1.4 Inspection and maintenance
1.5 Gross polluter program
  1. Voluntary buy-back
  2. On-road enforcement

A.2 Incomplete List for VMT Reduction Strategies

MODE CHOICE STRATEGIES
2.1 Short-term transit improvements
  1. Improve operations
  2. Reduction of headways
  3. Expanded feeder service
  4. Park and ride lots
2.2 Mid- and long-term transit improvements
  1. expanded rail system
2.3 Midday shuttles
2.4 Transit fare reduction
2.5 Provide bicycle facilities
2.6 Lanes/zones for common carriers

VMT REDUCTION STRATEGIES
3.1 Mandatory employer trip reduction
3.2 Ridesharing voluntary incentives
3.3 HOV lanes
3.4 Telecommuting
3.5 CBD traffic restricted zone
3.6 Mandatory no-drive days

LAND USE/LOCATION STRATEGIES
4.1 Short-term land use changes
  1. Pedestrian/bicycle redesign
  2. Regional centre densification

4.2 Mid- and long-term land use changes
1. Pedestrian-friendly design
2. Jobs-housing integration
3. Reduced parking requirements
4. Increased housing density

4.3 State/regional pricing strategies
1. Tax-base sharing
2. Fair-share housing policies
3. Economic development policies
4. Transportation impact fees

4.4 Redirect employment growth

PRICING STRATEGIES

5.1 Gas tax increase
1. Subsidise transit
2. Clean fuels
3. Gasoline price increase

5.2 Highway toll policies
1. Reduced or no tolls for HOVs
2. Increased SOV tolls
3. HOV toll bypass lanes

5.3 Parking fees (suburban)
1. Office parking fee
2. Industrial parking fee

5.4 CBD parking tax

5.5 Transit subsidy in lieu of free workplace parking

Appendix B

B.1 Derivation of Elasticity Model

Using a typical multinomial logit mode choice model such as:

$$P(A) = e^{-\alpha T_A - \beta C_A} / [e^{-\alpha T_A - \beta C_A} + e^{-\rho T_B - \delta C_B}]$$

where,

T_A (T_B): time related costs for mode A (B)
C_A (C_B): out of pocket costs for using mode A (B)

We can simplify the above model by defining that:

$$e^{-\alpha T_A - \beta C_A} = e^{-A}$$

$$e^{-\rho T_B - \delta C_B} = e^{-B}$$

Then,

$$P(A) = e^{-\alpha T_A - \beta C_A} / [e^{-\alpha T_A - \beta C_A} + e^{-\rho T_B - \delta C_B}] = e^{-A} / [e^{-A} + e^{-B}]$$

If we take a partial derivative of $P(A)$ with respect to the cost variable, i.e.:

$$\begin{aligned}
\delta P(A) / \delta C_A &= -\beta e^{-A}(e^{-A} + e^{-B}) + \beta e^{-A} e^{-A} / (e^{-A} + e^{-B})^2 \\
[\delta P(A) / \delta C_A][C_A / P(A)] &= \{-\beta + \beta e^{-A} / [e^{-A} + e^{-B}]\} C_A \\
&= -\beta \{1 - e^{-A} / [e^{-A} + e^{-B}]\} C_A \\
&= -\beta [1 - P(A)] C_A
\end{aligned}$$

Thus,

$$\Delta P(A) = -\beta [1 - P(A)] P(A) \Delta C_A \qquad \text{Q.E.D.}$$

CHAPTER 11

SPATIAL IMPACT OF ROAD IMPROVEMENT IN THE SUBURBAN RECREATIONAL AREA

Kenji Doi and Naohisa Okamoto

1. Introduction

In suburban areas near the metropolis, transport improvement such as provision of highways brings a drastic change in land use and the environment. It is often observed that savings in travel time through highway construction contribute to the increase in recreational development and travel demand. Since land use control is not yet implemented strictly in Japan, highway construction also triggers environmental problems, such as the increasing emission generated by road traffic of tourists and the depletion of natural green areas.

Supposing that transport improvement will be necessary in suburban recreational areas under the current trend of increasing leisure time, the evaluation of environmental impact due to the construction of transport facilities is urgently needed. However, the environmental impact in the suburban areas has rarely been analysed mainly because of the limitation in available data sources concerning land use and the environment in these areas. In the viewpoint of transport analysis, the short-term impacts of transport improvement and consequent increase in road traffic on the emission level in adjacent roadside areas has attracted much attention, while the long-term impact, such as the change in green areas and ecosystem has attracted quite few attention until recently.

The purpose of this study is to develop a method for quantifying the impacts of accessibility change on land use development and the consequent

Y. Hayashi and J. Roy (eds.), Transport, Land-Use and the Environment, 213–230.

decrease of green areas in suburban areas. A method of confirmative factor analysis is adopted for identifying the causalities among accessibility change, recreational attractiveness and the change of green conservation level. Also, a theoretical model which includes the travellers' demand for natural green and developers' strategy related to conservation/development of green area is proposed and is adopted for assessing the impacts of transport improvement.

2. Highway Improvement and its Impact on Green Conservation

2.1 Data

The Boso Peninsula Region, situated in the southeast part of the Tokyo Metropolitan Area is selected as a study area (see Figure 1). In this region, recreational development has increased drastically due to highway network improvement. The data used in the analysis are shown in Table 1.

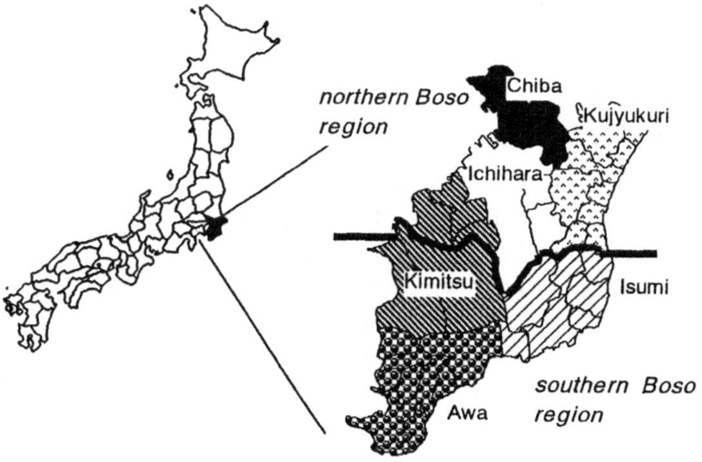

Fig. 1. Study Area and Zoning

Table 1. Outline of Data

Road Network Data - road traffic census: 1990 - travel time - 12-hour traffic volume counts
Recreational Development - survey on recreational facilities by Chiba Prefectural Office - annual survey on recreational visitors
Land Use and Zoning - 1 km^2 grid land use census: 1976, 1991 - aggregated land use data by municipality - published land price survey 1991

2.2 Recreational Supply and Demand

Recreational resources can be classified into two categories, namely outdoor recreational resources and indoor recreational resources, as shown in Table 2. From recent trends of recreational development and its composition shown in Figure 2, it is clear that developed area has rapidly increased since 1970 and this increase is supported mainly by an increase in outdoor recreational development. Regarding the outdoor recreational development, golf rinks have shown a marked increase in their site area. The total area increased from 1,200 has. in 1970 to 7,700 has. in 1980. This rapid increase, supported by a golf boom since the middle of the 1970's made up 65% of the total increase in site area of outdoor recreational development.

Table 2. Classification of Recreational Facilities

Indoor Recreational Resources
aquarium, museum, amusement park, zoo botanical garden, stadium, trade fair, etc.
Outdoor Recreational Resources
golf rink, hot spring, stock farm, auto-camp field, swimming beach, lake, natural park, etc.

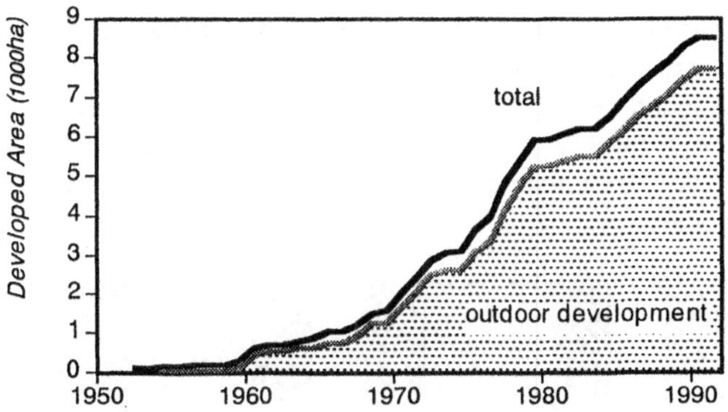

Fig. 2. Recent Trends of Recreational Development

Figure 3 shows trends of recreational visitors by zone since 1970. It is shown that recreational demand has increased much in the northern area of the Boso Region since 1980. However, the total growth rate is relatively less than that of recreational development. It is suggested that recreational supply is higher than the demand in the Boso Region.

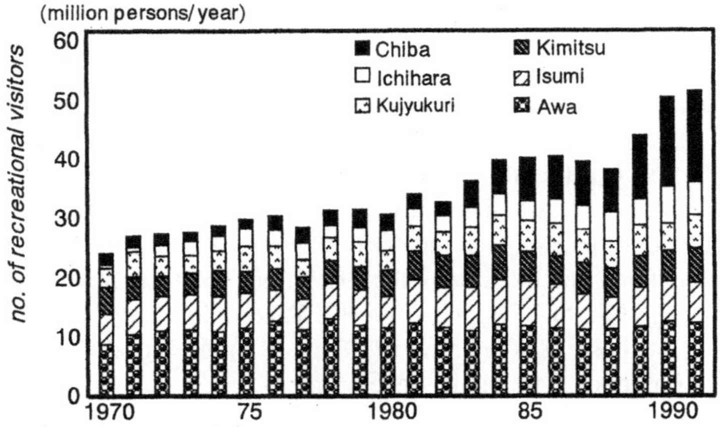

Fig. 3. Recent Trends of Recreational Travel Demand

2.3 Land Use Change

Figure 4 shows the change of land use configuration of the whole area and at roadside areas within 1 km from trunk roads. In 1991, forest and farm land made up 51% and 32% of the total land area, respectively and natural forest made up 48% of the total forest area.

Regarding the land use change between 1976 and 1991, although the area of forest and farm land has not decreased much with respect to the total area, it has occurred significantly in the roadside areas. In roadside areas, the average decline of forest and farm land are 4.3% and 6.6%, respectively. Deforestation and diminution of farm land are caused mainly by: a) sprawl development in urban fringe, and b) clustered development for recreational purposes in country side areas.

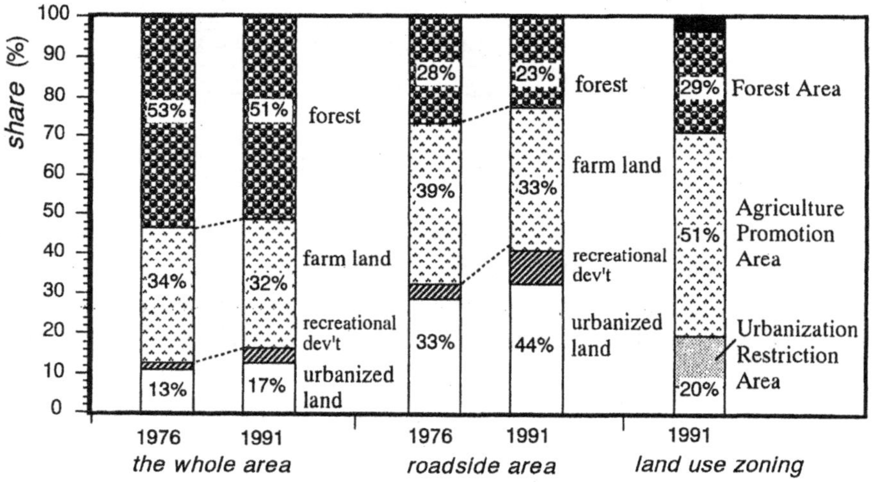

Fig. 4. Land Use Configuration and its Change

3. Causality Analysis of Land Use Change

3.1 Index of Natural Land Use

In measuring the natural use level of the area which includes various kinds of land use, the following index based on land use configuration is proposed:

$$NU_i = \sum_k g^k \frac{a_i^k}{L_i} \tag{1}$$

This is a composite index which assigns weights to each of the land use categories by its own degree of conservation g^k, a_i^k indicates the area of land use category k in zone i and L_i is the area of zone i.

In this study, the following 7 categories are predefined, namely: a) forest land, b) grassland and farm land, c) suburban recreational land, d) urban recreational land, e) residential land, f) industrial land and g) commercial land. In expressing the degree of green conservation of each land use category, land price index is applied because it is postulated to reflect the degree of urban use of land. Here, land price index is normalised between 0 and 1 as follows:

$$g^k = 1 - \frac{P_{current}}{P_{highest}} \tag{2}$$

where $P_{current}$ denotes land price in its current use and $P_{highest}$ is the price in its highest level of utilisation, which is regarded as commercial use in this study. Figure 5 shows the distribution of natural use level g^k which is derived from the land price data in the Boso Region.

The relative natural use level of farm land, suburban recreational land and residential land are calculated as 0.927, 0.912 and 0.693 of that of forest, respectively.

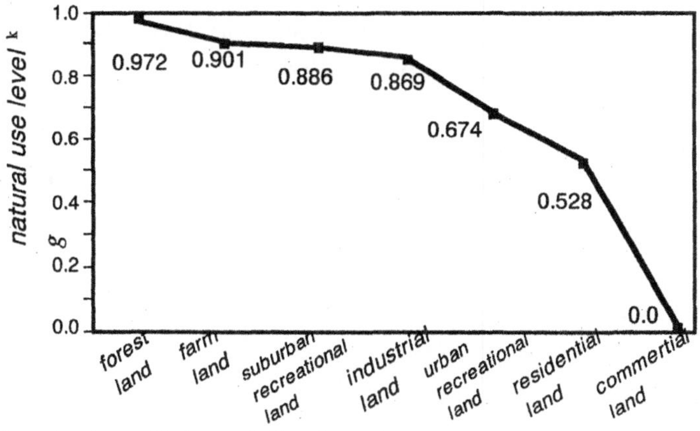

Fig. 5. Distribution of Natural Use Level

Figure 6 shows the relationship between the change in natural use level and that of accessibility from city centre to each of the 6 large zones of the Boso Region. This means that in general, the reduction of travel time from city centre contributes to the diminishing of the green conservation level and the promotion of development. There may be a time lag between the changes, where the change in accessibility is followed by the change in natural use level with an estimated five year lag. Since relatively large time savings in *Kimitsu*, *Isumi* and *Awa*, which are located in the southern part of Boso Region, did not contribute to the decrease in natural use level, it may be assumed that accessibility improvement does not always trigger a decrease in natural use level without the significant influence of urbanisation pressure.

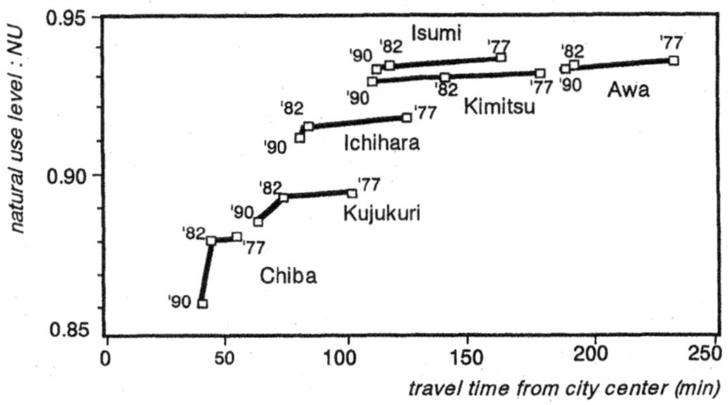

Fig. 6. Accessibility Improvement and its Effect on Natural Use Level

3.2 Causality Analysis Based on LISREL

Based on the evidence of the previous section, the diminution process of natural use level are assumed as follows:

In order to examine the assumption of the causal structure shown in Figure 7, we apply the comfirmative factor analysis (LISREL) which is composed of the following system of equations:

Measurement equations: $X = \lambda_x \xi + \delta$ (3a)

$Y = \lambda_y \eta + \varepsilon$ (3b)

Structural equation: $\eta = B\eta + \Gamma\xi + \zeta$ (4)

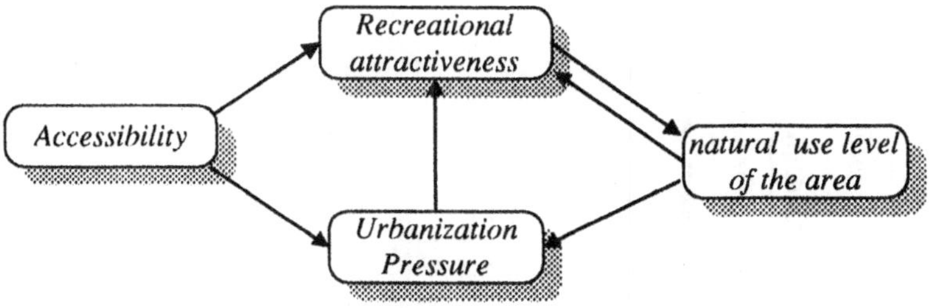

Fig. 7. Assumed Causal Structure

where X is a vector of exogenous observed variables, Y is a vector of endogenous observed variables, ξ is a vector of exogenous latent variables and η is a vector of endogenous latent variables. λ_x and λ_y are vectors of regression effects of X and Y. B is a matrix of structural (causal) effects among the endogenous latent variables (η) and Γ is a matrix of regression effects of the exogenous latent variables (ξ). ζ, δ and ε indicate vectors of disturbance or residual terms.

Causality analysis is conducted based on roadside land use data aggregated in each 1 km^2 grid, collected in 1991. Table 3. shows the list of observed variables X and Y and we set the "recreational attractiveness", "urbanisation pressure", "suitability to natural use" as latent variables.

Table 3. Observed Variables

Observed exogenous variables	
X_1	travel time from CBD (minutes)
X_2	regional recreational resources (0,1)
X_3	traffic volume in the trunk road (1000 vehicles/12hr)
X_4	land use zoning (share of designated area to the total area)
Observed endogenous variables	
Y_1	number of recreational visitors (1000 persons)
Y_2	recreational development area (ha)
Y_3	natural use level (non-dimensional)
Y_4	urbanisation of surrounding areas (non-dimensional)

The causal structure estimated by maximum likelihood procedure is depicted in the flow diagram of Figure 8 and summarised as follows:

a) accessibility is an essential part of recreational attractiveness,
b) higher recreational attractiveness lowers the suitability to natural land use and therefore accessibility improvement is assumed to have a negative effect on natural use level, however,

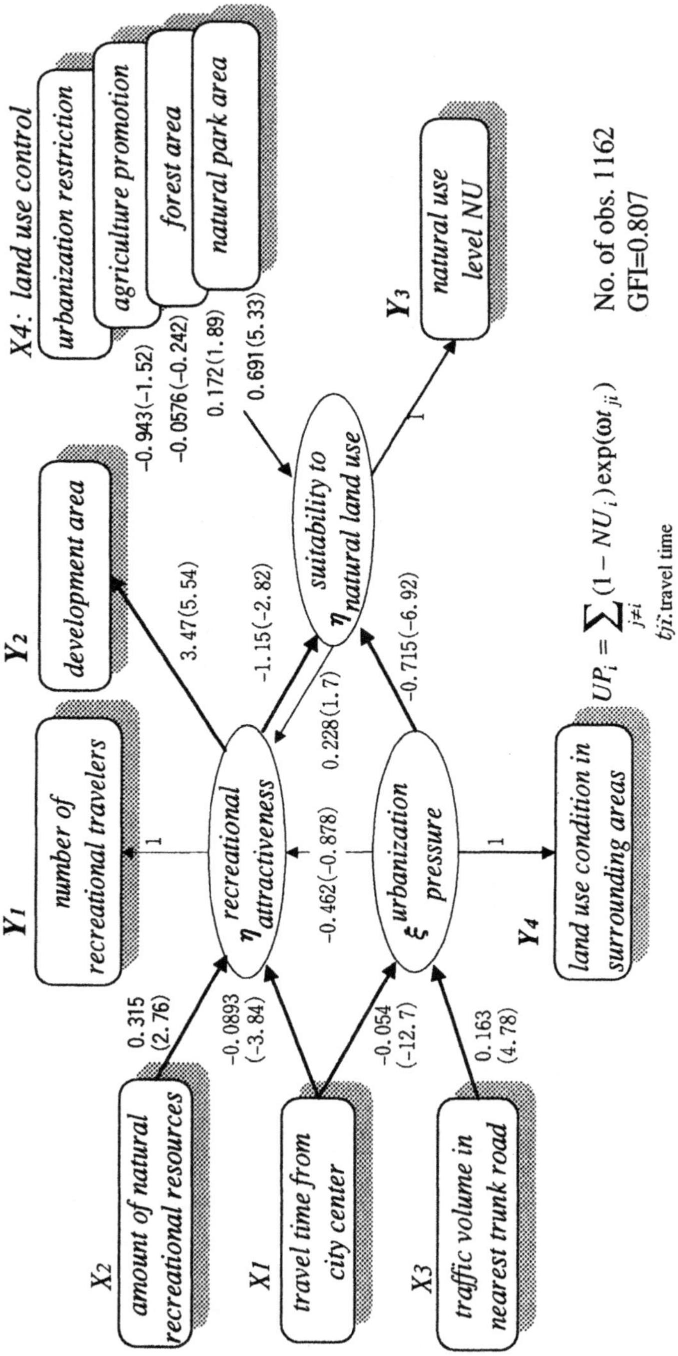

Fig. 8. Estimated Causal Structure

c) higher level of natural use has a positive effect on recreational attractiveness,
d) accessibility improvement also has a negative effect on natural use level through the promotion of urbanisation pressure from surrounding areas.

4. Modelling the Recreational Travel Demand and Developers' Strategy

In the previous section, the interaction between recreational attractiveness and suitability to natural land use is identified and it is suggested that accessibility improvement triggers the decrease in natural use level of the region. This section aims at proposing a theoretical framework to quantify the impact of transport improvement on natural land use pattern, based on the modelling of travellers' demand for green area and developers' strategy related to conservation/development of natural land.

4.1 Developers' Strategy

For describing the mechanism of deforestation and the decrease in the other green areas due to transport improvement, we first model the developers' supply of recreational land in suburban area in which the share of green area is dominant. Supply of developed recreational land is given by

$$S = S(p, w) \tag{5}$$

where p denotes the price of developed land for recreational use and w is the cost relating to land transaction and development. From the profit maximising criteria, the following conditions must hold:

$$\frac{\partial S}{\partial p} \geq 0, \quad \frac{\partial S}{\partial w} \leq 0 \tag{6}$$

Developers' demand for natural land, which will be utilised as an element of production, is expressed in the following function:

$$X = X(p, w) \tag{7}$$

where

$$\frac{\partial X}{\partial p} \geq 0, \quad \frac{\partial X}{\partial w} \leq 0 \tag{8}$$

Condition (8) shows that the higher the price of developed land and the lower the cost of land transaction and development, the larger is the developers' supply of recreational land. Furthermore, under the assumption that developers determine the price of developed land according to the recreational travel demand, p included in function (5) and (7) can be replaced by recreational travel demand RT.

It is obvious that developers' demand for natural land X increases with the increase in recreational travel demand RT and consequently, the amount of conserved green (natural land) decreases. Figure 9 illustrates the relationship between the amount of conserved green GC and recreational travel demand RT based on the following *green conservation function*:

$$GC = GC(RT, w) \tag{9}$$

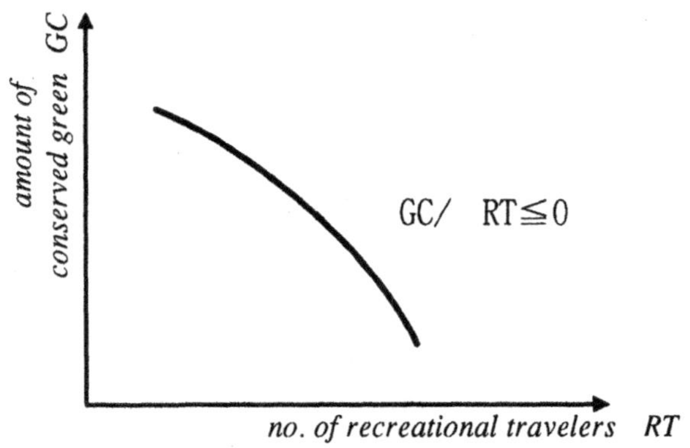

Fig. 9. Green Conservation Function

4.2 Recreational Demand Considering the Attractiveness of Green Area

The amount of conserved green is considered as one of the fundamental factors influencing suburban recreational activities. We express the recreational travel demand attracted to each area by using the concept of the gravity model,

$$RT = RT(GC, RA, AC) \tag{10}$$

where GC is amount of conserved green used as an index of environmental quality, which is one of the measures of attractiveness and RA indicates other measures of recreational attractiveness and AC denotes the accessibility in terms of savings in travel time from city centre to the area.

In this gravity model, if RA depends on the amount of developers' supply of recreational site, the effects of GC and RA on recreational travel are mutually related because an increase in RA causes a decrease in GC and vice versa. Therefore, the total effect of GC is shown as follows:

$$\frac{dRT}{dGC} = \frac{\partial RT}{\partial GC} + \frac{\partial RT}{\partial RA}\frac{\partial RA}{\partial GC} \tag{11}$$

Figure 10 shows probable functions of travellers' evaluation of green conservation level. The first case is a monotonic function of increasing travel demand with green conservation level *GC* (d*RT*/d*GC*0) and the second case is the monotonic function of decreasing travel demand (d*RT*/d*GC*<0) and the third case depicts a convex function of travel demand. The actual pattern will be confirmed later through model estimation.

Based on the above formulation, the realised amount of conserved green GC^* and travel demand RT^* are determined from the equilibrium of eqs. (9) and (10).

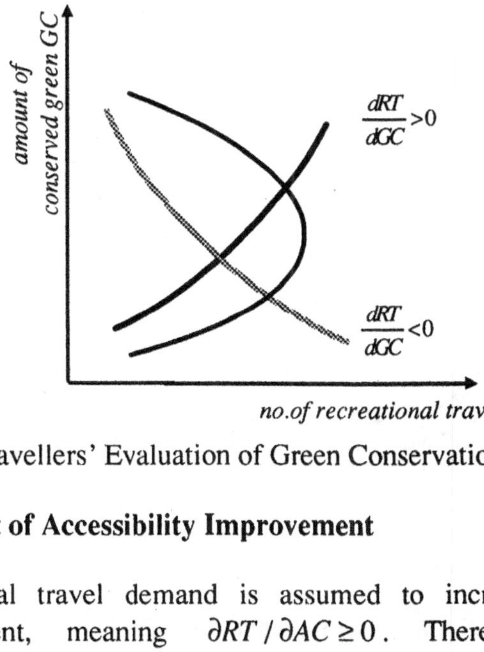

Fig. 10. Travellers' Evaluation of Green Conservation Level

4.3 Impact of Accessibility Improvement

Recreational travel demand is assumed to increase with accessibility improvement, meaning $\partial RT / \partial AC \geq 0$. Therefore, supposing that $dRT / dGC > 0$, accessibility improvement increases recreational travel and on the other hand decreases green conservation level as shown in Figure 11. Supposing that $dRT / dGC < 0$ and *RT* has a low sensitivity with the change of *GC*, accessibility improvement also increases recreational travel and on the other hand decreases green conservation level. Only if $dRT / dGC < 0$ and *RT* has a has sensitivity with the change of *GC* as shown in Figure 12, accessibility improvement increases green conservation level. However, under the strong assumption of monotonic functional form of recreational travel demand, an increase in the number of recreational travellers is not compatible with an increase in green conservation level.

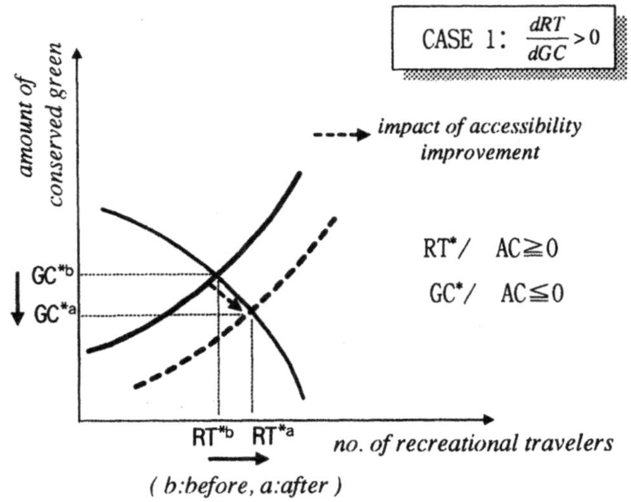

Fig. 11. Impact of Accessibility Improvement: Case 1

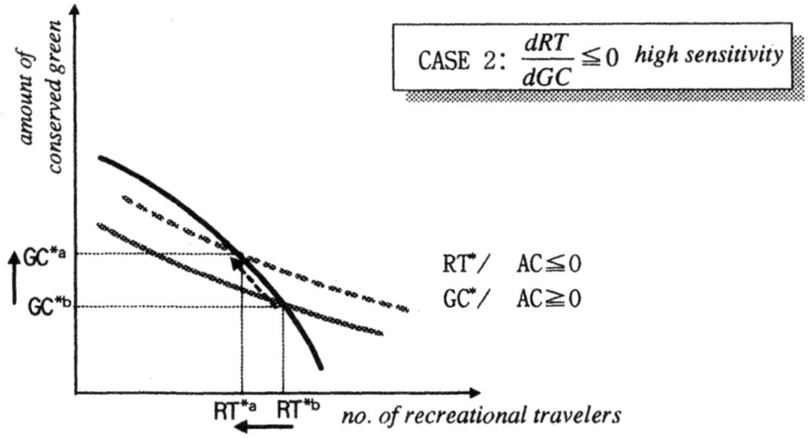

Fig. 12. Impact of Accessibility Improvement: Case 2

4.4 Model Specification

Under the assumption of Cobb-Douglas technology of developers' production, both eq. (5) and eq.(7) are derived as log-linear models. Furthermore, by using the ratio of conserved green to the total area L, eq. (7) is expressed as follows:

$$\ln X = \ln(1 - GCR)L = \alpha_0 + \alpha_1 \ln RT + \alpha_2 \ln w \tag{12}$$

where GCR is the conserved ratio of green area ($0 \leq GCR < 1$) and α_0, α_1 and α_2 are parameters. Conditions $\alpha_1 \geq 0$ and $\alpha_2 \leq 0$ must hold from the requirement of eq. (8). Also, by assuming the log-linear function (namely, Cobb-Douglas function) for recreational travel demand, eq. (10) is shown as,

$$\ln RT = \beta_0 + \beta_1 \ln GCR + \beta_2 \ln RA + \beta_3 \ln AC \tag{13}$$

where β_0, β_1, β_2 and β_3 are parameters. Recreational attractiveness *RA* is assumed to depend on the amount of natural recreational resources and developers' supply of recreational sites,

$$\ln RA = \ln NRA + \gamma \ln(1 - GCR) \tag{14}$$

Since it is difficult to derive the reduced form of equilibrium solutions GCR^* and RT^* in the non-linear equations system, we employ the following condition: $\alpha_1(\beta_1 + \beta_2\gamma) = 1$.

Consequently, equilibrium ratio of green area is derived as a function of the amount of natural recreational resources *NRA*, accessibility *AC* and the cost of land transaction and development *w* as follows:

$$GCR^* = \frac{1}{1 + e^{-\frac{1}{\alpha_1\beta_1}(\lambda_0 + \lambda_1 \ln NRA + \lambda_2 \ln AC + \lambda_3 \ln w + \ln L)}} \tag{15}$$

where

$$\lambda_0 = -\alpha_0 - \alpha_1\beta_0,\ \lambda_1 = -\alpha_1\beta_2,\ \lambda_2 = -\alpha_1\beta_3 \text{ and } \lambda_3 = -\alpha_2 \tag{16}$$

The derived model (15) is similar to the well known Logit model. Furthermore, travellers' evaluation of green conservation ratio is analysed by examining the elasticity of *RT* with *GCR* shown as follows:

$$\frac{\frac{dRT}{RT}}{\frac{dGCR}{GCR}} = -\frac{1}{\alpha_1}\frac{GCR - \alpha_1\beta_1}{1 - GCR} \tag{17}$$

The sign of eq. (17) will differ depending on the sign of α_1 and magnitude of $\alpha_1\beta_1$ and since the condition $\alpha_1 \geq 0$ must hold,

$$\frac{\frac{dRT}{RT}}{\frac{dGCR}{GCR}} \begin{cases} > 0 & & GCR < \alpha_1\beta_1 \\ = 0 & \text{if} & GCR = \alpha_1\beta_1 \\ < 0 & & GCR > \alpha_1\beta_1 \end{cases} \tag{18}$$

and therefore,

$$\frac{\frac{dRT}{RT}}{\frac{dGCR}{GCR}} \begin{cases} > 0 & \\ \geq 0 \ \text{or} < & \text{if} \\ < 0 & \end{cases} \begin{array}{l} \beta_1 \geq 0,\ \alpha_1\beta_1 \geq 1 \\ \beta_1 \geq 0,\ \alpha_1\beta_1 < 1 \\ \beta_1 < 0 \end{array} \tag{19}$$

The above conditions show that a) for negative value of β_1, recreational travel demand *RT* always increases according to the increase in green conservation ratio *GCR*, b) for non-negative value of β_1 and $\alpha_1\beta_1 > 1$, *RT* always increases according to the increase in *GCR*, and c) for non-negative value of β_1 and $\alpha_1\beta_1 \leq 1$, *RT* increases according to the increase in *GCR* if the ratio *GCR* is not larger than $\alpha_1\beta_1$ and decreases if the ratio *GCR* exceeds $\alpha_1\beta_1$. Figure 13 shows the case of non-negative value of β_1 and $\alpha_1\beta_1 \leq 1$.

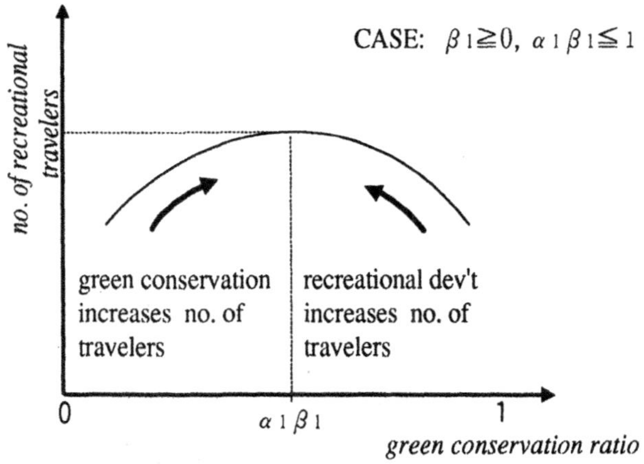

Fig. 13. Effect of Green Conservation Ratio on Recreational Demand

4.5 Estimation

For model estimation, the following two-stage procedure is applied, namely a) estimation of parameters of induced models, and b) estimation of parameters of structural equations. At first, parameters λ_0, λ_1, λ_2 and λ_3 of eq. (15) are estimated based on the maximum likelihood method as follows:

$$\max \prod_i \left[\frac{1}{1+e^{-\frac{1}{\alpha_1\beta_1}(\lambda_0+\lambda_1 \ln NRA_i+\lambda_2 \ln AC_i+\lambda_3 \ln w_i)}} \right]^{\overline{GCR_i}} \tag{20}$$

where $\overline{GCR_i}$ is the observed *GCR* in area i.

Table 4 shows the estimated results by using land use census data, number of recreational visitors and distribution of regional recreational resources in 1991. The study area is divided into two regions, namely the northern region and southern region, and model parameters are estimated in each region. In the table, land use zoning is used as a substitute of land acquisition and development cost because development control employed in each zoning area influences the cost. The following zonings, namely 'urbanisation restriction area: *URA*', 'agricultural promotion area: *AA*', 'forest area: *FA*' and 'natural park area: *NPA*', which are aimed at restricting development, are employed.

Table 4. Estimated Results

		Northern Region		*Southern Region*	
Variables		parameter	t-value	parameter	t-value
Constant		1.39	6.05	0.715	8.51
No. of regional recreational resources	*NRA*	-1.87	-0.946	-3.04	-2.56
Accessibility: 1/(travel time)	*AC*	-6.55	-4.29	-1.92	-1.37
Land Use Zoning:	ln*w*				
Urbanisation Restriction Area		-0.457	-1.64	0.072	0.443
Agriculture Promotion Area		-0.289	-0.545	0.564	0.281
Forest Area		1.21	2.28	3.57	1.90
Natural Park Area		--	--	14.1	6.85
Likelihood ratio		0.211		0.258	
No. of obs.		372		304	

In the northern region, all of the variables except the designation for *URA* and *AA* possess expected signs and it is shown that accessibility and number of regional recreational resources has a negative effect on green conservation ratio. Regarding the land use zoning, although the designation of *FA* contributes to the increase of the green conservation level, designation of *URA* and *AA* have a negative effect on it. The goodness-of-fit of the model shows likelihood ratio of 0.211.

In the southern region, all parameters have the expected sign, but accessibility and land use zoning except the *NPA* are not significant enough. As same as the northern region, the influences of accessibility and number of regional recreational resources on green conservation ratio are estimated to have a negative effect. The goodness-of-fit of the model shows likelihood ratio of 0.258, which is higher than that of northern region.

Table 5 shows the estimated results of structural equations (12) and (13). Regarding the developers' demand for natural land, parameter of number of recreational visitors *RT*, which indicates the elasticity of *X* with *RT*, is estimated as a positive value in both regions and the value in the northern region is higher than that in the southern region. This result shows that in the southern region, the increase in recreational demand causes more diminution of green area. The effect of land use zoning is similar to the result shown in Table 4.

Regarding the travellers' demand, it is shown that green conservation ratio, regional recreational resources, development ratio and accessibility have all positive effects in both regions. However, the contribution of green conservation ratio is relatively low and it is suggested that the abundance of green area does not significantly influence the travellers' evaluation of recreational attractiveness.

Table 5. Estimated Results of Developers' Demand Function

Variables		*Northern Region* parameter	*Northern Region* t-value	*Southern Region* parameter	*Southern Region* t-value
Constant	α_0	-0.613	-14.8	-0.247	-10.5
No. of recreational visitors	α_1	1.25	5.42	1.58	3.29
Land Use Zoning:	α_2				
Urbanisation Restriction Area		0.116	1.83	-0.0547	-0783
Agriculture Promotion Area		0.0773	0.691	-0.371	-0.295
Forest Area		-2.07	-2.59	-2.35	-2.12
Natural Park Area		--	--	-8.68	-6.93
R^2		0.646		0.725	
No. of obs.		372		304	

Table 6. Estimated Results of Travellers' Demand Function

Variables		*Northern Region* parameter	*Northern Region* t-value	*Southern Region* parameter	*Southern Region* t-value
Constant	β_0	0.249	8.52	-0.147	-5.08
Green conservation ratio	β_1	0.201	0.673	0.393	1.46
No. of regional recreational resources	β_2	0.382	4.26	1.27	13.9
Development ratio	γ	1.57	--	1.89	--
Accessibility	β_3	1.46	0.518	0.782	1.63
R^2		0.591		0.624	
No. of obs.		372		304	

Substituting the estimated values in Table 5 and 6 into eq. (17) yields the elasticity of recreational travel demand with green conservation ratio. First, since the sign of β_1 is nonnegative and calculated value of $\alpha_1\beta_1$ is not higher than 1, it is suggested that travellers' demand shows a convex change with the increase in green conservation ratio. In the northern region, the value of $\alpha_1\beta_1$ is estimated as 0.242, which is a little lower than the current value of green conservation ratio *GCR* of 0.274 as shown Figure 14. Also, in the southern region, the value of $\alpha_1\beta_1$ is estimated as 0.463, which is lower than the current *GCR* of 0.621. From the comparison between estimated $\alpha_1\beta_1$ and the current

GCR, it is suggested that a slight increase in development ratio (not higher than a few percent) in the northern region and an increase in green conservation ratio in the southern region will attract additional recreational visitors, respectively.

In Figure 14, accessibility improvement will shift travellers' demand curve to the right and consequently equilibrium point with developers' demand curve will move toward the lower right with the accessibility improvement, meaning the increasing travel demand and decreasing green conservation level in both regions. To prevent the decrease in green conservation level, upward shift of developers' demand curve by the increase in development cost or reinforcement of development control is necessary. This countermeasure will also lead to the increase in recreational visitors in the southern region.

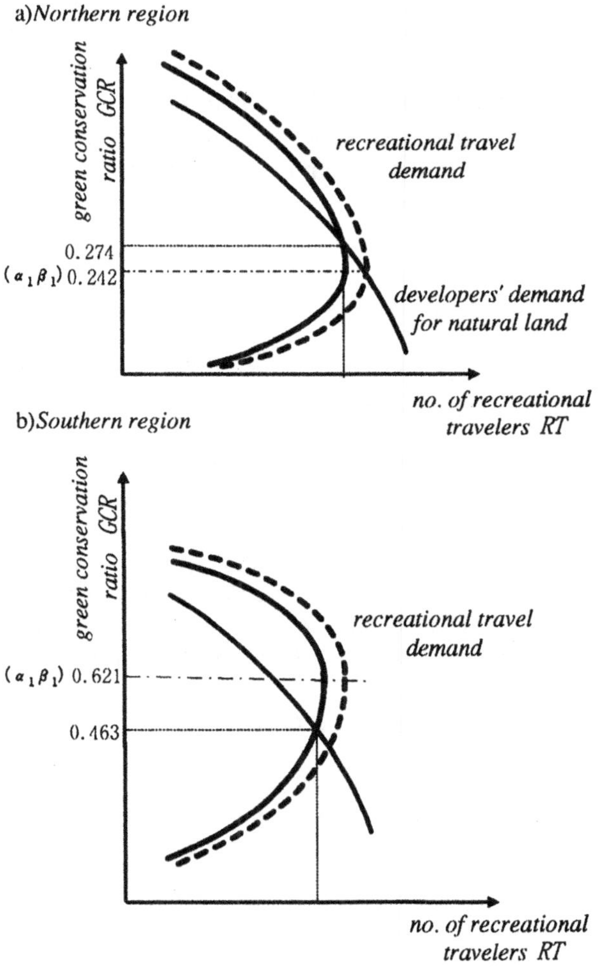

Fig. 14. Green Conservation Ratio and Number of Recreational Travellers at the Equilibrium

5. Conclusion

In the suburban areas, the abundance of natural resources is essential for recreational activities. However, it is often observed that the increase in recreational travellers causes the degradation of natural environmental quality, such as the decrease in green area. This study provided a fundamental framework for exploring the change in green conservation level due to transport improvement in the suburban areas.

Based on the model system including travellers' demand for the abundance of green area and developers' strategy related to conservation/development of green area, it is confirmed that a) higher level of green conservation is one of the measures of recreational attractiveness and has a positive effect on the attraction of recreational travellers, but, b) under the current travellers' evaluation of green conservation ratio v.s. development ratio and effectiveness of development control, accessibility improvement results in the diminution of natural green area in suburban recreational areas.

The methodology developed here seems to provide the potential capability of the long-term assessment of environmental change related to transport projects and suggests the perspective for sustainable development.

References

Takizawa, K.,Tokumaru, H. and Tsutsumi, K. (1985) Historical review of golf rink development in Kanto Region, *Proceedings of Japan Society of Real Estate Science* 1, 117-120

Nishizawa, N. (1992) Recreational city, Urban Planning 176, 40-42

Hirano, H. (1991) A study on urbanization of forest area, *Proceedings of Japan Society of Real Estate Science* 7, 21-24

Morichi, S., Hyodo T. and Okamoto N. (1992) A study on touring activity models for one-day car trip, *Infrastructure Planning Review* 10, 63-70

Morey, E. (1981) The demand for site specific recreational activities: A characteristics approach, *Journal of Environmental Economics and Management* 8, 345-371

Mendelsohn R. (1987) Modeling the demand for outdoor recreation, *Water Resour. Res.* 23, 961-967

Englin J. and Mendelsohn R. (1991) A Hedonic Travel Cost Analysis for Valuation of Multiple Components of Site Quality: The Recreation Value of Forest Management, *Journal of Environmental Economics and Manage-ment* 21, 275-290

Part C.
Alternative Modelling Approaches

CHAPTER 12

ANALYSIS OF URBAN ENVIRONMENTAL POLICIES ASSISTED BY BEHAVIOURAL MODELLING

Francisco J. Martínez

1. Introduction

Air pollution phenomena are essentially dispersed in the urban space, affecting population in a relatively homogeneous way. That is, differences between zones in a given city are so small in comparison with the phenomenon magnitude that is difficult for individuals to perceive them. This inaccurate perception of pollution introduces severe difficulties to analyze, as it is usual in cost-benefit analysis, the net benefits of projects and policies out of the demand reaction to pollution levels, i.e. economic evaluation through revealed preferences.

Besides, although air pollution is technically clearly defined, it is difficult to establish a measure that can be associated with people's perception, normally called "smog". Although scientists would promptly explain that what is visible of the smog are particles, the population is also aware of the existence of various types of gases which become clearly perceptible when they are in high level of concentration, for example in Santiago`s CBD.

These two arguments, dispersion and unclear measurement, hamper the use of a traditional method for the evaluation of policies based on people's perception (analysis of utility). Hence, the traditional cost benefit approach finds strong difficulties, which is the reason why herein an alternative method is proposed. This approach is based on measuring the impact, in terms of final pollution

Y. Hayashi and J. Roy (eds.), Transport, Land-Use and the Environment, 233–257.

levels, that different environmental policies may produce. Therefore, the method does not pursue an economic evaluation of policies, instead it attempts to evaluate their efficiency in achieving environmental goals compared with its implementation (plus external) costs.

In order to quantify such impacts, it is required to identify the type and location of mobile and fixed sources throughout the city, the level of emissions and the immission process and to calculate their expected change as a result of the policy under analysis. This estimation requires to simulate the change in the performance of the city, which is given by the location of households and firms, called the activity system and denoted by A, and the changes in the transport system, denoted by T.

This article describes this alternative method, which requires the use of an integrated land-use transport economic model. It begins describing the integrated approach proposed, followed by a discussion of the microeconomic framework of the model. In the next two sections, it continues with a brief description of the model, starting with the urban land market and followed by the transport sub-model. The explanation of this model is then completed by tackling a highly important issue: the interaction between the activity and the transport models, which allows to analyze, comprehensively, the final impacts generated by an environmental policy. Section five contains an analysis of different policies that had been proposed to improve air pollution conditions in Santiago, which affect systems A and T, describing the mechanism to forecast the impacts of that policies upon both systems. The reader has the option to review technical details in the Annex. This article does not include any kind of case results, which exceeds the aim of this paper.

1.1 An Integrated Approach

Experience on environmental studies shows that it is an area that refuses to be studied from any traditional point of view. In the case of public and private transport emissions (mobile sources) for instance, intuition points out that transport studies should be able to identify and forecast them. However, the cause of any transport movement is the existence of activities distributed in the space. This results in a long term relationship between systems A and T, which are inescapable in the analysis of the environment.

In the case of the city of Santiago, environmental pollution originated from mobile sources has been tackled with measures that, since they assume system A as exogenous, are called short term measures, whose objective is to optimize the transport function controlling emission levels. This includes the improvement of the network capacity by means of more efficient operations of highways and junctions, the modernization of traffic lights and, more recently, policies that regulate the management of transport services, as in the case of buses, where downtown services have been subject of a licensing scheme. Another policy proposed, is a road pricing system that would eventually replace the actual scheme of vehicle restriction (which prohibits circulation of two number plate's last digits every working day). Even though such measures seem to be efficient in terms of transport management, it is important to consider the impact of such policies on the activity system, which is the expected reaction of the city in terms of its degree and quality of development, and consequently, the long term environmental efficiency of that policies.

In short, although mobile emissions are produced by elements of T, vehicles that provide movement of people and things, the actual cause for this movement is found in the structure of the system A.

From the point of view of the activity system, the question arises on the potential environmental improvement that may be achieved via the application of urban policies designed for this purpose. An example, is a system of taxes directly associated to the external cost of urban land at each location, or a direct restriction for the land use (zoning laws), which may orientate the spatial dimension of urban development according to environmental long term objectives. In addition to the relocation of activities in space, hence the relocation of fixed emissions, such policies will inevitably modify transport flows patterns according to the modified activity system, therefore also affecting mobile emissions. This justifies the need to incorporate in the analysis all impacts, not only those on the activity system but also on the transport systems, if a comprehensive long term assessment is pursued.

Moreover, the impact of fixed emission sources upon the population, depends on the location of firms and households, which in turn depend on the transport system in the form of the accessibility available at each location.

Hence, the above arguments describe a clear long term relationship between systems A and T which is unavoidable in the analysis of urban environment. In fact, environment is, from a technical and value judgement point of view, a long term issue, where measures will have a delayed effect and final results will depend upon the dynamics of the diverse functions of the city.

1.2 The Microeconomic Framework

It seems necessary, then, to understand in some detail the unclear relationship between the transport system and the use of urban land. When I say unclear, I specifically refer to economic cause-effect mechanism that describes this relationship, in opposition to the more symptomatic relationship based on the interaction between physical factors, well treated by spatial interaction models, which is the most dominating modelling approach so far. The advantage of an economic framework is, obviously, that it is based on people's decisions, that is to say, on the description of individual's (or collective) behaviour as a result of their subjective valuation of alternatives. Within this approach, stimulus from very different origin can be compared by simulating the importance that people gives to each one of them, as is the case of the highly dependent transport and land use choices. The underpinning assumption is that individuals will always choose their best option under a given set of restrictions, regardless of the sort of decision faced.

This approach is particularly useful when urban planners concentrate on setting specific signals that can restrict or induce land uses, letting people's reaction to those stimulus to produce the final impact on the city; in other words, allowing the market mechanism to operate as the means for the application of environmental policies.

The purpose of this article is to describe a model of the relationship between A and T systems under strict economic basis. This will permit to analyze and to quantify the impact that environmental policies may have on both systems. Quantification is justified since it makes possible to compare on both systems, different levels of answer to stimulus that are produced by different policies, which is obtained by means of using a common economic framework. Certainly, every quantification is unavoidably exposed to some sort of errors, however what really matters is that these errors do not mistake decision takers on environmental policies given the information available. Nevertheless, the richness of this model lies not only on the possibility of quantification, but also on the explanation for the phenomenon studied -the interaction between transport and land use- transforming intuitive notions into a consistent mechanism that contains explicit relationships or laws.

Before introducing the model, it is worth mentioning that a model is understood as a mathematic entity that, on one hand, is fed with observations of those characteristics that describe the system under study, called "attributes", which is contained in the observed data. Secondly, it is also fed with a set of laws that rules the phenomenon -in our case they are economic laws of the A and T market- which describe the cause-effect mechanism of the phenomenon.

2. The Urban Activities System

2.1 Land Use Economics

The city has a structure of activities, defined by their distribution in space, which is the result of a number of processes that took part at different stages of its development. However, in the field of policy analysis, what matters is to predict the most likely future development of the city, according to the available information, rather than to fully understand its genesis.

For this purpose, we will concentrate on the decision that each activity, household or firm, faces to locate itself in the urban space; that is, we shall model people's decisions. Since the location of a single activity will modify somehow the city, individual location choices will inevitably affect others' decisions, which is a form of location external economy. This means that location choices across the population are interdependent, so only from their joint analysis one can define the demand for urban location.

The economic assumption of individual's behaviour is one of rational decisions, i.e. they maximize their individual utility. However, in contrast to other economic goods, the urban space is considered here as "quasi-unique", since its location is associated with a very specific surrounding that is only comparable with its closest neighbours. That is, a given land lot has a value directly associated to its relative location in the city with respect to the location of the rest of activities. This property justifies Von Thunen's (1863) and Alonso's (1964) best-bid rule, or the simulation of a bid-auction process to analyze the final spatial distribution of urban activities.

In this sense, the urban land market behaves similar to markets of unique goods, like pieces of art whose worth comes from the artist fame. In the case of a land lot, its value is given by its (quasi) unique location. In both cases, the natural way of trading is an auction. Thus, the best-bid rule properly describes the behaviour of the land supply, because it assumes that the owner will obtain the highest possible price, i.e. the maximum profit.

These characteristics of the urban land market: quasi-uniqueness of a land lot, consumers who maximize their individual utility and owners who sell to the best bidder, have been developed in two economic type of models or schools. The first one, the so called Alonso`s school, uses the best-bid rule to find the equilibrium in the spatial distribution of activities and has been (unfearly) described as a rent theory. The second one, follows the random utility theory (Domencich and McFadden, 1975) and develops a demand type of models for land (see McFadden, 1978 and Anas, 1982).

2.2 The Bid-Choice Theory

After having revised and reformulated these two approaches, the Bid-Choice model is proposed (Martínez, 1992a). This is a single theory that integrates them, demonstrating that both lead to identical results under the same basic assumptions. This is achieved by means of deriving a new version of the maximum utility model, which is based on the consumer`s willingness to pay for a land lot and replacing the maximum utility assumption by its equivalent, the maximization of the consumer`s surplus derived from their location choices (see Annex, A.1).

This theoretical model focuses on the analysis of the use of land, which is the element that is specifically related to location, in opposition to the problem of its property. Certainly, land users (residents and firms) are those who define the neighborhood, not the owners. Additionally, dwellings and other types of buildings are the result of investments on land, called the building market, which behaves as any other competitive market of goods or services. Of course it is possible to extend the analysis in order to include the dwelling choice in the Bid-Choice theory, but its conceptual contribution to the location problem is limited.

Now, to build a model of the activity system, more specifically, of the process of the distribution of activities on the urban space, consists on creating a mathematic representation of the assumptions previously indicated of the market behaviour. Such representation allows us to simulate in the computer the mechanism of economic forces in the urban system, both at present time and for the future. That makes feasible to predict changes on the use of the urban land that may result from future policies and projects. Furthermore, it allows us to compare future situations and to evaluate Government and private decisions according to economic criterions.

The main conclusion out of the Bid-Choice theoretical analysis is that the land use equilibrium can be described by means of two elements. First, the willingness to pay for land of households and firms, later denoted as WP. Secondly, consumers and suppliers laws of behaviour, which may be summed up by the best-bid rule. From the theoretical analysis an interesting corollary is derived: if a given consumer is observed to be the best bidder for a given land lot, is because this lot is also his/her best (optimal) location. To know more details of this theoretical issue in urban economy, see Martínez (1992a).

From the modeler point of view, the most important result is to confirm that the equilibrium mechanism of the urban land market can be solely described in terms of the willingness to pay of each consumer for a land lot. Here, WP functions are defined as the maximum price that the consumer (household or firm) is willing to pay for a given location, that amount will provide the consumer the maximum utility achieveable given market prices, income and tastes.

In opposition to other markets, the role of the supplier (the land owner) is passive, which is a direct result from the quasi-unique property of the urban land. In other words, since land and the attributes of a given location can not be produced, the value of a land lot is not associated with a productive process, i.e. there are no relevant costs involved. Therefore, the Bid-Choice model postulates that the only market force participating in land use equilibrium is the set of WP functions, defining one for every possible land purchaser.

2.3 The Empirical Location Model

Consequently, the modeler task is reduced to estimate WP functions for every customer, which are functions of land lot`s attributes, income and taste parameters. This would be relatively straight in a world with perfect information. Since that is not the real case, one must consider the fact that estimations of WP may be somehow uncertain, in other words, they may have some statistical errors. A simple way of handling this estimation (econometric) problem is to follow the discrete random utility theory in order to achieve simple expressions, like the multinomial logit form (see Annex, equation 4).

The estimating method requires observations of the location of households and firms in the urban space, in addition to the set of attributes that describes their chosen locations. Individuals' WP for each relevant attribute can be determined by observing how different consumers choose different combinations of attributes. For instance, it is normally important to estimate the extra value that each household is willing to pay for the following attributes: access, socioeconomic level of the surrounding area, etc. The WP is an aggregate value that results from the individual's valuation of each attribute.

2.4 Land Rents

The price of a land lot, is by definition, given by the best-bid, which is the maximum WP that every possible consumer -household and firms- is willing to pay for it. Hence, the only explanatory element for rent formation are, again, WP functions. Therefore, the Bid-Choice model is able to solve location and rent problems simultaneously, by estimating WP functions. A simple example is the case study of Santiago City, described in Martínez (1991), where the estimation process and the use of data are explained.

Thus, the Bid-Choice model allows the estimation of land equilibrium prices at any stage of the A and T systems. Certainly, in the case of the stochastic model, this value is given by the known formulae for expected maximum value of WP from alternative households and firms, which is directly calculated within Bid-Choice model (see Annex, equation 5). Thus, as the rent model is completely defined by consumer's WP functions, there is a direct relationship between WP, rents and the lot attributes. In urban economic literature these rent functions are known as a type of hedonic price models (see Rosen, 1974)[2].

The rent model, allow us to predict the impact on land values that may result from a policy that affects the activity system or the transport system. This is achieved, by specifying adequate access measures in the land lots's attributes vector, making the model sensitive to the lot's access, consequently, to the state of the activity and the transport systems (see Annex, A.3).

---

[2] Hedonic, or implicit, price refers here to value of different land attributes which, aggregated according to the WP function, generate the consumer´s willingness to pay for a piece of land. Note that the known hedonic theory argues that there is no theoretical functional form for the rent model, hence, it seeks one curve that fits best with the observed rent data, for instance, by using Box-Cox method (Halvorsen and Pallowsky, 1981). That is, it seeks the best empirical answer. Conversely, the Bid-Choice model proposes a clear functional form for rents, given by the so called logsum of WP (see equation 5 in the Annex).

In short, there is a theoretical model that supports a probabilistic logit model, which makes possible an estimation of the WP functions and land values. Certainly, the quality of the estimated functions will depend on the set of land attributes, the functional form specified by the modeler for WP and the data quality. Apart from this, the calibration of the model depends, of course, on how well this theoretic assumptions reveal the mechanism actually ruling the land market.

3. The Transport System

The transport system is also complex. To describe the movement of people and things it is essential to distinguish two prevailing dimensions: space and time; i.e. trips. Indeed, trips are performed between pairs of origins and destinations (space) in a given moment (time). These characteristics determine the trip performance conditions. In other words, a given produced trip must be consumed instantly -no possibility of stock- and between a given origin and destination (no spatial substitution). So, it is easier to analyze this system from the point of view of the user[3], that is, the demand, and to understand it as a group of decisions that the user has to make subject to transport supply conditions. Given the origin spatial location of the user, the required decisions will be the following:

3.1 Number of Trips and their Purpose

The decision of making a trip depends on whether there exists or not a benefit in making "contact" between the origin activity (e.g. residential) and another distant activity; this benefit depends on the trip purpose. Now, contact is verified by trips, whose cost must be lower than the benefit obtained. However, the final decision on making trips depends on the best use that the user can give to two basic constrained resources: time and income. Then, a trip requires a rational decision, obtained out of the assessment of benefits and costs. This issue is known as users' mobility and concerns to the <u>mobility sub-model</u>.

3.2 Destination

Contact with another activity has several feasible alternatives, as many as the number of places available where that activity exists. Nevertheless, each alternative is different, since they offer different levels or qualities, and impose different generalized transport costs (including time, fares, comfort, etc.). Then, the consumer faces a second decision: the trip destination. In this regard, he/she

---

[3]User: is the person who makes the relevant decisions related to the travelling process, which is not necessarily the traveller.

compares alternatives in terms of the economic net benefit (benefit-cost) offered by each one and, by assumption, chooses the best one. Therefore, alternatives are defined by destination -depending on level and quality of the alternatives for each case- and by the generalized transport cost. This is known as the trips distribution sub-model.

3.3 Mode Choice

Once destination has been decided[4], it will be necessary to choose one of the various alternatives of transport mode available for that time period and destination. In this case, the user evaluates the mode in terms of fares, time of travelling (walking, waiting and in the vehicle), comfort, security, etc. This issue is treated as the mode split sub-model.

3.4 Route

In the case that the chosen mode offers several routes to reach the destination, it is necessary to choose the best route alternative. This is modelled as the equilibrium assignment to a transport network. Note that, it is in this submodel, where trips demands are faced with transport supply (or network). Trips assignment to the network follows equilibrium conditions, e.g. Wardrop's (1952) equilibrium laws: for a given origin-destination pair of locations and for a given period, every used alternative route has the same transport cost. Thus, final vehicle flows in an equilibrated network reproduce the balance between: congestion, operation constrains and policies, on the supply side, and demand for transport on the other.

Each one of these trip decisions has led transport planners to design specific submodels, creating different versions of the known 4-stages model. However, for the purpose of this article, together these submodels may be seen as a way of expressing the demand for transport, considering the decisions process herein described.

Then, trips demand depends on the following elements:

a) Availability and level of the activity that the user may visit at each alternative destination.
b) Transport cost to reach contact with each activity, by destination, mode and route.
c) User utility obtained from the contact with another activity.

From an economic point of view, a trip allows the contact between two different activities, producing a change on the actual performance of each one of

[4] It has been established that, for some trips purposes, mode choice may be previous or simultaneous with the destination choice.

them, which can be expressed in terms of activities' net benefit. Here, benefit is defined as change on the utility function (measured in monetary units), in the case of household, or as the profit, in the case of firms, which may be positive or negative; these benefits can be measured directly at the origin and destination from the relevant activities demand and supply curves.

Once the transport demand model is known, i.e. specified in detail, it is also possible to derive these measures of benefits indirectly form the transport market, by calculating the known marshallian consumers' surplus of transport users. The consumer surplus can also be understood as a measure of the degree of access. From that interpretation (see Martínez, 1995) two access measures can be derived (see Annex, A.3):

a) Accessibility: measures the net benefit obtained by the user at the activity that generates the trip. This benefit is directly perceived by the user of the transport system and it is denoted as "acc".
b) Attractiveness: it is the benefit obtained by the visited activity as the result of being visited; it is denoted as "att" and beneficiaries are non transport users. Accessibility may be estimated directly, while attractiveness is obtained in an indirect way, in both cases by means of calculating transport user's benefit integrating the transport demand curve. The result of such analysis is that acc and att are calculated from standard parameters of the trip distribution model, which are generally available as an output of the 4-stages model. In other words, the transport model provides direct information of access, both acc and att.

These access measures have the property of being able to synthesize the transport supply system according to perceptions of each type of user and depending on the specific location. They represent a synthesis of users' best options of destination, mode and route to get contact with different activities.

To analyze how useful these access measures are, note that, in the case of a change in the activity system, acc and att measures are directly modified through the change on activities available at different destinations. Conversely, if a change occurs on transport supply, e.g. on fares, travel times, modes available, etc., this is also captured by acc and att measures through generalized costs. Therefore, acc and att are sensitive to changes on both transport and activity systems. These changes may be directly measured on the net benefit that users and non-users obtain when they visit (acc) or they are visited (att) by other activities.

Then, access measures are variables that "properly accuse" the impact that environmental policies on land use and/or transport may induce on the population. This allows the analyst to model people's reaction in terms of their decisions; i.e., it makes possible to model the demand on the land use and transport interaction consistently.

4. Interaction between Systems

Activity and transport systems interact in the following way. Transport supply defines the (generalized) cost of "contact" between distant activities, where distant refers to the spatial location of activities defined by land use. This approach is detailed in Martínez (1992b), where the 5-stages land-use transport model is described (5-LUT), which integrates the Bid-Choice model with the 4-stages transport model.

The basic components of access measures are the activities' level, their relative spatial location, transport costs and a set of weighting values associated with the utility change or benefit. Access measures are in charge of transferring information between systems A and T, in a way consistent with consumer's behaviour.

In the Bid-Choice model, the consumer's willingness to pay for land depends on a set of land lot attributes, whose components may be different between cities and consumers. Therefore, it is part of the "art" of modelling to include those relevant attributes of the city being studied for each household or firm. However, at the moment of deciding the location of any activity, a usually dominant attribute is the lot access. Depending on the kind of activity being located, access may be best represented by accessibility, as in the case of residences, or attractiveness, as in the case of firms (the typical example being the retail activity). More generally, the location of an activity is related to one or both notions of access. Therefore, the set of land attributes must include access which, in opposition to other physical attributes, like square meters of land, it measures the consumer's perception. This indicates that the set of attributes of a land lot is generally inherent to the consumer, which can be expressed in a vector format as follows:

$$Z_{hi} = (Z_{h1i}, Z_{h2i}, \ldots, acc_{hi}, att_{hi})$$

where Z_{hki} indicates attribute k as it is perceived by consumer type h, and i denotes a land lot.

Moreover, in Section 2 it is shown how the WP function depends on these attributes, i.e. on access and other attributes[5]. In this sense, the Bid-Choice model is very flexible, since it does not impose any restrictions on the kind of attributes that the modeler may consider reasonable to test, including social, cultural, environmental elements, etc. Certainly, defining vector Z is a task where it is important to deeply know reality and to listen the opinion of experienced people on the urban sector. On the other hand, the process of deciding which attributes

[5] The role of WP variables, particulary access, is well defined in Jara-Díaz, et. al. (1994)

must be included in vector Z is crucial, since it will determine which type of changes in the land use and transport system will be the model sensitive to. Consequently, these are the only changes able to represent policies which will be feasible to be analyzed by means of using the model.

For instance, in the case of environmental policies, access attributes permit to analyze the impact of policies that are focused on both systems A and T. Certainly, any policy that affects the transport system will be perceived by the consumer as a change on trips generalized cost. Consequently, that change will affect accessibility (acc) and attractiveness (att) measures that are associated with a given location. On the other hand, policies orientated to affect the activity system, by means of a rearrangement of land use, should alter access measures and any other land attributes, so as to affect rents through consumers' WP. This is the only way of obtaining a change on land use motivated by market incentives, while guaranteeing the natural equilibrium in land prices.

5. General Interaction Mechanism

We have briefly presented a model that allows the analysis of the interaction between the use of urban space and transport, which has been divided into 5-stages (four for transport analysis plus one for the location-rent model) with a consistent microeconomic approach. This is achieved by a definition of access, acc and att, which is essential in this approach, since it provides an economic link between transport and land use, allowing an adequate transference of information between them. Therefore, it permits to analyze the sequence of expected impacts that may result from any disturbance on the observed state of the city.

To fix ideas on how the market mechanism operates see Figure 1, which describes the impact sequence or interaction between the land use and the transport systems within the model. At the upper part of the figure, one can see state variables which describe the city system and have a direct economic interpretation: transport cost (c_{hij}, c_{hijm}, c_{hijmr}), willingness to pay for a land lot i (WP_{hi}), and the land rent on that zone (p_i)[6]. At the lower part of the figure, physical variables are displayed, like location of activity at each zone (a_j) and demand for trips, which is divided by origin/destination (V_{hij}), mode (V_{hijm}) and route (V_{hijmr}). The sequence of arrows and variables describe the mechanism of interaction. Circles highlight those economic elements that may represent, within the model, environmental policies based on economics incentives. On the other hand, policies that involve restrictions may be represented by location of activities and trip flows, i.e. physical variables.

---

[6] Indexes denote: h for type of household or firm, i-j for origin and destination zones of the trip, m-r for mode and route of a trip respectively.

This figure also shows two conditions of equilibrium. On the left extreme is the land market equilibrium, which is ruled by the best-bidder mechanism, and on the right extreme is the transport network equilibrium, which in the case of private vehicles (variable route), for instance, is ruled by Wardrop's conditions (1952). Both conditions must be accomplished simultaneously to guarantee a total equilibrium. On the network equilibrium, mobile sources of air pollution and vehicular congestion effects are detected, as a direct result of vehicle circulation and speeds on urban roads. In the case of fixed sources, pollution impacts will depend on the spatial location of activities (a_j), which is associated with the land market equilibrium. In sum, the difficulty in determining the production of pollutants is to know the location and characteristics of fixed sources and the operating conditions (flows) of mobile sources; **all this information is delivered by the model.**

Another feature of Figure 1 are two circular arrows (left side) that indicate the existence of circuits or loops of impacts. The one on the far left describes the fact that the location of an activity (a_j) affects willingness to pay for land by affecting the nieghbourhood of that zone (socioeconomic level, for instance), which in turns will affect its land rents, therefore, affecting activities location. The loop on the right shows that location (a_j) affects the structure of access measures (acc,att) which, in turn, alter willingness to pay for land, and land use. These loops reveal the complex equilibrium of the land market, where there are endogenous elements that make even more difficult the analysis (See Annex A.4). Other loops can also be identified in Figure 1 following avows in the transport system.

Figure 1 also shows the role of access variables located precisely where economic interaction between transport and land use is verified. It is particularly important to note that any impact on the transport system is transmitted to the activity system only by means of these measures. Conversely, information of changes in the activity system may also be transmitted through demand for trips. An important issue here is the equilibrium between transport changes and land use impacts. The model may be used in a dynamic version allowing lags between transport changes and their effect on land use, as well as between land use changes and transport impacts; in this case, simulations are performed for short time intervals (say two years). The alternative version is a simultaneous general static equilibrium in the complete system A+T, without lagged impacts but allowing longer time intervals (say every five to ten years), thereby permiting mayor interaction effects to take place. In the following section some examples of environmental policies, proposed both by transport and urban planners, are revised.

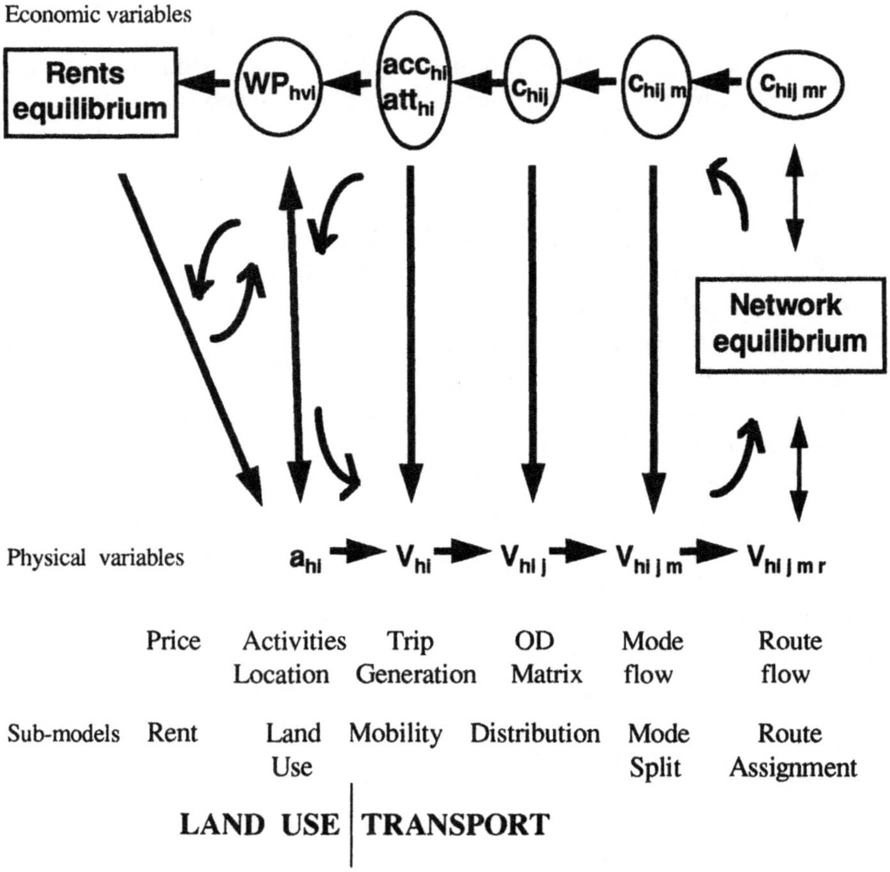

Fig. 1. Impact Sequence

6. Analysis of Environmental Policies

6.1 Road Pricing

It consists in charging, through any mechanism, the use of urban roads in order to control external costs, mainly congestion and pollution although there may be others, like accidents. This kind of policies is becoming increasingly accepted by public opinion and governments in many countries.

A monetary charge or toll may be modelled as a direct increase of the cost for using a route (c_{hijmr} in Figure 1). This will alter the network equilibrium, i.e. roads flow, as well as travel costs on each mode. It is expected that cars will be the main mode affected, decreasing its demand which switches to public transport alternatives according to the mode split model; this is a short term impact and, probably, the most important. This change in mode split provides also a new network equilibrium, with consequences on congestion and pollution that can be estimated.

Medium and long term impacts are changes on trip destinations, specially on trips that do not have an obliged purpose (shopping, amusement). This generates an impact on trips accessibility and attractiveness that produces changes in land values and land use. It is worth noting though, that the higher the trip cost, the lower the network congestion (due to changes in mode split), plus an increase in access levels and a change in the urban surrounding of the tolled zone. Then, the net effect on land is not easy to predict without using the model.

These impacts on the activity system depend in part on the pricing mechanism used. For example if it only affects some zones -tolled area- a greater impact on that zones is expected. Other mechanisms that embrace the whole urban area, like an automatic toll on main roads, have a more distributed impact in space. It is possible to analyze the final impact on congestion and pollution of different fare mechanisms. Besides, the model allows to forecast the magnitude of impacts on both A and T systems, for different fare levels.

In sum, the model analyses the road pricing policy to predict its impacts on congestion and traffic pollution, as well that on changes on the use of urban space (fix sources), at any level of the pricing mechanism.

6.2 Subsidies to Public Transport Supply

These type of policies pursue the improvement on the quality of public transport, for instance, on frequencies, security, comfort, information, services integration, etc. These type of policies may be applied as complementary to road pricing, in order to provide clear incentives to the use of public transport; moreover, they may be financed by means of collecting road tolls.

In terms of the modelling process, the improvement on transport supply is represented as a change (decrease) on the generalized transport costs of public transport modes (c_{hijm} in Figure 1). The direct expected impact is an increase on the use of public transport modes and a decrease on the use of private vehicles. The sequence of impacts in other levels of the T and A are similar to the road pricing policy, but in this case it positively affects medium and low income population groups, which are at present the main users of public transport and would be direct beneficiaries of a better service.

With regards to the impact expected on the use of urban space, the general tendency is that the downtown of the city will strength its role as a business and services area, while main corridors better served by public transport increase their location density. Tendencies of urban expansion can be restrained by a policy of this nature and its effects would be even greater if it is complementary to a road pricing policy.

It is important to indicate that this model has restrictions in analyzing some types of policies. These are cases involving improvements on the information given to the user or on the local operation of the bus system, if it improves the general perception of the system but it does not affect measurable variables used in the model.

6.3 Land Use Taxes

This policy is seen as a way of affecting the development of the city. It permits to internalize external costs of the suburbanization process, for instance, public services supply as roads, services, sewage, etc., which are usually financed by the Government. There are also external costs like pollution and road congestion. Certainly, the marginal cost of suburbanization is different to the marginal cost of land lots located in an area that has already been developed. There are even important differences among expanding zones, depending on the rural infrastructure or on the possibilities of using available interurban roads. This policy may be also considered as a way of capturing site values (location surplus value) resulting directly from general infrastructure and public goods.

The model can be used to analyze land tax structures -defined according to level and spatial distribution- in order to know the impact on the use of land as well as on the transport system and on the environment. In the case of Santiago, the main expected impact on the use of land is concentration and densification of households with medium income. As for low income groups, it would depend on the government specific regulations and on its capability to manage residential subsidies. Conversely, in higher socioeconomic levels, the impact may be even a tendency to migration towards the highly taxed periphery in search for better life conditions and status.

However, it is difficult to predict much without modelling. Certainly, the urban market mechanism is complex but also interesting, since it presents antagonic forces that tend to stabilize in what we call the equilibrium, which is precisely the situation that the model allows us to identify.

In terms of modelling this policy, it is important to determine the type of tax. Let us suppose that such tax corresponds to a percentage p% of the real increase of the land value, charged periodically (say monthly). This would be economically justified if p% represents the social external cost of the use of that piece of land lot for a specific purpose, which can be calculated by using the model.

Once we have determined that values we can assume that at each zone the price of a land lot exogenously increases in that amount. As a direct consequence, the consumers surplus (for residents and firms) will be reduced in taxed zones in some amount which accounts for this exogenous tax plus the effect on the endogenous rent. Low income groups will have less probability of purchasing in that zone and location demand for taxed zones is expected to decrease. Instead, demand for other zones will increase, with a consequent moderate raise in prices and a greater densification.

On the transport system, the new spatial distribution will modify trips' origins and destinations, hence decisions on modes and routes will also change, producing a new equilibrium in the network. All these processes can be simulated by the transport model. However, there are other second order impacts: the new equilibrium of transport network plus the changes on the urban distribution of activities will modify access levels (accessibility and attractiveness) producing a feedback impact on the use of land. Nevertheless, in the long term, there is a new equilibrium in the use of roads which results in a direct impact on congestion and air pollution produced both by mobile and fixed emission sources.

An alternative way of affecting the use of urban space is through an opposite method: subsidies or regulating incentives, to support densification and development of services in certain zones. The modelling process is similar, although it is oriented to different people. The final impact on congestion and air pollution will depend on the incentive type, level and distribution.

7. Conclusions

The analysis of expected impacts caused by environmental policies described on the previous chapter is rather speculative. It is based on the analysis of the main forces that rule the interaction between land use and transport systems. However, its speculative characteristic is on the magnitude of the impacts and not on the interaction -logical- mechanism among variables. Nevertheless, the purpose of this article is to demonstrate that a model like the 5-stages described earlier permits the calculation of these impacts.

Applications have been focused on a description of two main aspects: the way of translating environmental policies into variables that can be treated by this model, and the logical sequence of impacts that a variation on factors may generate on the activity and on transport systems.

However, a major objective of this article is to give a basic idea of how to make possible the analysis of environmental policies. This idea consists on modelling the impact that environmental policies have on land use and transport. It eliminates the necessity -required by alternative methods- to measure people's perception and valuation of levels and types of pollution and congestion. That is the case of hedonic price theory, which tends to detect the willingness to pay for a better environment according to the price of a lot -or to its variability[7]. The argument herein proposed is that environmental impacts are spatially distributed and difficult to detect by means of variations in the price of land. In the case of Santiago city, there is another element that increases the difficulty on using the hedonic prices method, which is the importance of the "socioeconomic level of the zone" at the moment of deciding the location of a residence. This characteristic makes more difficult to detect the role of other factors, such as access and level of pollution, which become less important or that are distributed in an almost homogeneous way.

Thus, the proposed method permits to detect impacts of environmental policies, which is particularly useful if the solution to the environment problem is pursued in terms of achieving a standard level of congestion and pollution. The method also permits to identify the degree of accomplishment of these objectives.

Finally, there is a question of how to model land use-transport interaction. This paper argues in favor of a consistent economic 5-stages model, linking transport and land use through well defined access measures. Consistency is obtained by modelling individual's behaviour at every decision, assumed to be done as rational beings which results in consistent choices in their location and transport decisions. Such behavioural approach provides an economic interpretation of the model throughout every stage and a direct link between transport and land use.

Acknowledgements

This research was partially financed by the Chilean National Fund for Science and Technology (FONDECYT), the Technical Department of Research of the University of Chile (DTI) and the Centre for Public Studies of Chile (CEP).

---

[7] The evaluation of environmental policies using hedonic price theory, is discussed, among others, by R. Palmquist (1988). More recently, Hopkinson et. al. (1992) propose the use of a stated preferences method for environment evaluation.

References

Alonso, W. (1964). *Location and Land Use*, Cambridge, Harvard University Press.

Anas, A. (1982). *Residential Location Markets and Urban Transportation*, Academic Press, London.

Ben-Akiva, M. and Lerman, S.R. (1987). *Discrete Choice Analysis: Theory and Application to Travel Demand.* The MIT Press, Cambridge.

Domencich, T.A. and McFadden, D.L. (1975). *Urban Travel Demand: A Behavioral Analysis*. North-Holland, Amsterdam.

Halvorsen, R. and Pollakowsky, H.O. (1981). Choice of Functional Form for Hedonic Price Equations. *Journal of Urban Economics 10*, 37-49.

Hopkinson P.G., Nash, C.A. and Sheehy, N. (1992). How much do people value the environment ? A method to identify how people conceptualize and value costs and benefits of new road schemes. *Transportation*, Vol. 19, Nº 2, pp. 97-115.

Jara-Díaz, S., Martínez, F.J. and Zurita, I. (1994). A Microeconomic Framework to Understand Residential Location. The 22nd European Transport Forum. Transportation Planning Methods, Vol.2, 115-129.

McFadden, D.L. (1978). Modelling the choice of residential location. In Karlqvist et. al. (eds). *Spatial Interaction Theory and Planning Models*, North-Holland, Amsterdam, 75-96.

Martínez, F., (1991). Transport Investment and Land Values Interaction. The Case of Santiago City. Proceedings 19th. PTRC Summer Annual Meeting, Seminar G. pp. 45-58, Brighton, U.K., 1991.

Martínez, F. (1992a). The Bid-Choice Land Use the Model: an Integrated Economic Framework. *Environment and Planning A*, Vol. 24 (6), 871-885.

Martínez, F. (1992b). Towards the 5-Stage Land Use-Transport Model, Land Use, Development and Globalisation. Selected Proceedings Sixth World Conference on Transportation Research, Lyon, Vol 1, 79-90.

Martínez, F. (1995). Access the Economic Link in Transport Land-Use Interaction. *Transportation Research B* (forthcoming).

Palmquist, R. (1988). Hedonic Methods. Working Paper, Department of Economics and Business, North Carolina State University, Raleigh.

Rosen, S. (1974). Hedonic prices and implicit markets: product differentiation in pure competition. *Journal of Political Economy*, Vol 82, Nr.1, 34-55.

Von Thunen, J.H. (1863). See english version in: *Von Thunen Isolated State* (ed). Peter Hall, Pergamon Press, London.

Wardrop, J.G. (1952). Some theoretical aspects of road traffic research. Proceedings of the Institution of Civil Engineers, Vol.1, 352-378.

Williams, H.C.W.L. (1977). On the formation of travel demand models and economic evaluation measures of user benefit. *Environment and Planning A*, Vol. 9, 285-344.

Annex

A.1 Bid-Choice Theoretical Model

Let us call WP_{hi} the consumer h's willingness to pay for a land lot denoted as i, and assume that it depends on a vector z that contains n elements which describe attributes of the lot (see Rosen, 1974). Then, WP can be written as a function of attributes as:

$$WP_{hi} = WP_h(z_{h1i}, z_{h2i}, \ldots\ldots z_{hni})$$

and one can define as many functions as consumers participate in the market. Element z_{hki} represents the value of the k-th attribute of lot i.

Index h represents land consumers, which may be households or firms. This is an important point, since it indicates that the model accepts that the urban land may be used for any kind of activity, except for some uses exogenously restricted. Other models distinguish areas for residential use, commerce or industrial use, etc. This segregation inhibits the adequate participation of all forces that may affect the equilibrium price in the urban space.

The market equilibrium is obtained as the spatial distribution of activities that accomplishes with two conditions simultaneously:

1.- Each consumer is located in a place that represents his/her maximum possible surplus (CS) at exogenous prices; that is to say, he/she chooses the best alternative available location. This may be expressed as:

$$\underline{\text{CONSUMERS}} \qquad \max_{i \in S} CS_{hi} = \max_{i \in S}(WP_{hi} - p_i) \quad \forall h \tag{1}$$

This equation must be verified by every household or firm h.

2.- In order to maximize the land owner profit, the consumer finally located in a given lot must be the one with maximum willingness to pay for it. This is the best bidder rule which makes sure that the owner obtains the maximum price p for a piece of land. This may be expressed as:

$$\underline{\text{OWNERS}} \qquad p_i = \max_{g \in H} WP_{gi} \qquad \forall i \tag{2}$$

which must be verified for each zone or plot i.

These two equations may be read as follow: i) each consumer chooses, among S alternatives, that location which maximizes his/her surplus. ii) each owner chooses the best bidder among H consumers.

The solution for the equations system (1) and (2), describes the equilibrium of the urban space market, which is obtained replacing (2) in (1). Then:

$$\underline{\text{EQUILIBRIUM}} \quad \underset{i \in S}{\text{Max}}\ CS_{hi} = \underset{i \in S}{\text{Max}}\ (WP_{hi} - [\underset{g \in H}{\text{Max}}\ WP_{gi}]) \quad (3)$$

Equation (3) represents the equilibrium equation of the activity system and is called the deterministic version of the Bid-Choice model. Writing one equation for each consumer, a system of equations of simultaneous solution in WP's parameters is obtained.

The structure of this equation is interesting. It has two levels of maximization over independent dimensions. The external maximization is across land lots offered $(i \in S)$, while the internal maximization is across consumers $(g \in H)$.

An important issue is that if the best bid for a land lot is submitted by consumer h, then from equation (2) the price is $p_i=WP_{hi}$; in that case the surplus of that consumer is null ($CS_{hi}=0$). Alternatively, the best bid may be submitted by another consumer g, that is $WP_{gi} > WP_{hi}$, in that case $CS_{hi}<0$. Therefore, the upper limit for CS_h is zero. In other words, the optimal consumer location is, by definition, where his/her surplus is null, which is exactly where the consumer is the best bidder.

Corollary: If a consumer is the best bidder in a given lot, then that is his/her optimal location. Other lots will, at the most, equalize the benefit obtained on that location.

But, what is most interesting of equation (3) is that equilibrium in this model is expressed only by one generic element, the consumers' willingness to pay functions (WP). In other words, owners behaviour does not have any direct influence, in that sense their roll is passive in the urban market.

A.2 Bid-Choice Empirical Model

There are two ways of expressing the theoretical Bid-Choice model. We will present here the Bid version, which is derived from the previous corollary.

The purpose of the empirical model is to estimate WP functions for the population of a city, based on the theoretical model that describes the market mechanism and observations of consumers' behaviour.

Let us call $WP_h^*(z)$ the total willingness to pay function, such that:

$$WP_h^*(z) = WP_h(z) + \varepsilon_h$$

where ε_h represents the random term and $WP_h(z)$ is the estimable function, that is, the systematic part of the total WP function under the assumption of imperfect information.

The problem now is how to estimate WP functions. Let us assume that we observe the location of a representative sample of consumers and the rent they pay for each land lot. We can deduct that the rest of consumers bid less than the consumer observed in each land lot. The price paid for the lot and the lot characteristics (vector z) is also observed. This information -collected for a representative sample of land lots offered in the urban area being analyzed- may be used in a probabilistic version of the theoretical model.

This version must assume some distribution of the error term ε_h. It is advantageous, in computing sense, to assume that such terms are identically and independently distributed Gumbel. In that case, according to the theory of the discrete choice, the probability that consumer h makes the best bid in a given lot i, called $P_{h/i}$, is given by the multinomial logit model:

$$P_{h/i} = \frac{\exp(\mu\, WP_{hi})}{\sum_{g \in H} \exp(\mu\, WP_{gi})} \tag{4}$$

with μ a scale factor.

Additionally, it is possible to demonstrate that the expected land price is given by:

$$p_i = \frac{1}{\mu} \ln\left[\sum_{g \in H} \exp(\mu WP_{gi})\right] + \frac{\gamma}{\mu} \tag{5}$$

with γ the Euler's constant (≈ 0.577)

This expression is very well known as the expected maximum value of WP among the discrete choice alternatives. In this case, it represents the rent model in the urban land market, which is endogenous in the Bid-Choice model. In other words, it represents the probabilistic version of price equation (2). Notice, however, that equation (5) may be seen as and hedonic land price fuction with attributes Z, but it is a special one in two ways: the functional form is defined from a theoretical background and, secondly, hedonic coefficients should also be consistent with location choices (equation 4). Therefore, there is a theoretical support for more constrained hedonic price functions than what is usual practice.

Equation (4) represents the location model, which has the well known multinomial logit form, described among others by Ben-Akiva and Lerman (1987), which is directly obtained from the assumption of the disturbance term. It is important to note that the term in parenthesis in equation (5) is identical to the denominator of equation (4), which allows us to rewrite the location model in a more simple way:

$$P_{h/i} = \exp[\mu(WP_{hi} - p_i)] \qquad (6)$$

The advantage of this last equation is that it can be expressed in a linear form if WP functions are linear, which is easier to calibrate.

Equilibrium conditions, stated in equation (3), require that equations (4) (or equations 6) are fulfilled simultaneously with equation (5); i.e., location and rents functions are satisfied simultaneously by a unique set of WP values. This can be achieved by searching the least squared root parameters of WP functions for the simultaneous non-linear system of equations, or by searching for the maximum likelihood parameters.

A.3 Access Measures

To understand what access measures are and the way they are generated (see Martínez, 1994), it is required to describe their formulation in more specific terms. From their definition as consumer surplus, these measures must be expressed in terms of the same factors than the demand function. Then, for a given trip, access depends on: the consumer's value of visiting activities (α_h), the availability of activities (a_j) at the destination and the transport cost (c_{hij}). That is:

$$acc_{hij} = f_1(\alpha_h, a_j, c_{hij}) \qquad (7)$$

$$att_{hij} = f_2(\alpha_h, a_j, c_{hij}) \qquad (8)$$

where f_1 and f_2 make explicit the definition of access as a measure of the benefit, or consumer surplus, perceived by an activity h located in zone i. In the case of acc, that benefit is derived from making contact with another activity, located at j, with a travel cost c_{hij}. In the case of att, the benefit is obtained from being visited by the traveller. It is important to note, though, that travelling between i and j can be done through several modes, therefore, c_{hij} is the lowest possible cost of that trip -or the expected value of the lowest cost in the stochastic model.

Access measures (7) and (8) are considered as relative access measures, since they described access associated with a given destination. They may be added to consider either: all the possible alternatives of destination to make the same trip, or visitors from all possible origins; which are called integral access measures. For this purpose, it is required to find the destination that offers the maximum benefit or the best accessibility (acc), or the expected value of the best accessibility (analogously for att). Fortunately, such added measures are directly provided by transport demand models, e.g. balancing factors of the trip distribution entropy model.

The stochastic approach is recommended for the analysis of individuals choosing among discrete alternatives, according to the principle of maximizing individuals utility. That, because of the impossibility of knowing exactly the individual utility function, which is overcome adding a random disturbance term; this argument follows from the random utility theory (Domencich and McFadden, 1975). Within the stochastic framework, one must think in terms of expected benefits or expected access, as it has been formulated by Williams (1977) and applied in the Santiago's transport model called ESTRAUS.

Therefore, the integral (aggregate) measure of the expected value of the access associated to a given location are given by:

$$acc_{hi} = f_1'(\alpha_h, a_j, c_{hij}) \tag{9}$$

$$att_{gi} = f_2'(\alpha_h, a_j, c_{hij}) \tag{10}$$

where f'_1 and f'_2 are functions that depend on the specific stochastic model of transport demand. That is to say, they depend on the explicit expression of the transport demand model. Once that model is known, explicit versions of equations (9) and (10) can be obtained, which represents an aggregate measure of the options of h to contact other activities, according to the probability of choosing each one and their location in space. It is worth noting, however, that acc and att can be obtained without ambiguity from transport demand models, only if activities (commodities and services) location and prices remain fixed, called the short run case (Martínez, 1995).

A.4 Land Use Simulation

Once WP functions parameters have been calibrated, the model can be used as a simulator of land use and rents in the urban area for any year in the future. This procedure starts by defining and calculating changes in the subset of attributes exogenous to the model for the prediction year t, e.g. zonal land use regulations, government development policies, etc., which we shall call vector Z_{hi}^{t}. Additionally, new population and total productive and commercial firms are forecasted by an exogenous input-output model.

The rest of attributes are endogenous (X_{hi}^{t}) because they are calculated from residential and commercial location and rents, for example agglomeration of activities, zonal average income of residents, etc. This defines an internal loop in the location model: activities' WP depend on location of other activities, while location of activities and rents depends on WP for each location. Analytically, the internal loop can be expressed as:

$$WP_{hi}^{t} = (Z_{hi}^{t}, X_{hi}^{t}(WP^{t-1}), \beta)$$

where WP in the right hand side is the vector of WP functions for every household and firm in the previous prediction year or previous iteration; b is the vector of WP parameters. This loop can also be explained in terms of locations probabilities using equation (4): location (probability) is a function of other activities location (probability).

The internal loop is a problem known as fixed point, whose solution exists depending on the actual functional form of WP function. If a solution exists, which represents the location equilibrium, it can be found iteratively by starting from an exogenously given location pattern, normally taken from the previous period t-1.

Finally, land use equilibrium should be constrained to land availability, which requires adjustments on land values (rents) and changes in land use tending towards densification of the city. Such changes in rents should be originated by changes in consumers' WP.

CHAPTER 13

AN ANALYSIS SYSTEM FOR INTEGRATED POLICY MEASURES REGARDING LAND-USE, TRANSPORT AND THE ENVIRONMENT IN A METROPOLIS

Kazuaki Miyamoto and Rungsun Udomsri

1. Introduction

Growing concern about environment is strongly urging planners and policy makers of a metropolis to more explicitly care the impacts of policy alternatives on various aspects of environment. Since most aspects of urban environment can be regarded as externalities of land-use and transport, it is indispensable to forecast and evaluate changes in environment which seem to occur when proposed policy measures related to land-use and transport, to say nothing of environment, are implemented. In addition, the policy measures should be an integrated set of instruments regarding land-use, transport and environment, because more effectiveness can be expected by integrating such policy measures. In other words, integrated planning and implementation regarding land-use and transport as well as environment is most necessary for metropolises, particular in developing countries where dramatic changes in urban structure are occurring (Miyamoto and Udomsri, 1994).

To make the integration possible, it is requisite to formulate an institutional set-up which is really functional. However, without effective tools for analysis, it is impossible for related agencies to discuss policies and their implementing measures sufficiently, because options are so various and complex in their

Y. Hayashi and J. Roy (eds.), Transport, Land-Use and the Environment, 259–280.

interactions. An analysis system covering land-use, transport and the environment is expected to provide the related agencies of a metropolis with a forum in which they can discuss policies and their implementing measures substantially.

Many kinds of land-use models and transport models as well as their integrated models have been developed to forecast their changes for various purposes, and some of them can be applicable to forecast as far as environmental changes. However, most of them have little compatibility with other models. It means that each model development aims to build its own simulation model only, and not to utilise existing program modules. In other words, there have been few ideas of standardisation in simulation programs of land-use, transport and environment models. Since existing program modules have various ways of input and output of data, they cannot be used as standard parts which constitute a large simulation system.

When we prepare an analysis system for integrated planning and implementation, the system should consist of modules of land-use, transport and the environment. It would take a long time and much cost if we try to develop all the simulation models only by ourselves. Therefore, establishment of a standard for other modules is as important as providing existing stock of program modules with interface, to connect them organically to the simulation system. For this purpose, we should set up a general framework for an analysis system that allows easy incorporation and replacement of modules of land-use, transport, environment and, moreover, any sector of a metropolis.

The objective of the present study is to develop a pilot system to evaluate integrated sets of policy measures related to land-use, transport and the environment in a metropolis, which will contribute to establish a forum of related governmental agencies for their integrated planning and implementation and to make it actually effective and substantial.

In the following part of this paper, we discuss some of the basic concepts for the development of the analysis system. In the latter part, we propose an integrated model of land-use, transport and the environment for the analysis system by improving a land-use model based on Random Utility/Rent-Bidding ANalysis (RURBAN) (Miyamoto *et al*, 1992).

2. Requirements for the Analysis Tool

2.1 Outline of the System of Land-Use, Transport and Environment

Before building an analysis system, it is necessary to identify the issues regarding land-use, transport and the environment in a metropolis. In this study, the system related to land-use, transport and the environment is grasped as shown in Figure 1. In this figure, land-use as well as transport are composed of two elements; markets and governmental agencies. It is assumed that environment is an externality of both land-use and transport and that it does not constitute a market by itself. Environment can be regarded as nothing but a situation of environmental qualities most of which land-use and transport determine. Therefore, the treatment of environment in the system is different from those of transport and land-use. However, in a wider sense, it can be said that land-use, transport and the environment constitute a market, in spite of that there are a variety of externalities in it. The metropolitan market has interactions with national, internationally regional and as far as global economies and environment. Particularly for the case of environment, pollutants as well as land cover changes in the microscopic level, such as emissions from vehicles and deforestation, affects the global environment.

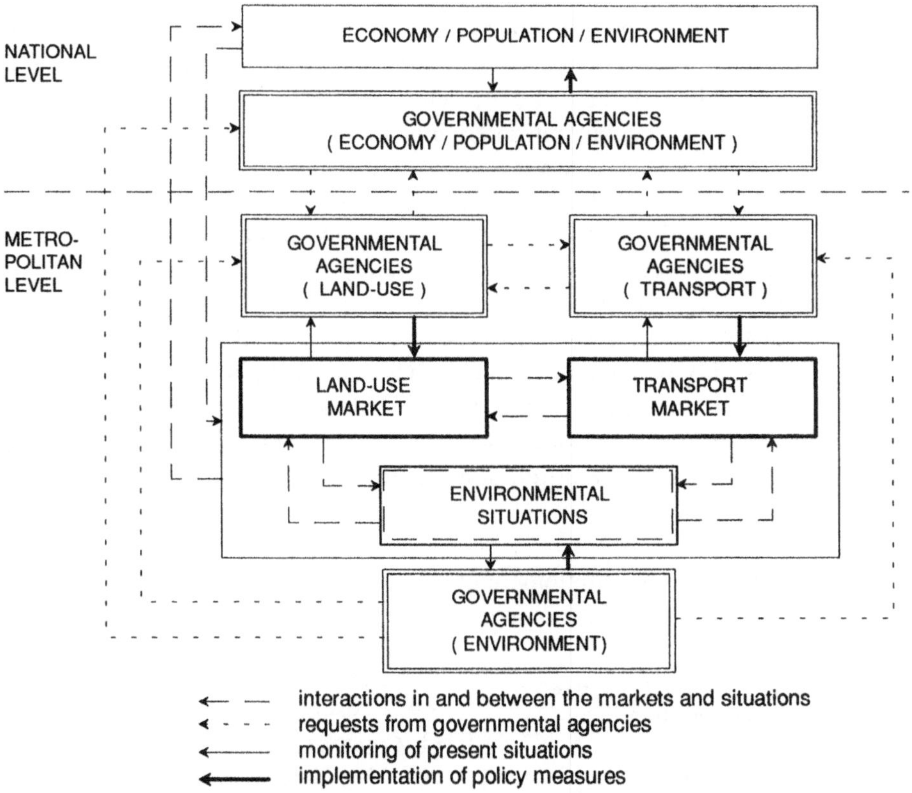

Fig. 1. Configuration of the System of Governmental Agencies, Land-Use and Transport Markets, and Environmental Situations in the Metropolis
Source: Miyamoto and Udomsri, 1994

The governmental agency in charge of the market is monitoring its situation through surveys and studies. Based on the studies, it makes a plan and selects policy measures for the implementation of the plan under its responsibility. If the interactions between the agencies are well coordinated, they would be able to work efficiently as if they constitute a single planning and implementing organisation. But this is not the case in most metropolises. In addition, even in the case of agencies in charge of the same market, either land-use or transport, it is seldom the case that they are well coordinated, to say nothing of the case of coordination between land-use and transport agencies. Therefore, there are such variety of problems that the most effective policy measures are out of options because they are under other the responsibility of another agency and that implemented measures don't work effectively because they contradict each other.

2.2 Integrated Sets of Policy Measures and Analysis Tools

Based on the above-mentioned system identification, one of the most important issues is how to adopt appropriate measures or instruments that facilitate the implementation of plans. Only with appropriate implementation measures, the plan can be translated into reality. Usually, each of them is regarded as a specific measure for either land-use, transport or environment, although it can also be a measure, sometimes a very effective one, for the rest through the interactions. In addition, a set of policy measures in integrated planning would be a combination of these policy measure elements which will be prescriptive for the metropolis. The concept of the integrated policy measures of this study is same as that of the integrated strategies approach (May, 1991) and/or management approach in transport planning except in that the approach of this study covers as far as land-use and the environment explicitly. The selection of policy measure elements should be made by taking the interaction into consideration. Some examples of policy measure elements are listed in Table 1.

Table 1. Examples of Implementation Measure Elements of Land-Use, Transport and Environmental Policies in Developing Metropolises

[Regulation]	(2) Provision of Mass Rapid Transit
(1) Bus priority / exclusive lanes	(3) Land development / readjustment
(2) Unleaded gasoline	(4) Housing Development
(3) Land-Use zoning	**[Operation]**
(4) Building control	(1) Mass transit operation
[Taxation / Pricing]	(2) Area traffic control
(1) Vehicle import / purchase taxes	(3) Flexible working hours
(2) Fuel taxes	(4) Open hours of shops
(3) Land-Use taxes	**[Education]**
(4) Development charges	(1) Car pooling
[Investment]	(2) Ride sharing
(1) Provision of road network	(3) Promotion by mass media

Source: Miyamoto and Udomsri, 1994

2.3 Analysis Tools

It is well known that the interaction between land-use and transport should be taken into consideration in the planning process. Integrated land-use and transport models which deal with both land-use and transport as well as the interaction between them have been developed as described in Webster *et al* (1988). However, there have not been many cases which employed such integrated land-use and transport models in actual planning even in developed countries. An integrated model should be built in accordance with the objectives of its application. The model should ideally represent the universe of land-use and transport as briefly as possible, so far as it satisfies the requirements given by the objectives. Bigger models are not necessarily better models.

In addition, provision of an analysis model with understandable presentation tools using computer graphics, will promote coordination among governmental agencies in the stages of both planning and implementation. With the analysis tool, they can discuss on the integrated land-use and transport and compare possible options of policy measures with each other. It can be expected that this kind of technical development is to bring about a better institutional set-up. In addition, a user-friendly analysis system can make it possible for both planners and implementers to analyse long-range or action plans flexibly even in uncertain future framework of a developing metropolis.

Since it takes long time and huge cost to build an analysis system which covers land-use, transport and the environment, it is not feasible for each local government of a metropolis to originally develop it by itself. Therefore, it is worthwhile for researchers to establish an methodology to provide metropolises with a standard system which has enough flexibility to install existing stocks and future developments of land-use, transport and environment models as well as computer system functions.

3. Basic Concepts of Analysis System Building

3.1 Background of System Building

The authors have been developing a land-use/transport analysis model named RURBAN (Random Utility/Rent-bidding ANalysis) by which land-use in a metropolis can be simulated by considering small units of land. In addition, a personal computer support system is also being developed to analyse policy alternatives with the model (Miyamoto *et al*, 1992). The system is fully user-friendly and actually operational. The system employs graphics as much as possible both for input of policy alternatives and output presentation. Almost all operations are done by using mouse through conversation with the system. Although the present study is an advanced version in the course of RURBAN development, it employs new concepts for system building as well as model development.

3.2 Basic Concepts for System Development

We intend to provide a general framework for an integrated land-use, transport and the environment analysis system which is feasible to be built even in developing countries. The principles for the system development can be summarised as follows;

- to make a prototype of system that can be easily introduced almost everywhere
- to make the system able to deal with an integrated policy measures
- to make the system user-friendly
- to make the system flexible for existing stock of models
- not to stick to our own model but to provide it as one of alternative models

3.3 System Structure

The conceptual framework of the system is represented in Figure 2. The system is designed under the condition that it is built in the environment of Microsoft Windows 3.1. Graphical User Interface (GUI) and Dynamic Data Exchange (DDE) are the functions of Windows 3.1. The reasons why Windows 3.1 is selected for the system development are as follows; (1) application programs can be shared with a standard input/output interface, (2) it is one of the most popular operation systems, (3) it can be operated under personal computers which are available at any places, (4) the system has high possibility for further development.

3.4 Functions of the System Parts

Graphical user interface

Since land-use and transport as well as environment are all essentially spatial matters, it is indispensable to fully make use of advantages of graphical user interface. Graphical User Interface is now readily available in some Geographical Information System (GIS) and also easily developed originally under Windows system. Even without such tools, original GUI can be relatively easily provided (Miyamoto *et al*, 1992).

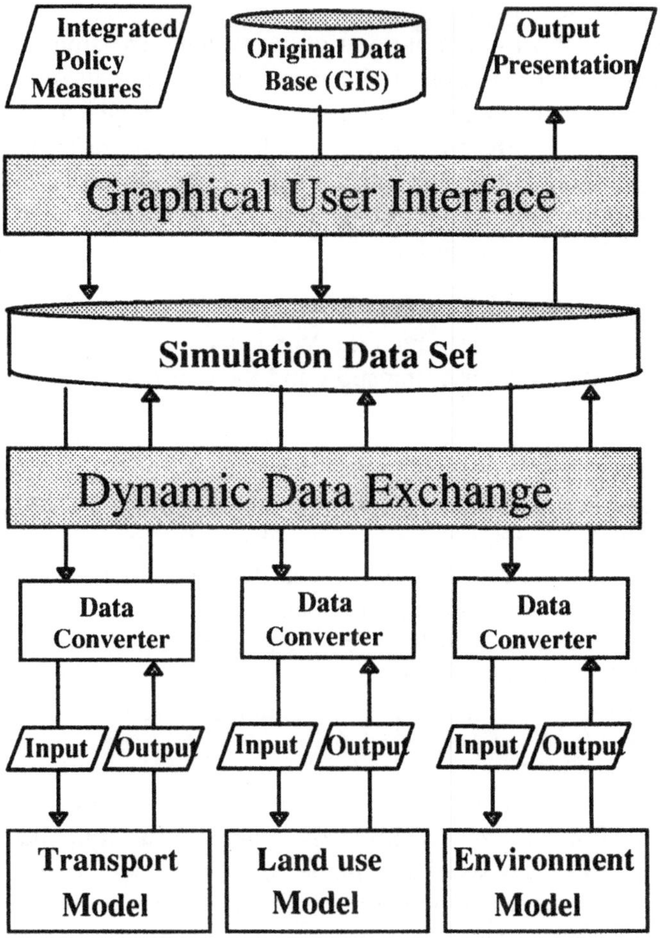

Fig. 2. The Conceptual Framework of the Analysis System

Integrated policy measures
Various policy measures should be integratedly analysed in the system. The input method should be also user friendly with GUI.

Original data base
Original data base can take any form; from established GIS data base to a file of existing data. The form can be selected by taking into consideration the availability of both hardware and software.

Output presentation
The same as input of policy alternatives, output presentation should be also user-friendly with the help of GUI.

Simulation data set
This is one of the most important part of the system. The simulation data set means "a model of the metropolis for simulation". The data set represents a simulation world of the metropolis. It contains all data which are necessary for land-use, transport and the environment at a level of analysis unit. Each model will make "the interactions through the simulation metropolis". The data of land-use, transport, and the environment by year are always updated through Dynamic Data Exchange function.

Dynamic data exchange function
This is one of the functions which the Windows system originally provided. DDE always update land-use, transport and the environment data when they are changed by models.

Data converter
Data converter has a function to connect simulation data set and application models. Existing program modules have their own input and output data layouts. In the case of the land-use model, data converter gets land-use data and explanatory variables for land-use changes, and process them to the input data layout of the model. After a calculation, the model outputs land-use changes, and the changed land-use data are transferred to the converter. Then, the changed data is again transferred to simulation data set.

Models
Existing program modules can be incorporated in the system without any modification. The system is completely independent from the application models. Therefore, any available model can be added to the system. In addition, plural number of models which simulate the same subject, for example some different land-use models, can exist in the system. The selection of the models for simulation is made by the simulation definition batch file.

Simulation batch file
Simulation batch file is an executive file to control whole simulation. It defines simulation periods and steps, selection of modules, sequence of module execution, judgement of simulation end. The operator can define in the file how land-use and transport are interacted in the simulation. Time lags and leads are also defined in the statement of the file.

4. Brief Explanation of the RURBAN Model

4.1 Assumptions and Basic Concepts

In this part, a land-use model based on Random Utility/Rent-Bidding ANalysis (RURBAN) (Miyamoto *et al*, 1992) is built is briefly explained.

This study discusses land-use and transport in a limited metropolitan area, which is hereafter called the "study area". The study area is assumed to be a closed city, which means that the demand for location is given from outside the model. To deal with the land market simply and conveniently, every zone in the study area is assumed to be owned by its own imaginary landowner and that every locator pays the rent to the landowner.

To segment the demand side in the land market, locators, which are travellers in the case of transport modelling, are classified according to their characteristics into a limited number of locator groups. These groups represent discrete options in the random rent-bidding analysis. The supply side of the land market is segmented by aggregating individual sites into zones based on locational conditions. The zones are regarded as discrete options in the analysis of location choice with random utility.

In this study, the land market is grasped from two viewpoints of locators and sites. If a locator chooses a certain site, it implies that the site must give the locator the highest utility compared with alternative sites. On the other hand, it also indicates that the locator must bid the highest rent among alternative locators at the site. At the level of aggregated locator groups and zones, the market can also be similarly explained, although probabilistic consideration should be introduced to represent the coexistence of a number of locators of various groups in a zone which consists of a number of sites. Locators belonging to a group are distributed in zones in proportion to the probability of each zone to give the group the highest utility. The area share by locator group in a zone is also proportional to the probabilities that the locator group bid the highest rent at the zone. These probabilities are obtained by logit models in RURBAN. The introduction of probabilistic terms in RURBAN follows the assumptions which most applications of the logit model employ (Ben-Akiva and Lerman, 1986). At this level of modelling, "the rents in all zones" and "the levels of utility of all locator groups" are indispensable in the former and the latter explanations, respectively.

In this study, it is assumed the existence of a state of general equilibrium of land market within the study area. The state of general equilibrium can be obtained through either equilibrium rents of all zones or equilibrium levels of utility of all locator groups, since the determination of the former brings about the settlement of the latter and vice-versa as is explained in 4.5. This general equilibrium is defined in this study as the case that the demand for location of a locator group in a zone is equal to the land supply of the zone to the locator group for all pairs of locator group and zone within the study area. The land used by a locator group in a zone is called demand, and land offered by a zone for a locator group is called supply. They represent not only newly generated or flow values but include the total distribution of locators or stock values. Therefore, such areas where housing units are built but nobody is living are regarded not as residential areas but as vacant or non-used areas in this model.

The model deals with only actual use of land.

4.2 Demand for Land Derived from Random Utility Analysis

The amount of location of a locator group in a particular zone depends on the corresponding utility in that zone which is represented by a "representative" indirect utility. The indirect utility, which is the maximum utility the locator can attain in that zone, is assumed to be distributed randomly around a represented utility by following the assumption of the logit model. All locators belonging to a group are assumed to locate themselves in the study area with an equal level of utility. The same level of utility is obtained as a probabilistic expectation of the maximum utility, which is given by so-called logsum function, for all zones. The demand function of a locator group for a zone is then defined as a probabilistic expectation of the demanded area which is given by the number of locators of the group allocated to the zone multiplied by the amount of land used by a locator of the group in the zone. The former is given by the logit model and the latter is endogenously obtained by an equation based on the Alonso model (Alonso, 1964).

4.3 Supply of Land Derived from Random Rent-Bidding Analysis

In the RURBAN, the total area of available land in the study area is exogenously given to the model. It means that the supply of land is rigid. In addition, the supply of either floors or buildings is not explicitly considered. The supply of land of a zone to a locator group in the RURBAN is determined according to its bid-rent compared with other groups' bid-rents as follows. At each site in the study area, the existing locator is bidding the highest rent which becomes the actual rent. This means that the imaginary landowner supplies the site for the maximum bidder at the maximum bid-rent. However there are a number of sites in each zone, and their characteristics are not necessarily the same within the whole zone. Therefore, it is assumed that the land in each zone is supplied to locator groups according to their "representative" bid-rents at the zone. The supply function of a zone for a locator group is given as an expectation obtained from the probability that the locator group is the highest rent-bidder in the zone. The probability is given by the random rent-bidding analysis (Ellickson, 1981).

4.4 Structural Equations of RURBAN

The followings are the structural equations of RURBAN. They are derived under the condition that the demand and the supply explained in the previous parts are equal (Miyamoto and Kitazume, 1989).

$$\mu U_{IS} = \mu\alpha_I X_{IS} - \omega B_{IS}^* \tag{1}$$

$$q_{IS} = \theta_I \exp(-B_S^*) \tag{2}$$

$$L_{IS} = \frac{A_S}{q_{IS}} \tag{3}$$

$$\omega B_{IS} = \mu\alpha_I X_{IS} - \mu U_I^* \tag{4}$$

$$U_I^* = \frac{1}{\mu}\ln\sum_S \exp(\mu U_{IS} + \ln L_{IS} + \ln w_{IS}) \tag{5}$$

$$B_S^* = \frac{1}{\omega}\ln\sum_I \exp(\omega B_{IS} + \ln N_I + \ln w_{IS}) \tag{6}$$

where,

I : the locator group
S : the zone
U_{IS} : the systematic part of random utility of locator group I in zone S
B_{IS} : the systematic part of random bid-rent of locator group I in zone S
q_{IS} : the amount of land used by a unit of locator group I in zone S
X_{IS} : location conditions (except rent): $(X_{IS1}, \ldots, X_{ISk}, \ldots)$
α_I : parameters for locator group I: $(\alpha_{I1}, \ldots, \alpha_{Ik}, \ldots)^t$
L_{IS} : the number of available sites for the use of locator group I in zone S
U_I^* : the level of utility of locator group I
B_S^* : the representative rent of zone S
μ : a positive scale parameter of indirect utility function in location choice
ω : a positive scale parameter of bid-rent function
N_I : the number of individual locators belonging to locator group I
A_s : available area of zone S
w_{IS} : the measure of heterogeneity of individual locators in locator group I and individual sites in zone S
θ_I : a parameter of locator group I

Equation (1) represents an indirect utility of a locator group I in a zone S. Equation (2) gives the amount of land used by a unit of the group in the zone which is inversely proportional to representative rent of the zone. This function implicitly represents multistoried uses of land in a high land price area. Equation (3) shows the number of available sites for the use of the group at the zone. The number of optional sites in the zone affects the probability of the group to choose the zone. Equation (4) represent bid-rent of the group at the zone. This equation is a kind of dual equation of equation (1). Equation (5) gives the level of utility of the group in the whole area. It is a logsum function of all utilities of the group in the study area. Finally, equation (6) represents the representative rent in the zone which is also a logsum function. The latter two functions give key values which determines the general equilibrium of the land market as explained before.

4.5 Structure of RURBAN

Figure 3 shows the general structure of the RURBAN model. The RURBAN model has two partial equilibrium parts; location choice of locator groups based on utility analysis which is represented in the upper part of the figure, and locator choice of zones based on rent-bidding analysis which is represented in the lower part of the figure. The given values in the upper partial equilibrium are the rents at all zones and in the latter are the levels of utility of all locator groups. The levels of utility are obtained from the utility analysis and the rents are derived from the rent-bidding analysis. The equilibrium levels of utility and the equilibrium rents are obtained as converged values after iteration between two partial equilibrium parts. In the convergence, the area of locator I in zone S obtained by random utility analysis, which is represented as Φ_{IS}^{U} in the figure, should become equal to the area Φ_{IS}^{b} obtained by random rent-bidding analysis for all pairs of I and S. Then, they can be regarded as those in the state of general equilibrium.

In order to obtain both of them, the land market should be modelled with two partial equilibrium aspects. In addition, when either of the general equilibrium values of levels of utility or rents are obtained, the other values naturally become those that are in the state of the general equilibrium.

As Figure 3 shows, the structure of RURBAN consists of two kinds of single constraint entropy models. In the end, that is in the state of general equilibrium, RURBAN becomes a doubly constrained entropy model.

So far as the authors know, such idea as that both random utility and random rent-bidding are simultaneously considered in the land-use model was firstly proposed in a former version of RURBAN in Miyamoto and Yagi (1987), although there was inappropriate interpretation regarding the joint probability between utility and rent-bidding analyses. In addition, recently have applied a few models such as Hayashi *et al* (1989) and Martinez (1992) which employ very similar scheme to that of RURBAN with different ways of formulation. The most particular difference of RURBAN from these two models is that the amount of area used by a unit of locator in each zone is endogenously obtained in the RURBAN.

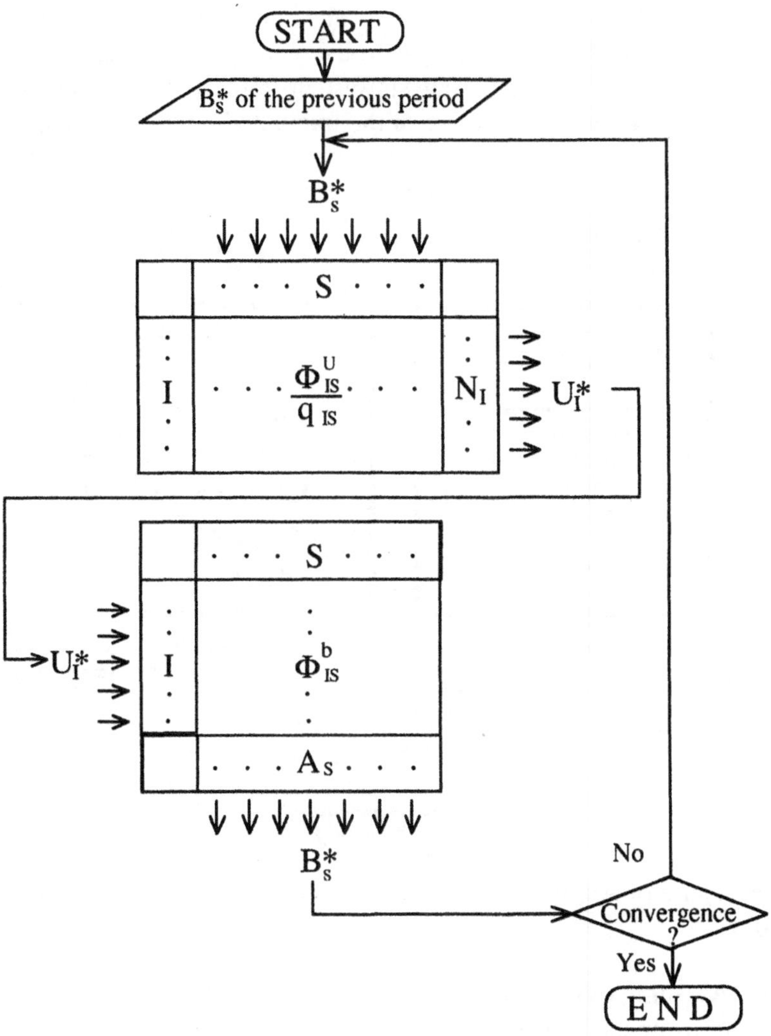

Fig. 3. The Partial Equilibrium in Location Choice of Locator Groups, the Partial Equilibrium in Location Choice at Zones and the General Equilibrium of the Land-Use Market in RURBAN

5. An Integrated Model of Land-Use and Transport with Environment

5.1 The Approach

The main part of the analysis system is an integrated land-use and transport model. In this study, RURBAN model has been improved to represent transport more explicitly. In this improvement, it is intended to keep consistency between land-use and transport. The improved model can be operated not only by directly integrating land-use and transport but also by separating them under the proposed analysis system framework as discussed in section 3. Figure 4 shows the structure of the integrated land-use, transport and environment model. It corresponds to the system configuration of land-use, transport and the environment in a metropolis shown in Figure 1.

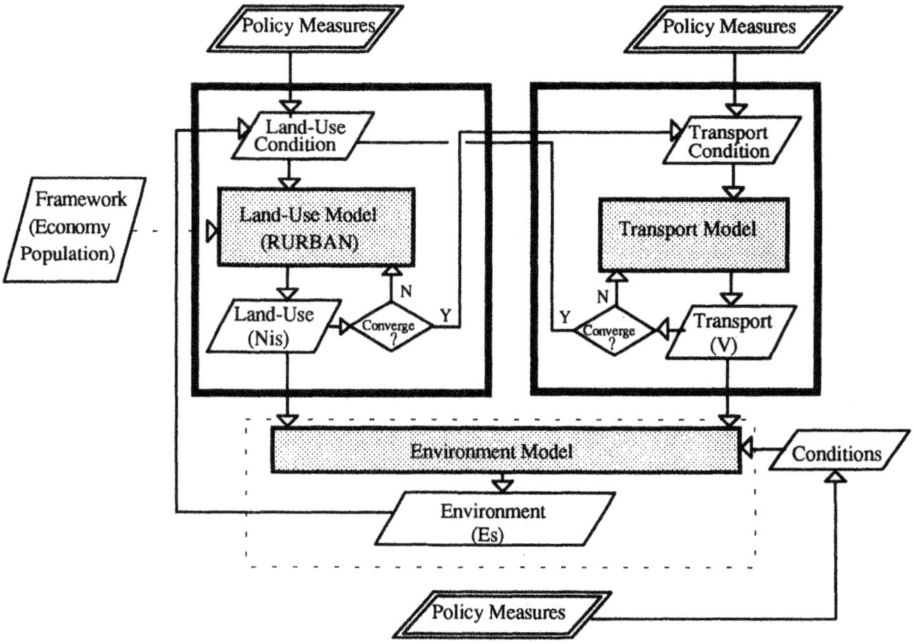

Fig. 4. The Structure of An Integrated Land-Use, Transport and Environment Model

The RURBAN employs aggregate logit model structure as described before. The improvement is being done along with the same structure. The reasons why we decided to improve the RURBAN for developing an integrated model of land-use, transport and the environment are explained as follows. Firstly, the model is able to fully and consistently integrate transport choice steps within the location choice. Therefore, land-use and transport are modelled within a single model framework. Secondly, the model has mechanisms to represent market equilibrium both in land-use and transport markets. Thirdly, the logit model or the nested logit model, which is the theoretical background of the RURBAN, can consistently deal with the cases in which units for analysis are either aggregated or disaggregated.

The choices in location and trip are viewed as outcome of a probabilistic choice process. The process is simply described by four levels of choice hierarchy in decision-making chain starting from location choice and destination choice in land-use level, to mode choice and route choice in transport level. The basic concept has been also employed in Martinez (1992) and others. Figure 5 and Figure 6 show how this process can be described in the hierarchical concept. In the case of residential location, destination choices mean "choices of working place, school and shopping places". The hierarchical structure represents that "a site which is convenient for commuting" means that it is close to large working places. It is also explained that the site has better accessibility to large working places compared with other sites. The structure covers both behaviour of aggregated locators and travellers.

5.2 Choice Tree and Location Utility

Figure 5 shows the location conditions or explanatory variables for the utility function of a locator group *I* in a zone *S*. For explanation, let *I* denote the residential locator group. There are four categories in location conditions. The first category covers "accessibilities" which represents, for example, work, school and shopping trip conditions which are expressed by trip purpose *p*. They are represented by both the attractiveness of available destinations and transport conditions, which are also explained by the nested tree structure shown in Figure 6. The second category consists of the characteristics of the zone itself which are foot-tight conditions. The third category includes all aspects of the environment which are determined by land-use and traffic of the neighbouring zones. The forth category indicates the price of the land which is named "the representative rent" of the zone in the RURBAN.

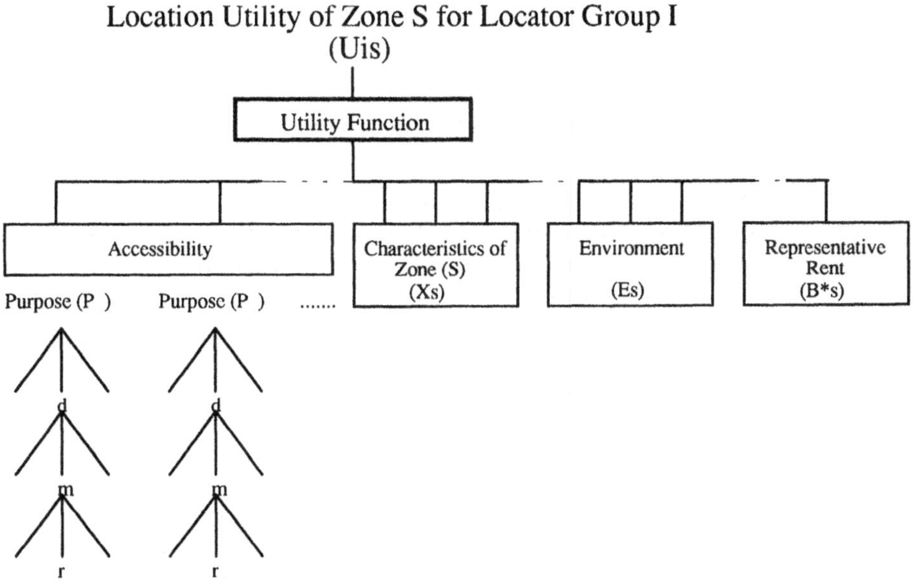

Fig. 5. Location Utility and Factors (Explanatory Variables)

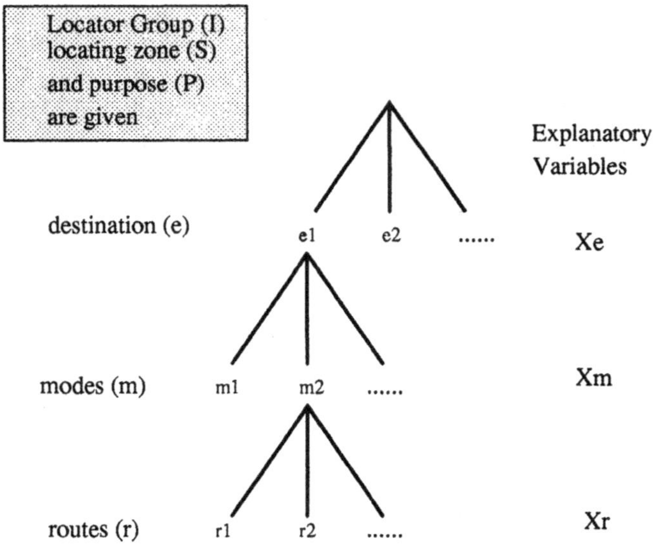

Fig. 6. Accessibility of Zone S for Purpose P and of Locator group I

The choice tree structure of transport shown in Figure 6 consists of a destination, mode and route choice hierarchy. In the case of commercial location, the top choice level is not destination but origin choice, because the accessibility means the size of the population in the hinterland which are discounted by the transport condition. However, for the convenience of explanation, the word of destination is used in the choice of accessibility, because reverse-direction trip can substitute for it.

In the level of mode choice, it also includes type of vehicles as well as other type of communications, for example telecommunications. It is indispensable to classify the type of vehicles, for example passenger cars or heavy trucks, for the evaluation of environmental changes caused by the mixed traffic.

Regarding route choice, each route consists of some links as shown in Figure 7. In addition, each link is generally shared by some routes. In order to obtain the traffic volume on a link, it is necessary to sum up trips of all routes which have the link as a part. The general service level of the link is obtained by its service function which contains its conditions such as capacity and length. Moreover, the service level is determined by the composition of different vehicle type in the total traffic volume. The service level will also determine the efficiency of energy consumption as well as pollutant emission which are estimated by the environment model.

5.3 Utility Function and Rent-Bidding Function

Also in this improvement of the RURBAN, the basic concept of the model consists of both random utility of the locator group in location choice and random rent-bidding in locator choice of the zone. The top level of behaviour is location choice. The bid-rent function can be derived from the utility function since they have dual relation between each other. The structure of locator choice is shown in Figure 8.

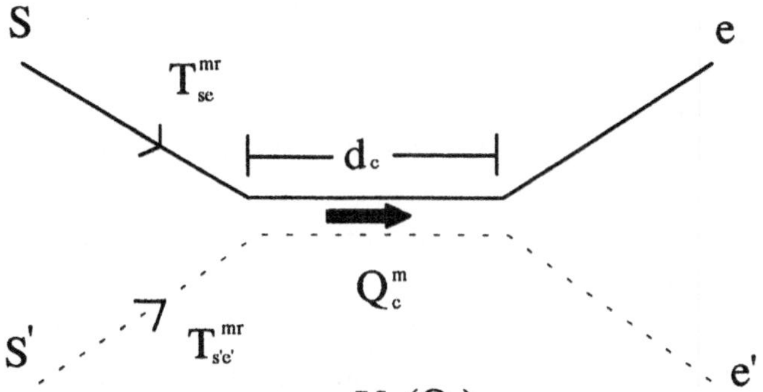

$$v_c = V_c\,(Q_c)$$

$$Q_c = (\,Q^1_c\,,\ldots\ldots, Q^m_c,\ldots.\,)$$

$$Q^m_c = \sum_S \sum_e \sum_{r(c \in r)} T^{mr}_{Se}$$

Fig. 7. Accessibility (Q_C) and Service level (v_C)

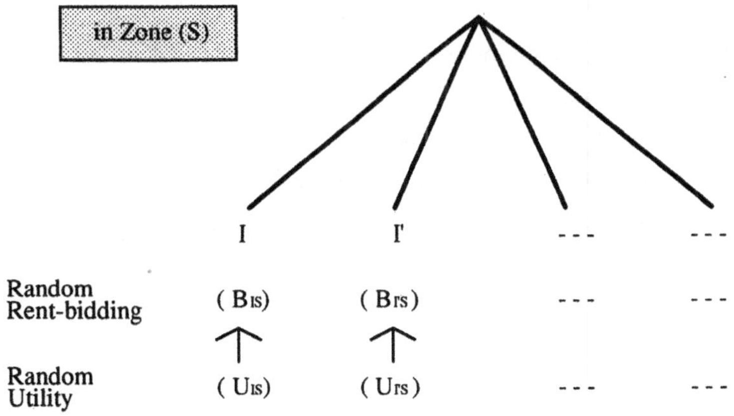

Fig. 8. Locators Choice in Zone S

5.4 Formulation

Based on the above-mentioned explanation, the following equations are derived to represent land-use, transport and the environment. Denotations are as follows,

I : the locator group
S : the zone
c : link
r : route
m : mode (including type of vehicle)
e : destination
p : (trip/communication) purpose
U_c, U_r, U_m, U_e, U_s : utilities for link, route, mode, destination and locating zone
$\mu_r, \mu_m, \mu_e, \mu_s$: positive scale parameters of route, mode, destination and locating zone choice
$\alpha_m, \alpha_e, \alpha_s$: parameters for mode, destination and zone choice
X_m, X_e, X_s : explanatory variables for mode, destination and zone choice
$C\,(\,)$: cost function
v_c : service level of link c
d_c : condition of link c
${}^p_I g_S$: trip production of locator group I for purpose p from zone S
${}^p_I \pi$: trip production rate of locator I for purpose p
n_{IS} : number of location of locator group I in zone S
${}^p_I t^{mr}_{Se}$: assigned trips locator group I for purpose p for route r
T^{mr}_{Se} : total assigned trips for route r
$V\,(\,)$: service level function
Q^m_c : traffic volume of mode m in link c
E_S : environmental indicators (vector) in zone S
$E\,(\,)$: estimation function of environmental indicators.
$v = [v_c]$
$n = [n_{IS}]$
$Q = [Q^m_c]$
$Q_c = [Q^m_{(c)}]$

Regarding the choice tree, it is understandable to trace from the bottom to the top. The utility of route r can be expressed by the summation of the utility of each link c which constitutes the route as shown equation (7). The utility function of the link c can be represented by the cost function of service level and link condition as equation (8).

$$U_r = \sum_c U_c \tag{7}$$

$$U_c = C(v_c, d_c) \tag{8}$$

The choice probability of route r is expressed by equation (9), under the condition that locator I, locating zone S, purpose of trip and destination p/e, and mode m are given.

$$\text{Prob}\,(\text{I,S,p / e,m} \mid \text{r}) = \frac{\exp(\mu_r U_r)}{\sum_{r'} \exp(\mu_r U_{r'})} \tag{9}$$

Following the choice tree, the choice probability and utility of mode choice are expressed as equation (10) and (11). The utility of mode m contains a logsum function of utilities of routes which belong to it.

$$\text{Prob}\,(\text{I,S,p / e} \mid \text{m}) = \frac{\exp(\mu_m U_m)}{\sum_{m'} \exp(\mu_m U_{m'})} \tag{10}$$

$$U_m = \alpha_m x_m + \frac{1}{\mu_r} \ln \sum_r \exp \mu_r U_r \tag{11}$$

In the level of destination choice, it depends on the trip or communication purpose p. Same as the previous derivation, equations (12) and (13) are derived.

$$\text{Prob}\,(\text{I,S} \mid \text{p / e}) = \frac{\exp(\mu_e U_e)}{\sum_{e'} \exp(\mu_e U_{e'})} \tag{12}$$

$$U_e = \alpha_e x_e + \frac{1}{\mu_m} \ln \sum_m \exp \mu_m U_m \tag{13}$$

In the top level of choice behaviour, the choice probability of locating zone S by locator group I is expressed by equation (14) which is same as the original RURBAN. In this improved version, however, the original utility function (1) is substituted by the following equation (15). In this case, logsum functions which represent the "accessibility" are summed up over purposes.

$$\text{Prob (I|S)} = \frac{\exp(\mu U_{IS} + \ln L_{IS} + \ln w_{IS})}{\sum_{S'} \exp(\mu U_{IS'} + \ln L_{IS'} + \ln w_{IS'})} \quad (14)$$

$$U_{IS} = \alpha_I x_{IS} + \sum_p \frac{1}{\mu_e^p} \ln \sum_e \exp \mu_e^p U_e^p - \omega B_S^* \quad (15)$$

As for the calculation of travel demand, trip production is calculated by equation (16). Then trips assigned to route r is calculated by equation (17). By summing up the trip production over purposes and locator groups, assigned traffic volume for route r is obtained by equation (18).

$$ {}_I^p g_S = {}_I^p \pi \, n_{IS} \quad (16)$$

$$ {}_I^p t_{Se}^{mr} = {}_I^p g_S \text{ Prob (I, S, p / e, m, | r)} \quad (17)$$

$$T_{Se}^{mr} = \sum_I \sum_p {}_I^p t_{Se}^{mr} \quad (18)$$

The service level of each link is determined by the total traffic volume in it which is obtained by equation (19) (see Figure 7). The service level of link c is determined by the function of equation (20).

$$Q_c^m = \sum_S \sum_e \sum_{r(c \in r)} T_{Se}^{mr} \quad (19)$$

$$v_c = V(Q_c, d_c) \quad (20)$$

In the final stage, the environmental indicators are estimated by the environmental model function of (21).

$$E_s = E(v, Q, n) \quad (21)$$

5.5 The Structure of the Integrated Model

Once again see Figure 4 which shows the whole structure of the integrated model. The land-use distribution (n_{IS}) is not only output of the model but also input as the attractiveness of destination. In the land-use model (RURBAN), iteration is needed to get to a convergence. Also in the transport model, transport service level (v_c) is both input and output of the model. Some iteration are necessary before reaching to a convergence. In addition, as a whole model system of the integrated land-use, transport and environment model, iteration processes of a larger circle are necessary for a convergence of the total system.

The iteration of the whole system should start from the equilibrium situation, that is the point of convergence, of the previous period.

6. Concluding Remarks

This paper briefly describes the basic concepts of the analysis system which we are now developing. Since the conceptual system building goes first, we may be facing several issues which we have to resolve in the development of the pilot system. In addition, regarding the development of the integrated model, some devices will come to be indispensable to adjust the theoretical model with actual situations of land-use, transport and the environment. Data availability may urge the development to add some assumptions and special treatment of parameter estimation.

References

Anas, A. and Duann (1983), Dynamic Forecasting of Travel demand, Residential Location and Land Development: Policy Simulations with the Chicago Area Transportation / Land Use Analysis System, Paper prepared for presentation at the International Symposium on New Directions in Urban modelling.

Ben-Akiva, M., and Lerman, S. R. (1986), *Discrete Choice Analysis : Theory and application to Predict Travel Demand*, (The MIT Press).

Ellickson, B. (1981), An Alternative Test of the Hedonic Theory of Housing Market, *Journal of Urban Economics*, 9, pp. 56-79.

Hayashi, Y. and Doi, K. (1989), A Model for Analysing the Imputation of Commuters' Benefits to Land Property Value, *Proceedings of the Fifth World Conference on Transport Research*, Vol. 1, pp. 303-317.

Martinez, F.J, (1992), Towards the 5-Stage Land Use-Transport Model, *Proceedings of the Sixth World Conference on Transport Research.*

May, A.D.(1991), Integrated Transport Strategies: a New Approach to Urban Transport Policy Formulation in the UK, *Transport Reviews*, Vol. 11, No. 3, pp 223-241.

Miyamoto, K. and Yagi, S. (1987), A large-scale Land Use Model based on the Random Utility / Bidding Theory, *Proceedings of the International Symposium on Transport, Communication and Urban Form*, pp. 235-250, Monash University.

Miyamoto, K. and Kitazume, K. (1989), A Land-Use Model based on Random Utility /Rent-Bidding Analysis(RURBAN), *Proceedings of the Fifth World Conference on Transport Research*, Vol. 4, pp. 107-121.

Miyamoto, K. et.al.(1992), An evaluation method of transport projects with the aid of RURBAN (Random Utility/Rent-Bidding ANalysis) Model, *Selected Proceedings of the Sixth World Conference on Transport Research*, Vol. 1, pp. 55-66.

Miyamoto, K. and Black, J.A.(1992), Editorial Introduction, *Regional Development Dialogue*, Vol. 13, No. 3, pp. 1-4, United Nations Center for Regional Development.

Miyamoto,K. and Udomsri,R.(1994), Present Situations and Issues of Planning and Implementation regarding Land-Use and Transport in Developing Countries, *Journal of Infrastructure Planning and Management*, Japan Society of civil Engineers, No. 482 / IV-22, pp. 87-97, Japan Society of Civil Engineers.

Webster, F.V, Bly, P.H. and Paulley, N.J.(1988), Urban Land-use and Transport Interaction - Report of the International Study Group on Land-use/ Transport Interaction (ISGLUTI), Avebury.

CHAPTER 14

OPTIMISING TECHNIQUES IN ACTIVITIES-TRANSPORT MODELS

Tomás de la Barra

1. Introduction: the Need for Optimising Techniques

Thirty years ago, the possibility of building a mathematical model to represent a city or a region was at an experimental stage in just a few research centres. Today, after perhaps thousands of applications world-wide, it is a fairly common practice, and the field has had a significant effect on planning. Urban models are used to assess the effects of policies and investment proposals of a wide variety. After years of research, development and applications, urban models are becoming increasingly available and, of course, better and easier to use. The wide availability computer power and the emergence of low cost geographical information systems are only two elements that are going to increase the importance of urban models in the future even further. Also the range of applications will become wider than what it is today.

A large number of applications seem to support the view that we now have workable urban models with the ability to reproduce real situations, and simulate future ones conditioned to a range of policies and investment proposals, evaluate such proposals and provide enough information to make a sound decision. Nowadays, there is a plentiful supply of commercial transport packages that are both flexible and easy to implement in low cost computers. Integrated land-use/transport systems are less common, but probably increasing in the near future. It is also quite clear that these models have had

Y. Hayashi and J. Roy (eds.), Transport, Land-Use and the Environment, 281–296.

an important effect on planning and even academic research and teaching. It is probably true that urban and regional planners have been reluctant to use models in everyday practice, but for the transportation planner the use of models is commonplace. What can be asserted with confidence is that we now have sound theories and workable models that can be used in practice with relative ease.

Most of the urban models of today fall into the category of simulation models. The analyst provides a partial set of data, such as employment, land, transport supply, and so on, and the model is asked to reproduce the remaining parts of the urban structure, such as the location of induced activities, the use of land, land rent, and transport demand. Usually this leads to a cross-section type of dynamic simulation in which the model predicts future states of the system at discrete time intervals. This simple scheme has proved to be satisfactory, because it produces realistic and useful results with a reasonable amount of effort.

Now that we have efficient and well structured models and that we have increasingly powerful low cost computers, we should ask ourselves what else do we want our models to do, beyond simulation. This depends on what we are currently doing with our models, the intended future use and today's hottest planning issues, restricted by what our models can do. Also, it must be recognised that policies are becoming more complex in nature, and it may well be that our models are not prepared to cope with some of the current planning issues. Transport planning is a good example; the first generation of models was designed with the ability to simulate and evaluate highway proposals with automobiles in mind, which is a relatively straightforward task. With increased congestion due to a growing population of cars, interest has shifted to public transport, requiring a much more detailed process of simulation and evaluation. Today's generation of transport models can represent complex transit networks with the required degree of sophistication. In the last decade or so, interest has shifted further to more complex problems, and planners are required to assess the energy and environmental aspects of projects. Policies such as road pricing, emission controls, and particularly financial schemes are new requirements for our models, and it is in no way clear that these can be tackled effectively.

In many cases, these new requirements involve optimising certain aspects of the policies or projects, but the simulation approach may not lend itself towards this end. For instance, a model can be used to simulate the effects of a proposed toll road; in this case the model is not only asked to simulate the effect on users and the entity in charge of the road, but it is also asked to determine the optimum price of tolls for each vehicle, with the purpose of maximising revenues, and optimising user benefits at the same time. Currently this is done by repeated runs of the model, changing the price of tolls and looking at the corresponding indicators: revenues and benefits. This *brute-*

force practice, however, may run into problems in certain conditions, and the aim of this paper is to identify them and suggest possible solutions.

The potential problem is that proposals must me designed by the analyst outside the model in a non-systematic manner, usually by trial and error, a practice that is in sharp contrast with the rigorous way in which we turn them into data, feed them into our models and arrive at a recommendation. In the example of the toll road above, it may be that the optimum price scheme never gets tested, because there is no guarantee that the optimum has been imagined in the first place. This is made worse when the proposals are complex in nature, such as road pricing, in which the optimum scheme might involve different prices in each road section, discriminating by vehicle type; the same can be said about different toll levels, energy efficiency or environmental controls. There is also a practical problem: repeated tests of a policy to explore its range is error prone and time consuming. Finally, there are some theoretical issues that are raised; when a variable such as tolls, fares, controls, and the like are allowed to vary, the system will converge to a new equilibrium, and many other related variables will be affected. The evaluation of these new planning issues also pose important questions, such as what is the value of clean air or low energy, and up to which point these benefits balance higher user costs.

What is proposed here is that the potential solution to some of these problems is to develop methods to explore policy ranges in a systematic way, thus improving our confidence that the optimum policy has been tested and evaluated, and allowing for more complex and efficient solutions with less effort. This paper proposes an area of research towards this end: optimising techniques, as just one of many possible solutions that can help in making the policy design process stand on a more firm basis.

Optimising techniques means that we build into the models the ability to explore a policy variable in a systematic way, comparing it to one or more objective variables. This does not mean that the model is turned into an optimising model, because it does not assume that the behaviour of the real system will become optimum, as would be the case, for instance, in a linear programming model. It is still a behavioural model looking for an optimal policy.

Building optimising capabilities into the models opens a wide range of possibilities. For instance, the model could be asked to charge vehicles travelling along roads in order to keep emissions under certain pre-specified limits, or perhaps price fuel to keep system wide emissions under control or reduce energy consumption. The results of such optimising procedures should be taken with flexibility, because in many cases they would be used to explore ranges and limits of the policies being considered. It is only a suggestion and it is useful to explore the consequences and limitations of a particular policy. It may well be that the results of the analysis suggest that the policy is doomed, because the amount that should be charged turns out to be unrealistic, or not viable from a political point of view.

It is clear that the simulation approach with repeated runs of the model will not be an effective way of dealing with these complex issues. If, instead, these calculations are built into the model structure, the results may be obtained in a more efficient and reliable way. There are, however, considerable problems that must be solved, such as:

will the model always converge to a unique and correct solution?

which is the most efficient way of arriving at a solution?

is there a solution at all?

how many objective values can be optimised simultaneously?

This paper does not intend to provide definite answers to all these questions, that became part of a long term research project. Instead, it addresses the main topics and discusses its implications for modelling. The paper first describes briefly the TRANUS integrated activities-transport model developed by the author and colleagues, which is taken as a basis for the introduction of optimising procedures. Similar procedures, in fact, can be implemented in other models as well, but this particular model is taken as a reference. The paper then gradually introduces a number of additional concepts that are built into the model structure to provide it with certain optimising and goal seeking capabilities.

The first of these techniques is the use of *undefined values* to determine supply in the transport model. Elements such as transit frequencies are specified as undefined, and the model is asked to estimate the appropriate values according to demand conditions. Then optimising procedures are introduced. In this case, tariffs for particular transit operators are left undefined, and the model is asked to determine the optimum tariff, subject to a set of restrictions or objectives. In other cases, the model is asked to estimate the price that should be charged to, say, vehicles or households, to keep certain variables under predefined standards, such as energy consumption, emissions, congestion levels or some overall system performance indicator.

2. The TRANUS System

TRANUS is based on the concept that the interaction of activities in space give rise to transport demand; in turn, the accessibilities that arise from transport demand-supply equilibrium condition the way in which activities interact. The activity-transport system is conceived as a sequence of interrelated choices in the form of a decision chain. For example, a typical urban chain would be:

place of work→residence→shopping→transport mode→ transport route

The population of decision makers is distributed probabilistically among available options as a function of perceived costs. Based on random utility principles, TRANUS uses a set of Nested Multinomial Logit (NMNL) models to represent and relate each decision link along the chain, from the location of activities to land use and transport.

A dynamic structure relates the two main components of TRANUS: activities and transport. In the activity system, economic activities in space interact with each other generating flows; these flows determine transport demand, and are assigned to transport supply in the transport model. In turn, demand-supply equilibrium determines accessibility, which is fed back to the activity system, influencing the location of activities and the corresponding flows. This feedback is lagged: accessibilities in time period 1 affect the location of activities in time period 2. As a result, a change in the transport system, such as a new road or a mass transit system, will have an immediate effect on travel demand, but will only affect activity location and interaction one or even several time periods later. This dynamic structure is outlined in Figure 1, and the main sequence of calculations is shown in Figure 2.

2.1 The Activities Model

The activities model is a spatial input-output structure with a very general and flexible formulation. Production and consumption of economic sectors in zones are related through demand functions with price elasticities and substitutions. Consumption is assigned to production zones with a logit model.

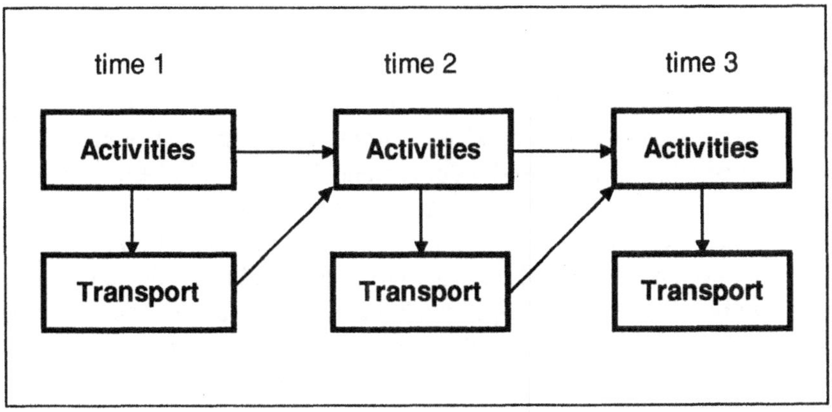

Fig. 1. Dynamic Structure in TRANUS

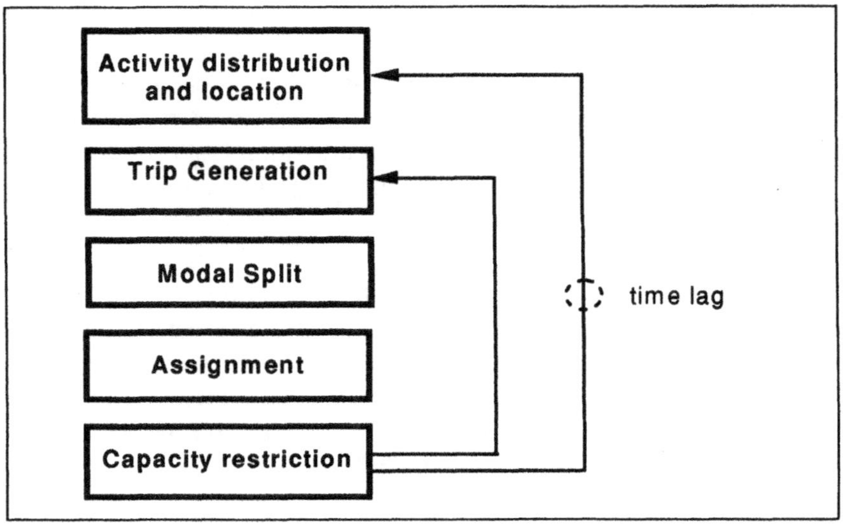

Fig. 2. Sequence of Calculations in TRANUS

The results of these transactions are flows represented in origin-destination matrices, which are later transformed into transport demand. The model keeps track of the costs involved in the transactions: transport costs are added to production costs and value added, and these are transferred along the production chain. If there are restrictions to the production of a sector in a particular zone, the model simulates an equilibrium price. Typically land and floorspace are restricted to existing stock and equilibrium prices represent land values or rents; these elements are fixed in space, that is, they must be consumed in the same zone as they are produced, and do not give rise to flows. Energy consumption by activities or floorspace can be represented as one or more sectors. An explicit representation of imports and exports is included in the model.

This general framework allows for the representation of cities or regions with different degrees of complexity. A simple urban application may consider only one type of population and employment. More detailed applications may require dissaggregation into several income groups, different types of employment, and several types of land and floorspace. The activity model allows for two level zones: when distributing demand to production zones, the model applies conditional logit probabilities similar to the well-known hierarchical modal split model. This feature is useful to represent parts of a city in more detail, or to combine urban and sub-regional levels.

The use of substitutes in demand functions may be very useful to represent a complex property market. Substitutes are treated probabilistically with logit distributions. Activities may be defined as consuming variable amounts of several types of floorspace, in which case demand is assigned probabilistically to supply in each zone. In turn, each floorspace type may be defined as consuming variable amounts of land of several types. This process will result on the simulation of complex combination of housing and land sub-markets with corresponding rent values. Value added may be used to represent construction costs.

2.2 The Transport Model

The transport model is also a nested MNL model, with a decision sequence of trip generation (trip choice), mode choice and route choice. The model first analyses the transport network and performs a multidimensional path search to find a set of *n-paths* connecting each origin to each destination by each transport mode. Each mode in turn may be divided into several transport operators and transit routes, allowing for multiple transfers. Detailed operating costs, tariffs and transfer costs are calculated for each path, together with perceived costs (travel and waiting times) and penalising factors. In TRANUS, a multi-path search procedure identifies several travel options for each origin-destination pair from the network data, and then assigns trips probabilistically with a logit model. One advantage of this scheme is that the perceived cost of paths can be aggregated up to the modal split level using a *log-sum* average. The probability of choosing a path is calculated as a function of the generalised cost, and a composite cost is aggregated from a path level to a mode level with another *log-sum* average. Similarly, costs are aggregated over all modes to obtain the average origin-destination cost as a measure of accessibility. Thus, if a new transport option is introduced, the *log-sum* calculation over paths ensures that the improved utility is passed on to the modal split and trip generation stages, and, eventually, to the activities model.

An important attribute of logit assignment is that it does not depend on congestion in order to disperse trips off the minimum path. Popular algorithms such as incremental or equilibrium assignment are known to behave as an all-or-nothing procedure when dealing with uncongested networks, as is the case in regional applications, small towns, or even uncongested parts of large urban areas. Because logit assignment distributes traffic probabilistically among options, it can be applied virtually to any scale. An added benefit of the probabilistic model in practical terms is that it is no longer necessary to work with a large number of zones in order to get realistic results.

The well-known problem of the independence of irrelevant alternatives in MNL models is particularly acute in assignment. This has been solved in TRANUS with a method called *overlapping control.* A measure of overlap among competing paths is calculated simultaneously with path search, and the probabilities of each path are compensated accordingly. The result is that path choices represent distinct options with very little correlation even in complex networks, a fact that substantially increases the realism of the results.

Another important feature of the model is the concept of *dual networks*. The user codes the network in the usual way, but the model creates a dual of the network, i.e., transforms the links into the vertices of the internal graph, and creates the links as possible connections. This process is transparent to the user, because the results are reported back in the original format. It has the advantage that prohibited turns are represented in a simple and error-free fashion, completely avoiding fictitious links. Whenever a prohibited turn is found, the model simply eliminates a link in the dual network.

A further characteristic of the model that has important practical implications is the idea of *multidimensional networks*. Each physical link has a number of characteristics, such as distance, capacity, speed, and so on. But the information also includes the types of vehicles that can use the link, such as cars, trucks, railways or transit lines. With this information, the model transforms, also internally, the single-dimensional graph into a multi-dimensional one, where each physical link is transformed into several virtual links, representing a combination of physical link with vehicle type. This has obvious advantages in practical terms, because the user can code a complex transport network with transit routes in a simple way. After assignment, all vehicles in each physical link are added to estimate capacity restriction.

There is a more important feature of this method to consider: mode constants can be applied to vehicle types. This means that in a combined trip between an inconvenient and a convenient mode, each component is penalised accordingly. This method has proved very useful in practical applications. Even in a third-world city like Caracas, buses, mini-buses, metro, metro-feeder-buses and jitneys, can be combined in thousands of different proportions. In regional applications commodities can also travel in combined modes, such as trucks, railroads and waterways.

In the literature and in practice, such complex networks are dealt with hierarchical modal split models. The main problem with such models is that penalising constants can only be applied to predefined study-wide combinations of modes. This is an important limitation when applied to complex transport systems, where the number of combinations becomes excessively large, making the estimation of constants impossible. The multidimensional model, by placing the constants where they really belong, solves the problem in a topological fashion: demand is assigned to combinations of links and modes simultaneously, multiplying travel time by the penalising constants on a link-by-link basis. In other words, the first level of the hierarchical modal split model is carried out in the usual way, but second and subsequent levels are treated at a network level, simultaneously with assignment.

Trip generation is calculated as an elastic function of the composite cost by zone pair. Elasticity in trip generation allows for the estimation of induced demand that results from the introduction of new or improved transport facilities. Modal split and assignment are performed with NMNL logit models, taking car availability into account if applicable. The model can estimate empty return vehicles, particularly useful for freight movements. Finally, a capacity restriction procedure adjusts travel speeds and waiting times as a function of congestion. The model iterates until equilibrium is reached.

2.3 The Evaluation Procedure

The evaluation procedure included in the TRANUS system compares the results of two strategies: base case and alternative, and estimates a number of socioeconomic, energy consumption and financial indicators within a multi-criteria framework. Current costs and benefits are arranged in three main accounts: users, operators and administrators. The latter are the entities in charge of maintaining the infrastructure and may perceive revenues from user charges. The procedure estimates indicators such as cost/benefit ratios, net present values and internal rates of return, considering economic or financial costs and the shadow price of energy.

2.4 Applications

This general model has been applied to a wide variety of cases, from local to metropolitan to nationwide. Over the past ten years, more than sixty applications have been carried out. At a national level, the model was used to simulate the movements of passengers and freight in Venezuela, and has been used to asses the effects of several toll road schemes, and to evaluate highway proposals, as well as railways and waterways. At an urban scale, the model has been used to simulate land use in several cities, such as Caracas, Maracaibo, Valencia, Maracay and many others. Most of the urban applications have been made to improve current transit systems or to design and evaluate rapid transit proposals and their land use implications. At a more detailed urban level, the model has been used to estimate the impact of estate developments on the surrounding transport system.

3. Undefined Variables

Taking the TRANUS system described in the preceding sections as a starting point, the first step is to introduce the concept of *undefined variables*. The principle of this method can be stated as follows:

- the user specifies one or more variables as undefined,
- the model is asked to estimate values for the undefined elements, according to pre-specified criteria.

The current version of the model already includes some of these types of variables in the transport model: link capacity and transit frequencies and routes. If the capacity of a link is declared undefined, the model assigns traffic to it and does not perform capacity restriction, i.e., speeds along such links will remain as free flow speeds, because the model assumes that capacity equals demand. This feature has been used on several occasions as a way to explore the capacity that would be required for a highway proposal; once the initial test is performed, realistic capacities can be re-introduced.

An improved procedure to estimate the optimum capacity of a proposed road scheme is shown in Figure 3. In this method, minimum and maximum capacities are given for each proposed road section. The model starts with the minimum values, calculates travel costs, estimates demand and assigns demand to the network. At the end of the first iteration, the level of service in the proposed links is evaluated and compared to a given criteria; if the result is under the specified criteria, the capacities of the corresponding links are increased, subject to the maximum capacity values. The process is repeated in subsequent iterations, checking for convergence.

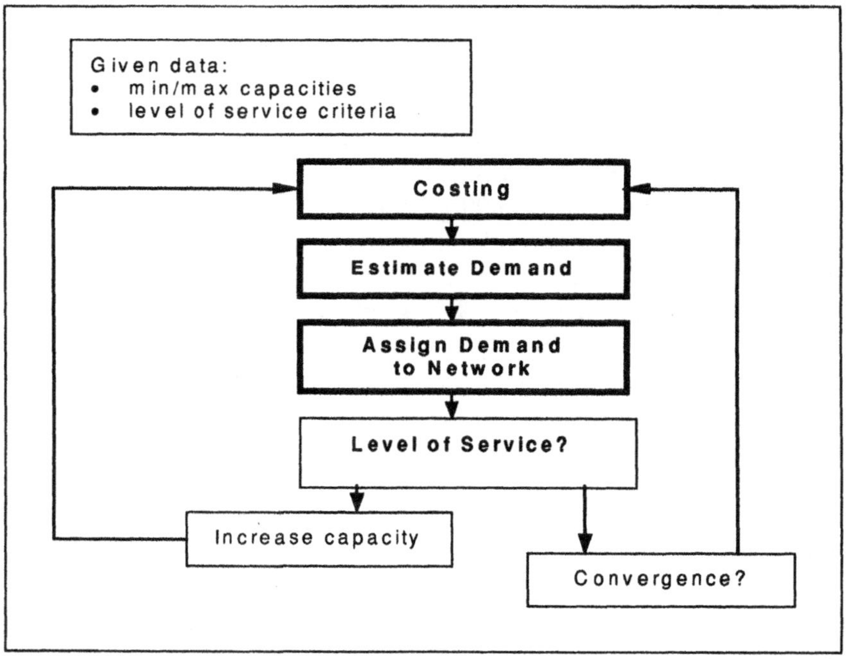

Fig. 3. Undefined Link Capacities Method

In this method, it is important to consider long term projections, to make sure that enough capacity will be determined to cover the projection period. It is also important that competing road schemes should not be introduced. Once the optimal capacity has been determined, fixed capacities can be re-introduced to the model, repeating the simulation.

Undefined frequencies are more complex. In this case the user declares some or all transit lines as having an undefined frequency. In the first iteration, the model assigns demand to such transit lines assuming a default waiting time. After assignment, the model calculates the number of vehicles as a function of demand and average occupancy, and this, in turn, determines the simulated frequency and corresponding waiting times. In second and subsequent iterations, the new frequencies affect demand, resulting in a new number of vehicles and new waiting times. At the end, the model outputs the resulting equilibrium frequencies.

An added freedom is that the analyst may specify public transit operators with undefined routes as well as frequencies; in this case the model is told which link types can be used by particular transit operators and will estimate the sequence of links according to demand and traffic conditions. Again this type of operators may be combined with fixed routes-undefined frequencies or fully defined routes and frequencies.

This feature has proved very useful in a number of studies. It is particularly applicable when long term projections are performed, because it is very difficult to know how transit lines are going to be organised in, say, 20 years time. The implied assumption here is that transit lines are going to be organised as a function of demand. But there is more to it than this. The feature can also be used to represent a policy of de-regulating public transport, or maybe parts of it. It may be that existing frequencies are coded, and turned into undefined values in future years simulations. In this case the model re-organises such parts of the public transport system and simulates what operators would do if they are left free to choose their frequencies. Further degrees of freedom can be introduced if the routes themselves are left undefined. These methods can be useful to evaluate de-regulation policies, or can be used to design bus feeder services to mass transit systems.

An interesting result from this process is that a market mechanism applies if transit routes and/or frequencies are left undefined. In the first iteration those origin-destination pairs where demand is large are assigned transit services with high frequencies, while those pairs with poor demand get services with low frequencies. In a second iteration, high frequencies attract more passengers and low frequencies result in less attractive services, shifting demand towards the private mode, or simply inhibiting trips. The result is that those services that got a high frequency in the first iteration, increase their frequency in the second iteration, and the opposite happens with the low frequency services. At the end, the system converges to a state in which transit services concentrate in high demand corridors with an over-supply, while low demand areas get very poor services. Since this probably means overcrowding along the high demand corridors, operating speeds deteriorate, passengers (as well as cars) take longer to reach their destinations and public transport operators have to increase the number of units, reducing their operating revenues. This is precisely what happens when public transport is de-regulated.

A similar procedure can be implemented in the land use model. Usually the analyst specifies fixed amounts of floorspace in each zone, and makes it grow in a controlled way for future year predictions, subject to land use regulations that specify maximum limits. These maxima may be relaxed by setting them to undefined values in some or all zones. In this case the model will simulate future developments assuming de-regulated land use.

4. Optimising Procedures

Optimising procedures take the undefined variable method as a starting point and take it a step further. Optimising procedures are at an experimental stage in TRANUS and have not been used in practice. In this case the model is given an undefined variable linked to one or more variables that have to be optimised. This method can be useful to estimate optimal transit tariffs, highway tolls, road pricing schemes, improvements in energy efficiency or emission controls. In the following paragraphs, some of these applications are explored.

Consider first the case of optimising transit operations. The method, shown in Figure 4, takes as starting values minimum and maximum frequencies, minimum tariffs equal operating costs and a minimum level of service requirements. With these elements, the model is asked to determine the optimum combination of tariffs and frequencies such that revenues are maximised, subject to a minimum level of service, probably measured in terms of demand/capacity ratio. To do this, the model uses the starting values in the first iteration and assigns demand. In each iteration, the model will have estimated demand and total revenues, i.e. the difference between income from tariffs and operating costs (zero in the first iteration) for the transit route being optimised. If the level of service criteria has not been satisfied, the model must increase frequency. Next revenues are checked; if revenues have increased with respect to the previous iteration, adjust fares up, down otherwise. Finally, check for convergence. At the end of the process, the transport system must have reached a state of equilibrium, maximising the operators' revenues, subject to a minimum level of service. Note that if frequencies are increased in any iteration, operating costs are also increased, reducing revenues; this, however, could be compensated by increased ridership.

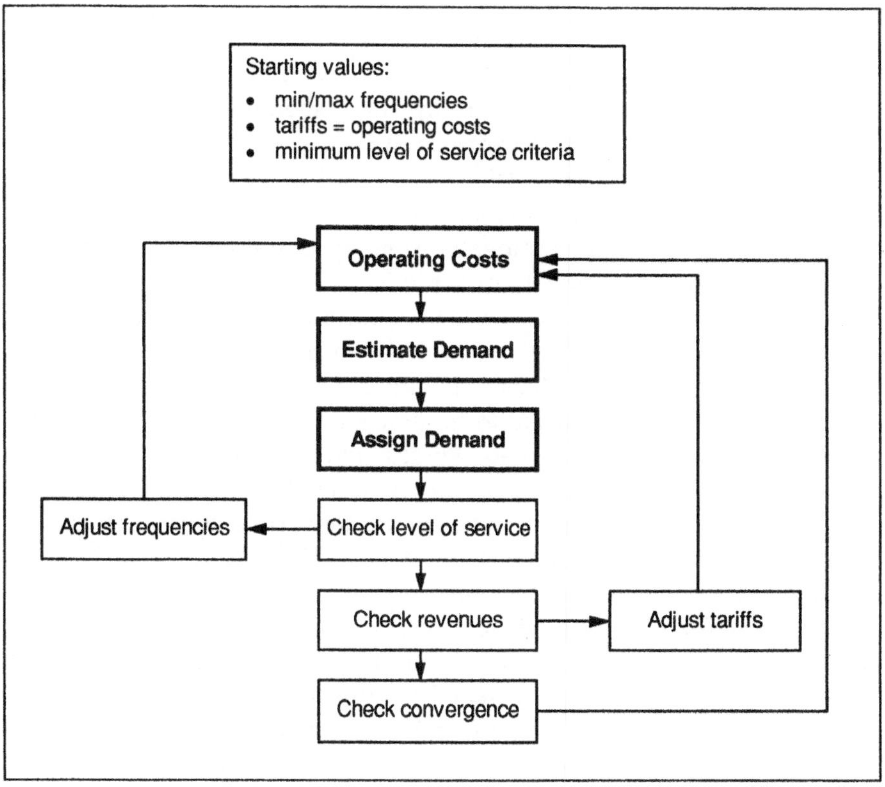

Fig. 4. Optimising Transit Operation

A similar approach can be used to optimise other elements of the transport system. Tolls is the most obvious choice, and in this case the model would be asked to maximise the administrator's revenues, subject to a utility target. The utility target can be a measure of user benefits, and the target can be defined as the level of utility greater than that obtained in a parallel run of the model without the toll scheme. Again the model is given a starting value for the price of tolls, in this case minimum maintenance costs. In the iterative cycle, the model calculates maintenance costs as a function of assigned traffic, and calculates revenues as the difference between maintenance costs and income from tolls. If such revenues are higher than those of the previous iteration, tolls are raised, lowered otherwise. Since tolls are probably different for each vehicle type, tolls can be adjusted in proportion to the marginal maintenance cost of each vehicle type. In every iteration, the utility target is also checked, limiting any increase in the price of tolls. Several runs can be made with different utility targets.

Road pricing can be treated in a similar way, as shown in Figure 5. In this case the criteria for determining the amount by which charges are raised within the model for each vehicle type will be different to that of tolls. Standard vehicle ratios and occupancy rates should be used instead of marginal maintenance costs, because the whole point of road pricing is to reduce congestion; hence those vehicles with low passenger/standard vehicle units should be charged more. Similarly, the variable to be maximised will not be total revenues from road pricing, but some measure of system throughput, such as average speed or total user disutilities; after all, road pricing is meant to be a policy to benefit users. The idea is that by charging users in specific sections of the network, traffic can be diverted in such a way that all users benefit, or at least, those that benefit could compensate those that loose. This is why the model must be asked to minimise total user disutilities. In every iteration, the model will raise road charges, and will check total disutility to make sure that the increased prices imply a benefit to users. There are no financial implications behind this strategy, i.e. there is no intention to use road pricing as a way to finance road improvements. This is, however, an important consideration that should be taken into account. To consider the possibility of improvements financed through road pricing, the model should be run to determine a first approximation of optimal road charges, and an estimate of the total amount that would be collected. Then this amount must be allocated to projects, such as new roads, better transit services or improved maintenance; the corresponding data is then fed back into the model to obtain a new set of road prices and disutility estimates. If money collected from road pricing is invested in improvements, the resulting prices should be, in general, lower. Note that it is possible that the first runs might show, in particular circumstances, that the resulting prices should be zero or too small to make it worthwhile. This may occur if there is not enough congestion to justify the policy of road pricing, or if users have enough alternatives to divert.

It is also interesting to consider the long term land use effects of these transport policies. In the short term, the optimised transport system with road pricing will converge to a particular set of transport disutilities. These, in turn, will produce long term effects on the land market and the location of activities. For instance, if a road pricing scheme is implemented in the CBD of a town, such area will become less attractive for residents, who will tend to move to the outskirts. This, in turn, would result on longer trips, compensating the intended gains of the original policy.

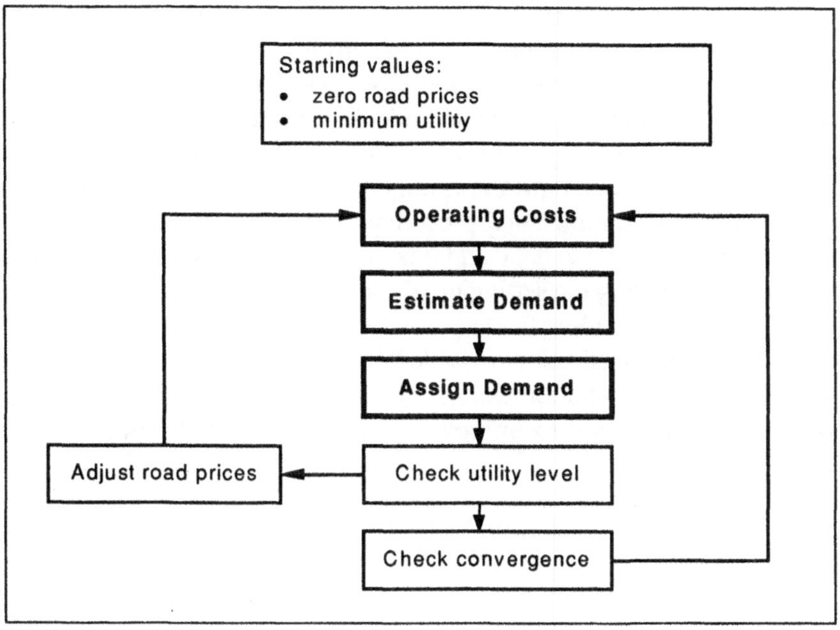

Fig. 5. Method for Calculating Road Pricing

5. Conclusions

In this paper the idea of complementing a land use-transport model with optimising procedures has been proposed and discussed. It has been argued that in this way the models can be used not only to simulate and evaluate a given set of policies, but can also help in defining the policies themselves. Several optimising procedures are explored, mostly to deal with elements of the transport system and related environmental policies. These range from optimisation of public transport, emission controls, road pricing and toll roads. Only a few of these procedures have been tested in practice.

Hence, a considerable amount of research and development lies ahead. First, there are some theoretical aspects that must be tackled; sound principles on which optimisation procedures must be designed are needed to avoid improper or inadequate use of the methods. Considerable research should be devoted to analyse the optimisation techniques themselves, in order to determine the most appropriate and efficient methods, to have a clear idea about the conditions in which some models will converge to a unique solution, to determine in which cases there is a solution or more than one solution, and to minimise the number of tests required. There is also the economic side of these procedures that must be taken into consideration, particularly when the policies imply direct payments, or when there are complex externalities involved as in the case of environmental controls.

Second, there is development work to be done. The proposed optimisation methods work as a jacket around the modelling system, and hence lie between the models and the analyst or user. It is important to design this interface in such a way that the techniques are immediately understandable and easy to use. After all, these optimisation procedures are performing non-trivial tasks on both the models and the data, and it is important to present them to the user in a clear way, covering the setting up of each procedure and the proper way to use and analyse the results.

Finally, real applications are needed to test each procedure thoroughly. There is little doubt that after testing, a feedback will be needed to solve both theoretical questions and to improve the procedures at the implementation level.

CHAPTER 15

MODELLING THE IMPLICATIONS OF NEW PUBLIC TRANSPORT TECHNOLOGY: AN APPROACH USING ARTIFICIAL INTELLIGENCE

Roger L Mackett

1. Introduction

This paper is concerned with the issue of improving urban public transport and so improving the quality of the environment. In particular, it is based on the premise that in Britain, public transport is not given the same emphasis as in continental Europe and that there is a need to learn from experience there.

In the next section the need for improving public transport is considered. The growth in car ownership and its implications in terms of energy consumption, pollution and traffic speeds are examined. It is argued that it is necessary both to discourage car use, for example, by road-pricing, and to provide an attractive alternative in the form of modern public transport. The provision of public transport in cities in Britain, France, Germany and the United States is examined in terms of patronage and finance.

The role of conventional modelling techniques in representing urban public transport is considered. These are found to be useful for estimating the effects, but not for actually deciding about the type of system which is most appropriate. This is because such decisions tend to use judgements and approximate data, rather than the exact values required by conventional

Y. Hayashi and J. Roy (eds.), Transport, Land-Use and the Environment, 297–315.

techniques. One way forward is to use the branch of artificial intelligence known as expert systems. These can be used to encapsulate knowledge of experts for use in situations similar to the one in which they have experience, in this case, deciding on the appropriate form of urban public transport system. Such a project, named UTOPIA (Urban Transport Planning and Operations using Intelligence Analysis) is being carried out at the Centre for Transport Studies at University College London. After a description of the project, the implications in terms of transport, land use and the environment are considered.

2. The Need for New Public Transport Systems

Car ownership is growing rapidly in countries around the world. Table 1 shows the number of cars per head in various countries in 1981 and 1991. Over ten years the number of cars per head in Great Britain has risen from 0.28 to 0.38, an increase of 36 per cent. The equivalent figure in Japan is 38 per cent. The United States had a growth rate of 7 per cent, but from a much higher base, suggesting that saturation is being approached. In some parts of the United States, for example Los Angeles County, there is now more than one registered car per licensed driver (Wachs, 1993).

Table 1. Cars per Head

	1981	1991	% change
Great Britain	0.28	0.38	36
Japan	0.21	0.29	38
United States	0.54	0.58	7

Source: Transport Statistics, Great Britain 1981 and 1993.

The levels of pollution emission in the United Kingdom in 1982 and 1991 are shown in Table 2. This shows how much transport contributes to the total of all emissions and how much of that is due to road transport, which is mainly cars, and that the proportions are increasing. A second environmental effect of the growth in road use is the demand for energy, as Table 3 shows. The total amount of energy used in the United Kingdom rose by 11 per cent over the period 1983 to 1992. However, transport energy consumption increased by 36 per cent. Of the overall increase in energy consumption, 68 per cent of it is due to increased demand for petroleum for cars.

These are all national figures, but much of the effects of car usage resulting from car ownership growth is in cities. Table 4 shows traffic speeds in London. In the central area these have fallen from 20.3 to 16.3 in the morning peak and from 19.4 to 16.8 over a twenty year period. The overall value for London has also decreased.

Table 2. Pollutant Emissions by each User in the United Kingdom

	Nitrogen oxides		Carbon monoxide		Volatile organic compounds	
	1982	1991	1982	1991	1982	1991
Road transport	870	1440	4360	6020	1090	1230
All transport	1060	1650	4400	6060	1160	1300
All emissions	2280	2750	5240	6740	2550	2680
Road transport as % of all transport	82	87	99	99	94	95
Transport as % of all emissions	46	60	84	90	45	49

Note: Units are thousand tonnes
Source: Transport Statistics, Great Britain 1993

Table 3. Energy Consumption by Transport Modes in the United Kingdom

	1983	1992	% change
Road transport - petroleum	11372	15545	37
Railways	441	408	-7
Other transport modes	2501	3494	40
All transport	14314	19447	36
All energy consumption	54043	60153	11
Energy used by road transport as a percentage of all transport usage	79	80	
Energy used by transport as a percentage of total	26	32	

Note: Units are millions of therms
Source: Transport Statistics, Great Britain 1993

Table 4. Average Traffic Speeds in London

	1968 - 70	1986 - 90	1990 -
Morning Peak			
Central area	20.3	18.4	16.3
All areas	29.0	25.6	na
Daytime off-peak			
Central area	19.4	17.6	16.8
All areas	34.1	30.2	na

Note: Units are km/hour
Source: Transport Statistics, Great Britain 1993

The continued growth in the ownership and availability of the motor car is likely to cause increasing problems. There are ways of reducing some of the effects, for example, fitting catalytic converters to cars can reduce pollution, but it will take many years before all the cars without them have ceased to be used. Furthermore, as Wachs (1993) observes, about 75 per cent of the daily pollution produced by modern cars is in the first few kilometres because a catalytic converter is less efficient when cold. Hence there is a need to eliminate car trips rather than just to shorten them if the objective is to reduce environmental damage. This suggests that land use policies aimed at reducing car journey lengths are not appropriate. What is needed is policies to reduce car use and to make the alternatives attractive. Reduction in car use can come about by policies such as road-pricing. In London the Department of Transport (1993) has commissioned a 3 million research project into road pricing in London, probably involving some form of electronic charging system.

However, it is also important to make the alternative public transport modes attractive. In many cities this means investing in new systems, because the existing public transport is provided by bus and suburban rail only. Bus tends to suffer from congestion caused by cars and suburban heavy rail tends to have poor spatial coverage because it is expensive to build and requires large flows to be economically viable.

Table 5 shows the characteristics of various urban public transport modes. The cheapest and quickest mode to introduce is the standard bus running in traffic because it needs little new infrastructure. However, it is subject to delays because of congestion and tends to have a poor image and so does not attract motorists from their cars. A guided bus system, such as that in Essen in Germany or Adelaide in Australia, permits high-speed running along radial corridors, thereby avoiding congestion, but retaining the flexibility to cover the suburbs by using ordinary roads. It is debatable whether such systems can overcome the prejudice against buses. Many cities in continental Europe have retained trams, which can provide efficient movement of passengers to the city centre. However, street running means that trams are delayed by cars, so in some cities, such as Vienna and Prague, tram routes are being removed as metro lines are being opened. Segregated light rail is really a modern form of tram, but running on separate corridors. Such systems carry large numbers of people at high speed. The disadvantage is the need to find land upon which to build. In some places, such as Newcastle-upon-Tyne in the north of England, the system goes underground in the city centre. This can increase the cost substantially, but it may be necessary to provide sufficient penetration of the city centre to attract car users. Higher capacity can be provided by a full-scale metro running underground. This system completely segregates the passenger from the surface, so that road congestion has no effect. The disadvantages are the high capital cost and the

Table 5. Costs and other Characteristics of Public Transport Modes

	Maximum capacity (1000pph/direction)	Commercial speed (km/h)	Operating cost per km per annum ($ x 10^6)	Capital cost for for twin lanes ($ x 10^6)	Total cost over 30 year life ($ x 10^6)	Cost per passenger-km in cents
Standard bus in traffic	7.2 - 9.6	15	0.5 - 0.7	0.4 - 0.5	5.7 - 8.1	0.8 - 0.9
Guided bus	19 - 29	15 - 25	1.2 - 2.1	1.1 - 2.6	14.7 - 26.7	0.8 - 0.9
Tram (street running)	9 - 25	15 - 25	0.3 - 0.9	6.7 - 13.3	10.7 - 23.3	0.7 - 1.9
Light rail (segregated)	9 - 25	30 - 40	0.3 - 0.7	3.3 - 6.7	6.7 - 14.0	0.5 - 1.1
Metro (underground)	35 - 70	30 - 40	0.7 - 1.3	20.0 - 43.0	26.7 - 60.0	0.5 - 1.3

Note: It is assumed that system is operating at 50 per cent capacity for 18 hours a day, 363 days a year over 30 years. The total operating costs over the 30 year life have been annualised at 8 per cent a year. The figures have been converted from to $ at an exchange rate of 1 = $1.50

Source: Modified from a table in a review of people mover systems and their potential roles in cities, by B H North, published in the Proceedings of the Institution of Civil Engineers. Transportation, Volume 100, pp 95-110

length of time it takes to build. These factors tend to mean that areal coverage is poor, particularly when a new system is built. Suburban rail can also convey large numbers along corridors, but city-centre penetration is usually poor.

In practice, a large city needs a combination of public transport modes, with buses in the suburbs where their flexibility can be exploited, a high-capacity rail-based system, along the radial corridors and an efficient distributor system in the city centre.

There are many types of urban public transport systems around the world. The nature of the present systems arises from a combination of political, economic, geographical and historical factors for each city. In Great Britain most public transport is provided by bus, except in the very largest cities. London has large metro and suburban rail systems. Glasgow and Newcastle-upon-Tyne have small metro systems. Manchester recently opened a light rail system with some street running. In fact many cities in Britain wish to build light rail systems, and are seeking funding. Bus services outside London were deregulated in 1986 and this has led to a major loss of patronage (White, 1990). This effect can be seen, in part at least, in Table 6. This shows public transport patronage in three urban areas in Britain: Edinburgh, Leeds-Bradford and London. (The cities have been selected on the grounds of data availability). Edinburgh and Leeds-Bradford are served by buses, with some suburban rail services supplied by British Rail. Britain had a small economic boom, which peaked at about 1989, and this is reflected in the patronage figures, but the overall trend is downwards. In London, which has not yet had bus deregulation, patronage grew, peaked about 1989, but has now declined as the economic recession has deepened.

Table 6 also shows public transport patronage in some American cities. It grew in New York to the year 1987/88, but has declined since. Boston shows a general downward trend. By contrast, Washington DC is showing growth, as is the Bay Area Rapid Transit System (BART) in San Francisco. The metro in Washington, which opened in 1976, is still being expanded, as is BART, which opened in 1972. It can be seen that there is growth in patronage on the modern rail-based systems rather than the bus.

Cities in continental Europe take a very positive view of public transport. Table 7 shows the public transport patronage in a selection of cities in France and Germany. None of the cities are showing declining patronage, and some are showing significant growth. Often this is associated with a new public transport system. A light rail system was opened in Grenoble in 1987, a fully-automatic metro in Lille in 1983, a metro in Lyon in 1978, which is still being expanded, a metro in Marseille in 1978, and Paris has expanded the regional metro (RER) system. A similar positive approach to infrastructure investment is taken in German cities. In Cologne, construction started on a high-speed metro (S-bahn) in 1985, and in Dortmund branches of the tramway system are to be connected by three cross-city tunnels. A similar set of tunnels is being built in Dusseldorf, and in Munich a metro was opened in 1971 and several expansions built since.

Table 6. Public Transport Patronage in British and American Cities (in millions per year)

	1985/86	1986/87	1987/88	1988/89	1989/90	1990/91
British cities						
Edinburgh	160.5	155.1	144.8	145.3	149.0	140.4
Leeds/Bradford	318	na	313	322	313	300
London	1832	1910	1941	2066	1995	2001
American cities						
Boston	232	na	na	195	198	192
New York	1517	1524	1574	1567	1559	1496
San Francisco						
- Muni	na	252	na	236	233	239
- BART	58.9	56.2	57.6	60.5	70.5	na
Washington DC						
- bus	136.8	136.4	138.5	139.5	na	na
- metro	116	127	135	144	na	na

Note: na - not available
The figures for Edinburgh are for buses operated by Lothian Regional Transport and SMT Omnibuses Ltd. Rail is not included, but is very small.
The figures for Leeds/Bradford are for the Passenger Transport Executive and contracted bus and rail services.
The figures for London and for London Transport Underground and bus services, including contracted services.
The figures for Boston are for the Massachusetts Bay Transportation Authority.
The figures for New York are for the New York City Transit Authority.
'Muni' is the San Francisco Municipal Railway, which operates buses, trolleybuses, tramways and cablecars, but not BART.
'BART' is the Bay Area Rapid Transit.
Source: Jane's Urban Transport Systems, 1989 and 1992-93.

Car ownership is continuing to grow in France and Germany which makes the growth in public transport patronage even more remarkable. The evidence from these tables is that investing in new rail-based systems can cause patronage to grow. Conversely, lack of investment is likely to mean a continuous decline in patronage. There is a major difference in the attitude to urban public transport of the government in Britain from that in France and

Germany. In France and Germany investment of public funding is encouraged. In Britain the government will allow investment, but only if it is from the private sector. Local authorities are not allowed to build new systems so they must rely on complex arrangements with private companies, whose only interest is making a profit.

Table 7. Public Transport Patronage per Year in French and German Cities (in millions per year)

	1985	1986	1987	1988	1989	1990
French cities						
Grenoble	37	35	38	43	46	49
Lille/Roubaix/ Tourcoing	76.2	75	77.9	80.5	89.2	94.1
Lyon	na	206.7	212.2	195	210	217
Marseille	128.5	140.7	158.5	157.8	164.5	na
Paris	na	2222	2227	2381	2388	2420
German cities						
Cologne	169.0	163.1	161.1	161.1	164.3	173.1
Dortmund	71.7	72.3	72.1	72.0	76.7	82.1
Dusseldorf	163.4	162.4	167.6	161	161	161
Munich	484.7	482.3	487.1	489.4	499.9	507.2

Note: na = not available
Source: Jane's Transport Systems, 1989 and 1992-93

The very different attitude to the funding of public transport is shown in Table 8. This shows the sources of funding to cover public transport operating costs in the cities discussed previously (except that Glasgow has been used instead of Edinburgh because of data availability).

In terms of the percentage that comes through the farebox there is no overlap between the British cities and the rest. In Glasgow over 80 per cent of the bus revenue and over 70 per cent of the metro revenue is from the farebox. The figure is nearly as high in Leeds-Bradford. In London less than a quarter of the operating costs comes from grants or subsidies.

In San Francisco and Washington DC, the new metro systems are more successful at covering their operating costs than the other modes in those cities but, of course, required massive capital investment. Public transport in Boston and New York requires much more subsidy than in the British cities.

With the exception of Paris, the French and German cities fall within a fairly narrow range of 41 to 57.1 per cent coverage of costs by fares. Subsidy is consequently quite high. In France much of this comes from a payroll paid by employers ('versement transport'). In Paris much of the funding comes from other commercial sources and in Dortmund from cross-subsidy from gas and water supply.

Table 8. Sources of Funding to Cover Public Transport Operating Costs (in percentages)

		Fares	Grant or subsidy	Other commercial sources	Pay-roll tax
British cities					
Glasgow	- bus	83.0	16.7	0.3	0.0
	- metro	70.1	28.3	1.6	0.0
Leeds/Bradford (Yorkshire Rider buses)		67	32	1	0
London		75.5	24.5	*a*	0.0
American cities					
Boston		30	66	4	0
New York		56	44	0	0
San Francisco	- Muni	30.1	68.8	2.1	0.0
	- BART	52.0	44.6	3.4	0.0
Washington DC	- bus	30.0	67.1	2.0	0.0
	- metro	64.4	28.9	6.7	0.0
French cities					
Grenoble		41	38	0	21
Lille/Roubaix/Tourcoing		50.6	49.4	0	0[b]
Lyon		51	34	0	15
Marseille		57.1	41.3	1.6	0[b]
Paris		32	37	15	15
German cities					
Cologne		47	53	0	0
Dortmund		49	22	29	0
Munich		45	45	0	0

Notes: [a]Included with fares
[b]May be included under grant or subsidy
Source: Jane's Urban Transport Systems 1992-93

Thus it can be argued that subsidy is lower in British cities than elsewhere, and this reflects a difference in the attitude of the government. In France and Germany subsidy is used to encourage public transport usage, to help reduce urban car use and to keep cities more civilised (Mackett, 1993a). In Britain, the government is strongly opposed to the concept of subsidy believing that it encourages inefficiency. There is some evidence for this. Bly *et al* (1990) argue that every extra one per cent of subsidy leads to a decrease in productivity of between 0.2 and 0.6 per cent. This is an argument for better control of the subsidy, for example by franchising specific routes rather than a blanket subsidy to cover losses. It is not an argument against using subsidy to achieve objectives such as improving public transport in order to reduce car use and so reduce congestion and pollution levels.

It has been argued above that cities need good public transport to attract people from their cars, and that this may require major investment in new infrastructure. There are a variety of technologies available and so decisions have to be made on what is appropriate for a particular city. It has also been shown that public transport is given a lower priority in Britain than in continental Europe. British cities need investment in public transport if the damaging effects of the car are going to be limited. Two key questions are: how does one decide what is the appropriate form of public transport technology to adopt; and how can the positive experience in cities in continental Europe be drawn upon? These questions are addressed in the rest of this paper.

3. Modelling Decisions about Public Transport

Conventional modelling techniques are very useful for showing the effects of various public transport modes. For example, in the work of the International Study Group on Land-Use Transport Interaction (ISGLUTI) (Webster *et al*, 1988) policies examined included building new cross-town metro lines and changing public transport fares. In fact, the effects of the new metro were small. However, when two of the models were applied to Leeds (Mackett, 1991), making public transport more attractive by reducing the fares not only increased patronage but had land use effects of reducing the loss of economic activity from the city centre while slightly increasing the decentralisation of population. It has been argued (Mackett, 1993b) that these land use effects arise because of the radial nature of public transport which means that if public transport use is encouraged, more trips will be made to the city centre, and fewer to suburban destinations.

Such trips are better served by public transport. This means that building a radial public transport system will become self-perpetuating in patronage terms, as illustrated by the growth in public transport in cities in continental Europe. This is more likely to happen with a fixed-track system such as rail or guided bus rather than buses running on conventional roads because the former demonstrate a state of permanence which may influence locational decisions. It is however, worth bearing in mind that there are dangers in such investment. Wachs (1993) argues that the dispersed nature of Los Angeles arose from the building of the suburban rail system rather than because of the freeways, which were a response to dispersal rather than its initial cause. This then became self-perpetuating. Another example is in Munich where Kreibich (1978) found that the new rapid transit system encouraged the dispersal of high-income people to suburban locations served by the new system. Such people are likely to buy more cars and then abandon use of the metro, possibly following the Los Angeles pattern unless positive action is taken to prevent it, for example road-pricing and continued improvement to the rapid transit system.

This suggests that integrated land use and transport models of the type used in the ISGULTI study have a useful role in the long-term analysis of the impact of such systems. Such models could be used to look at the impacts of different types of public transport technology, assuming that each can be represented in terms of speed, capacity, route coverage and so on. However, decisions about which technology to invest in involve judgements and approximate values that are not represented by this type of model. The following are characteristics of this type of model:

(a) They require precise data;
(b) They contain formal relationships expressed in precise mathematical terms;
(c) They produce precise answers;
(d) The range of solutions must be specified in advance;
(e) Parameters representing spatial differences are assumed valid for temporal forecasts;
(f) They have limited explanatory power.

What is needed is a systematic way of generating the alternatives to be considered and then deciding between them. Such decisions tend to be based on experience and judgement as much as formal modelling techniques. Such decisions are not made very often, so it would be very useful to be able to use the lessons from one city in another. To do this would require the encapsulation of the relevant knowledge from decisions about one city in order to transfer it to another. One way to do this is to use artificial intelligence, as discussed in the next section.

4. The Use of Artificial Intelligence Methods

One approach that can be used to encapsulate the knowledge and judgement of experts who have made decisions of the type being considered here, is to use the branch of artificial intelligence known as expert systems. Expert systems are computer programs that provide advice on solving a problem using the knowledge of experts. Ortolano and Perman (1990) explain that an expert system has the following elements:

(a) **The domain** which is the subject area;
(b) **The knowledge base** which is a collection of facts, definitions, rules of thumb and computational procedures applied to the domain;
(c) **The control mechanism** which is a set of procedures for manipulating the information in the knowledge base; this may be in the form of logical deductions from a set of facts and rules of the form 'if (premise) then (consequence)'.
(d) **The user interface** which will usually be a visual display unit and keyboard linked to the computer running the expert system, and the associated software.

The domain being considered here is the decision about the type of public transport technology to adopt to meet the travel needs of a city. The knowledge base is information on various types of transport technology, their characteristics and implications, plus information on the city, in terms of population, density, existing transport, and so on. The control mechanism is based on information from experts who have been involved in the design of such systems, mainly in continental Europe.

It is relevant to consider whether this is an appropriate topic for representation by an expert system. Ortolano and Perman (1990) have identified six criteria for deciding whether a particular task can be codified into an expert system:

(a) The knowledge needed task performance is specialised and narrowly focused;
(b) True experts exist, that is, people who know more than novices;
(c) The task is neither trivial or exceedingly difficult;
(d) Conventional computer programs are inadequate for the task;
(e) The potential pay off from an expert system is significant;
(f) An articulate expert is available and willing to make a long term commitment to helping to build the expert system.

The first five criteria seem to be met by the task being considered here: it is a specialised area, there are experts who have made such decisions, it is not a trivial task, nor is it exceedingly difficult, conventional models help, but cannot actually do the task, and the potential payoff is huge, given the scale of the investment. Whether or not the sixth criterion is met depends on the particular circumstances.

A project using expert systems for this task is discussed in the next section.

5. The UTOPIA Project

UTOPIA (Urban Transport Operations and Planning using Intelligent Analysis) is the name of a project being carried out in the Centre for Transport Studies at University College London, with funding of about £130000 from the United Kingdom Science and Engineering Research Council. It has the following objectives:

(a) To help produce more civilised cities by improving transport operations and planning;
(b) To transfer between cities, experience of decision-making about appropriate transport technology;
(c) To use artificial intelligence techniques to improve decision-making in the field of transport.

The idea behind the project is to talk to experts in cities in continental Europe who have been involved in the design of such systems, to encapsulate their knowledge in the expert system, and then to apply it to cities in Britain. Expert systems are not very good at representing complex mathematics so more conventional models will be used to calculate the impacts of the options generated by the expert system.

The methodology being used is shown in Figure 1. `The expert' is the person who has experience of designing an urban public transport system. His or her expertise will have been encapsulated by means of one or more interviews, supplemented by written documents on the background to the decision. There may be knowledge from several experts in the system, and each can be used in turn. `The user' is the planner or politician in a British city who is interested in designing such a new public transport system for his or her city. It will be necessary to input both the characteristics of the city and the objectives of constructing a new public transport system. These might cover the speed and capacity of the system and environmental features. These data will be input in response to questions. If the person does not know the answer the computer will search the knowledge base to offer a suitable value, for example from a similar city or a national value. When the data have all been input, the control mechanism in the expert system will make inferences to produce one or more possible solutions. These will then be applied using a model such as a simplified version of LILT (the Leeds Integrated Land-use Transport model)(Mackett, 1983). This will calculate the implications of the possible solutions in terms of revenues, environmental implications and so on. These will then be fed back to the expert system which make interpretations, for example to see if the predicted revenue produced is sufficient or the environmental criteria are met. These will then be displayed to the user to explain the proposed solutions. The user can then

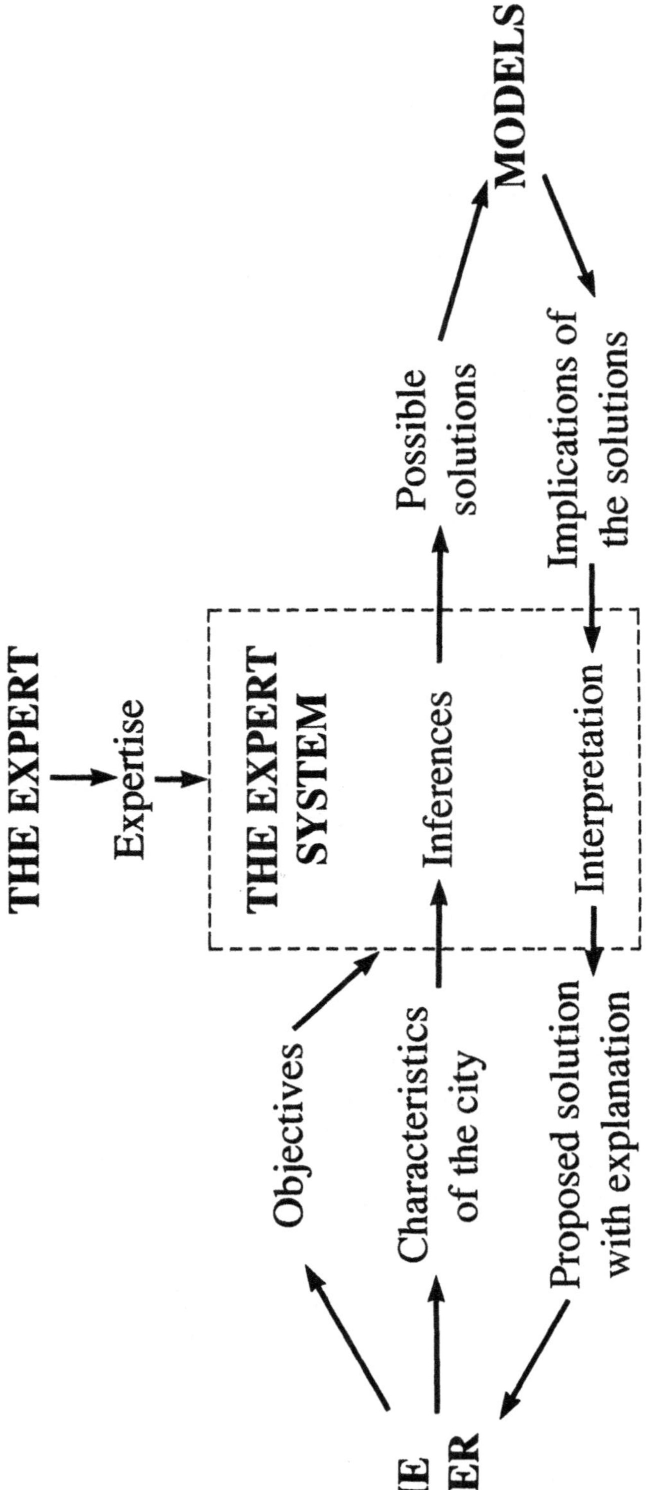

Fig. 1. The Methodology being Adopted in the UTOPIA study

change the objectives or use another set of expertise to explore a variety of options.

The UTOPIA project started in January 1993 and will last three years. So far the emphasis has been placed on identifying appropriate public transport systems to study, talking to various relevant people to build up knowledge and to help to identify experts for interview, taking to British experts and starting to develop the expert system. The work on the expert system has concentrated on the design of the `Intelligent Cities Database' (ICD). This will form part of the knowledge base of the expert system, It will also be used during the knowledge acquisition process and use of the model, allowing experts and users to enter data on their cities in a systematic way, responding to questions from the computer. It will also provide the most appropriate value for a particular city if none is available.

To date, one formal interview has been held. This was on the Manchester Metrolink project, which opened in April 1992. This is a light rail system which has taken over two former British Rail lines to Bury and Altrincham and linked them with street running though the city centre, focusing on Manchester Piccadilly station. The interview was with Bill Tyson, Director of Planning and Promotion of Greater Manchester Passenger Transport Executive (GMPTE) and Tony Young, Operational Planning Manager of GMPTE. Before the interview a set of questions was sent, at the request of Bill Tyson. These were as follows:

(a) What alternatives were considered?
(b) How explicit was the process of deciding between the alternative technologies?
(c) If an alternative technology was considered, would the design of the system have been different, for example, alternative routes, stopping points or interaction with other traffic?
(d) What factors were taken into account when deciding on the type of technology? (For example, capacity, speed, influence on demand.)
(e) Have there been compromises made because the vehicles run both off and on the streets? Was tunnelling under the city centre considered?
(f) To what extent have the level and method of funding influenced the design of the system?
(g) What would you do differently if you were starting now?
(h) Who actually decided on the type of system - politicians, managers or technical staff?
(i) What effects do you expect the system to have on Manchester in terms of, for example, employment patterns and car usage.

The interview was tape recorded and is being analysed. A list of the responses based on the author's notes is given elsewhere (Mackett, 1994).

The following is a list of some examples of the type of information obtained.

(a) The options considered were: closure of the rail lines, continuing the existing system with some investment, a street-running light rail system, a light rail system in tunnel, a heavy rail system in a tunnel, a busway and a guided busway.
(b) The criteria used to choose between the systems were: capacity, top speed, ability to use the existing rail lines without extensive additional costs, high level of reliability, acceptable environmental features, capability of expansion, ability to run on the streets (non-tunnelling options only), proven technology, ability to carry cross-town passenger movements.
(c) Tunnelling was rejected because of the high cost of passenger access and low visibility.
(d) Busways are an expensive option if you have to remove existing rail track.
(e) Busways are more likely to suffer from car congestion in the city centre because there is less incentive to segregate; buses are also more likely to suffer from roadworks because with a rail-based system utilities under the road such as gas and water are more likely to be moved; buses can be diverted to alternative roads, trams cannot.
(f) If the system includes street-running it can never be driverless; driverless systems need more secure lines and depend on unproven technology; there may be political problems with driverless vehicles in an area with high unemployment.
(g) If all the processes of design, build, operate and maintain are part of the same contract (DBOM), then the body commissioning the work, that is GMPTE in this case, should put a great deal of effort into the design specification, particularly at the interface with other bodies, for example, British Rail.

These are just a few examples of the type of information that has been obtained. Some of these can be put into the form `if (premise) then (consequence)'. Copies of the report produced at different stages in the decision process have been received, so these can be used to supply further data.

In due course interviews will be held with others involved in such decisions in Britain, for example Croydon Tramlink and Sheffield Supertram. Then trips will be made to talk to experts in cities such as Essen with its guided bus, Grenoble with a new tram system and Amsterdam with a light rail extension to the metro, once the appropriate experts have been identified. The knowledge gained will be encapsulated in the expert system and then applied to the various cities currently considering investing in such systems.

6. Transport, Land Use and Environmental Implications

It was argued in Section 3 that the growth of the use of the car was causing more energy use, pollution and congestion. The introduction of a new urban public transport should cause some shift from car to public transport, although it is possible that the suppressed demand for car travel is such that the car trips lost to the new mode are simply replaced by new car trips. There may well be an element of this, but nonetheless there should be some reduction in car use. Many of the trips on the new public transport systems may come from existing public transport modes, especially bus, and walking. While such trips do not have the effect of reducing car use, they do offer a better quality transport experience for the user and may lead on to land use changes. Such land use changes would reflect the perception of permanence that a new light rail, metro or guided bus system offers, which means that activities may relocate confident that users will be able to use the system in the future. It was argued in Section 3, based on evidence from modelling work (Mackett, 1993b), that the radial nature of such public transport systems means that economic activities will tend to be concentrated more in the city centre (or, in practice, city centre decline may be slowed down). Population may well decentralise more, partly because the new public transport system will enable people to commute from the outer areas served by the system and partly because land in the centre might be used by economic activity rather than housing. Greater concentration of economic activity in the city centre may well cause patronage to increase on the public transport corridors that serve it, for example heavy rail. These longer term land use effects would reinforce the shift from car to public transport. They should therefore lead to reductions in energy use and atmospheric pollution, or at least, reductions in the rates of growth. In the UTOPIA work all these effects would be estimated in the models which calculate the implications of the solutions generated by the expert system.

The UTOPIA expert system will include environmental characteristics of the various modes. For example, light rail systems and trolley buses are electric, so the atmospheric pollution would be produced at the power station, which may be coal, oil, gas or nuclear. The vehicles on a busway would normally be diesel, so the pollution will be local. Hence, in a city where the reduction of atmospheric pollution is an issue, the expert system can be used with environmental criteria to emphasise modes which use electricity. Similarly, noise characteristics of each mode could be included in the knowledge base and then used to provide advice, for example, on the appropriate mode along a corridor where a maximum noise level is specified. Thus, the UTOPIA methodology not only permits the estimation of the transport, land use and environmental implications of new public transport systems, but also allows the incorporation of system design criteria formulated in terms of these topics. Thus, for example, the solution generated by the UTOPIA system could be based explicitly on the objective of reducing atmospheric pollution. Hence the methodology being discussed in this paper could aid the co-ordinated planning of transport, land use and the environment, thereby improving the planning of cities.

7. Conclusions

This paper has shown why there is a need for better urban public transport systems. Many cities in continental Europe have modern systems and increasing patronage. In contrast, British cities have declining patronage. However, there is now considerable interest in the construction of light rail and similar systems in Britain. However, there is no systematic way of deciding on the type of system. It is intended that UTOPIA will fill this gap by enabling knowledge from experts who have been involved in such systems to be imported for use in British cities. While the work is still at an early stage it is showing great promise and generating much interest. It can lead to better co-ordination of the planning of transport, land use and the environment.

Acknowledgements

This work is being carried out as part of the project entitled 'An intelligent design model for multimodal urban public transport systems' funded by the UK Science and Engineering Research Council (Grant number GR/H 78481).

References

Bly, P.H., Webster, F.V. and Pounds, S. (1980) *Subsidisation of Urban Public Transport*. Supplementary Report 541, Transport and Road Research Laboratory, Crowthorne Berkshire, 1980.

Department of Transport (1983) *London Congestion Charging: a review of technology for road use pricing in London*. HMSO, London.

Kreibich, V. (1978) The successful transportation system and the regional planning problem; an evaluation of the Munich rapid transport system in the context of urban and regional planning policy. *Transportation*, 7, 137-145.

Mackett, R.L. (1983) The Leeds Integrated Land-use Transport Model (LILT) Supplementary Report SR805. Transport and Road Research Laboratory, Crowthorne, Berkshire.

Mackett, R.L. (1991) LILT and MEPLAN: a comparative analysis of land-use and transport policies for Leeds. *Transport Reviews, 11* 131-154

Mackett, R.L. (1993a) Why are continental cities more civilised than British ones? Paper presented at the 25th Universities Transport Study Group Annual Conference held at the University of Southampton, January 1993.

Mackett, R.L. (1993b) Structure of linkages between transport and land use. *Transportation Research, 27B*, 189-206

Mackett, R.L. (1994) Determining the appropriate public transport system for a city. Paper written for presentation at the 73rd Transportation Research Board Conference held in Washington DC, January 1994.

Ortalano, L. and Perman, C.D. (1990) Applications in urban planning: an overview, in T J Kim, L L Wiggins and J R Wright (eds) Expert Systems Applications to Urban Planning. Springer-Verlag, New York.

Wachs, M. (1993) Learning from Los Angeles: Transport, Urban Form and Air Quality. *Transportation 20*, 329-354.

Webster, F.W., Bly, P.H. and Paulley, N.J. (eds) (1988) *Urban Land-use and Transport Interaction* Avebury, Aldershot.

White, PR. (1990) Bus Deregulation: a welfare balance sheet. *Journal of Transport Economics and Policy, 24*, 311-332.

CHAPTER 16

GIS INTEGRATED SYSTEM FOR URBAN TRANSPORT AND DEVELOPMENT PLANNING

Takayuki Ueda, Hideo Nakamura and Eihan Shimizu

1. Introduction

1.1 Urban Issues

Growth of population and economic activities in metropolitan areas causes serious issues such as heavy congestion of transport, urban sprawl, and so on. Again with a background of more concern for environmental problems, these urban issues should be discussed not only from the point of economic efficiency, but also from the point of sustainability of a society, or quality of life. Therefore, urban planners or engineers are strongly required to present a set of effective policy options for solving the above problems, under more complex criteria and constraints than ever experienced before.

1.2 Need for a GIS Integrated Planning System

Analysis of complex urban issues requires a lot of time and cost, and many computer technologies have been utilised to reduce the time consuming steps in planning. In recent years, geographical information systems (GIS) are becoming a powerful tool in the fields of geographical analysis or spatial statistics, and furthermore in regional and urban planning. However, a GIS itself is usually a software package for data base and for geometric calculation, not a tool that is well-specified for complicated planning

Y. Hayashi and J. Roy (eds.), Transport, Land-Use and the Environment, 317–336.

functions. We can say, therefore, that need for a GIS integrated planning system which consists of a GIS itself and some subsystems developed for specific planning works stimulate the R&D not only by computer engineers but also by urban planners and civil engineers. (See Nakamura et al (1993).)

A GIS integrated planning system has the following merits;

i) We can make storage and retrieval of planning information more efficient than before, that is, improvement of information management.
ii) Under a specific constraint for time and cost, we can propose and analyse more alternatives than before, among which we can find more favourable ones. Even in the case where optimisation techniques are not so effective, we can make alternatives with trial and error.
iii) We can understand urban issues and policy impacts from a geographical view or a spatial view, more clearly, more quickly, and more easily.
iv) We can clarify the accountability of decision making because any steps in planning must be made clear and reorganised to be systematic in the process of system development.

This latter merit should be more emphasised. In practice, the process of decision makings is not always transparent, or the decision can not be always accounted for. By the system, the process of decision making can be traced in a backward direction from results to origins. Thus, the accountability of decision making can be improved by the GIS integrated system.

1.3 Urban Transport and Activity Location Model as a Core of the System

Since a majority of impacts of urban policies are revealed as changes in transport and activity locations, we must analyse such changes quantitatively. Therefore, the GIS integrated system that we intend to develop must have a function for analysing transport demand and activity locations. This function gives or receives variables and data sets to or from other subsystems in the system. This means that urban transport and activity location model plays a role of a core of the system.

The models for such a function, which are known as a land use transport interaction model, should posses both of accountability and operationality. In other words, it should be not only based on sophisticated location theories or urban economic theories, but also be computable in practical forecasting works. Therefore, our model, as will be explained later, is built within the theoretical framework of Walrasian multimarkets equilibrium, and made computable by formulating as a kind of modified mathematical programming.

2. Design of the GIS Integrated Planning System

2.1 Outline and Major Components of the System

The GIS integrated planning system that we are now developing has been designed as shown in Figure 1. The system consists of a GIS and subsystems. Furthermore, the GIS consists of i) a data base, ii) a data base management system, iii) a geometric data processing system, and iv) an user interface. The subsystems integrated in the system are organised to cover 3 phases of the planning process, i) model analysis, ii) preparation of policy alternatives, and iii) evaluation of policy impacts. For the phase of model analysis, we develop transport and activity location analysis subsystem, and for preparation of policy alternatives, transport planning subsystem, land use planning subsystem, and urban development planning subsystem. Since most of urban development projects in Japan are land readjustment projects, we have developed a computer-aided design system for such projects. Finally, for the phase of evaluation of policy impacts, monetary evaluation subsystem such as economic and financial evaluation works, and non-monetary one for environmental quality assessment are joined with the GIS.

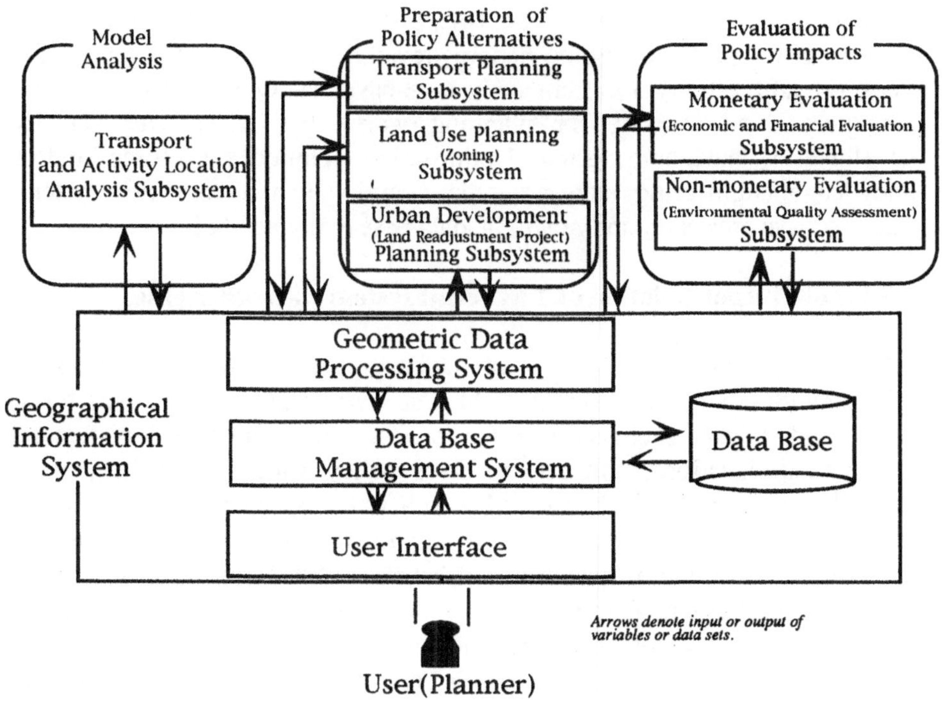

Fig. 1. Outline of the Integrated System

Any subsystem integrated to the system receives and gives variables or data sets from or to the GIS. Thus, a GIS is a base of our integrated system. Since the data structures of different types of GIS shells have been already compatible with each other and also with CAD systems, any computer system developed for a specific planning work can be joined with our integrated system in principle.

2.2 Brief Explanations of Components

In the data base, various kinds of geographical data such as population, employee, and transport networks, are accumulated. The data base should have not only the data sets representing status quo, but also those for the past and future ("future" means some results of forecasting to be reference.)

The role of the interface is, first, to facilitate the easy input of policy variables into the above two subsystems, and then to present their outputs. The former means user friendly ways for policy setting which are used in CAD systems. The latter includes various formats of graphic presentations, for example, map displays, or 3D bar charts, etc.

The transport and activity location analysis (TALA) subsystem is to be explained later. Monetary evaluation subsystem gives indicators for projects evaluation such as cash flows, IRR, B/C, FIRR, R/C, and so on. Some items of necessary data for assessing benefit, revenue, and cost are given by the transport and activity location analysis subsystem. Other subsystems are computer aided systems for planning works that have been already developed by Nakamura's group. Optimisation techniques are some kind of knowledge engineering ones embodied within such a computer system. These subsystems can jointly own the data base explained before.

2.3 Input-Output Relation of TALA Subsystem with other Ones

TALA subsystem, which is a core of our integrated system, has input-output relation with other subsystems ; i) TALA subsystem receive, as inputs for it, many kinds of data sets necessary for focusing the future transport demand and activity locations, and also receives policy variables set by planers through the user interface. ii) TALA subsystem gives changes in traffic volumes by route and by modes as inputs to the monetary evaluation subsystem for estimating cash flows of benefit, revenue, and cost. iii) TALA subsystem presents its outputs to planners through user interface. iv) TALA subsystem receives policy settings such as zoning or urban development plans from the other peripheral subsystems, and then, gives its outputs to them as frameworks for a plan such as future population in the development area.

3. Urban Transport and Activity Locations Model Embodied in the System

3.1 Introduction of the Model

In Japan, many types of so called landuse transport model have been developed and applied in practice. However, most of these models deal with only the land market. This is partly because data on floor use have not been available, and partly because their prototypes were developed in the U.S. or Europe, where building and land are traded in one set.

Contrast, it is a fact that land and building are traded separately in Japan. This is an indication that they are regarded as different types of goods, and therefore, there exist unsatisfied volume ratios or vacant lands even in the areas of high development potential.

With these backgrounds, we are now developing a new type of landuse model which can deal with not only land markets but also building markets explicitly. For the above purpose, our model is based on the theoretical framework of Walrasian (general) multimarkets equilibrium, and location choice behaviour theories described by Logit model. Therefore, we can say that our model has *"systemic consistency"* and *"behavioural consistency"*, (See Anas (1982)), when compared with the Lowry type landuse model.

Adding to the above concepts of our model, here, the importance of the modelling of "General Equilibrium of Land and Building Market" should be emphasised from the point of environmental challenges. Nakamura and Ueda (1993) summarised the mechanism where urban growth causes environmental issues, as shown in Figure 2.

The urban growth causes environmental issues not only through the expansion of urbanised area, also through increase of density and intensity of urban activities. The former can be recognised as land use changes. However, the latter must be accompanied with raising of high buildings, while conventional land use models are not well specified for the analysis of such 3 dimensional structural changes of urban space. Therefore, the model embodied in our system might be more useful from the point of environmental challenges.

3.2 Model

Outline of the model

Major assumptions and concepts in the model are summarised as follows.

1) There $i = 1, \cdots\cdots, I$ zones in the spatial coverage of our model. Each zone has one land market and one building market.

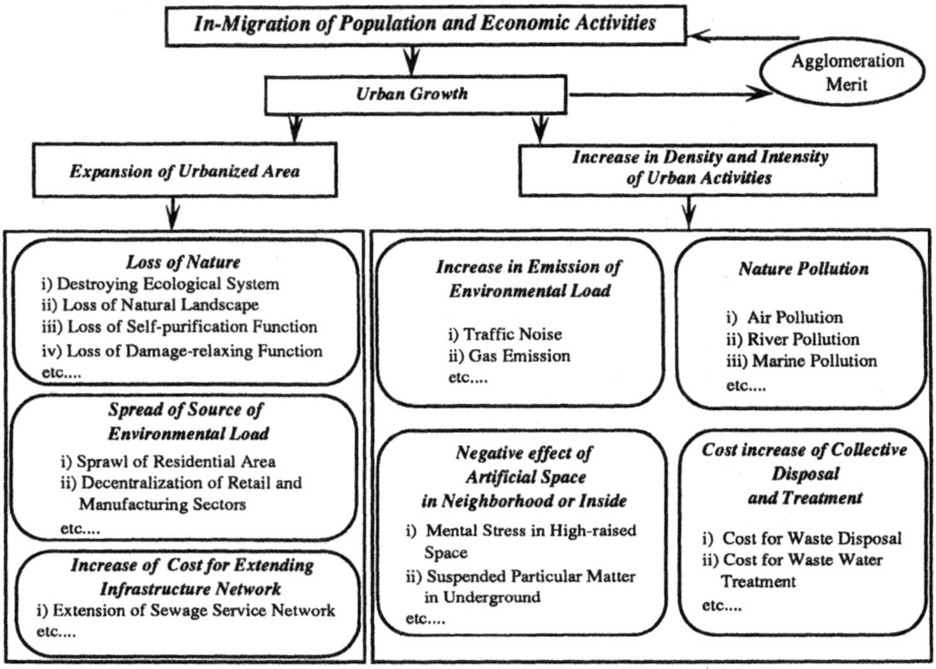

Fig. 2. Causality of Environmental Issues with Urban Growth

2) Economic agents considered in the model are, several types of locators, developers, and landowners. The locator population consist of $k = 1,\cdots\cdots,K$ types with N_{kT} locators. Total number of $k = 1,\cdots\cdots,K$ type locators N_{kT} is given. Hence, an urban system in our model is a closed city in a sense of urban economics. There is only one representative developer and one landowner in each zone. Each representative developer can supply floor service and demand for land service only in its own zone. Also, each landowner can supply land service in its own zone. Hence, subscript $i = 1,\cdots\cdots,I$ labels developers and landowners as well as zones.
3) Location choice behaviour of locators is modelled as logit model which gives choice probabilities as a function of variables denoting attractiveness of zones. The attractiveness of zone for locators is called *"location surplus"* in this paper, and it is defined as surplus of floor consumption.
4) In the framework of Walrasian (general) multimarkets equilibrium, all land and building markets should be cleared. All of locators should be allocated so that none of them can have an incentive to relocate. The former state is *"market equilibrium"*, and the latter *"location equilibrium"*. Furthermore, we call the state where both of them hold

"simultaneous equilibrium". We simulate this equilibrium in each period appropriately discreterised.

The outline of the model structure is sketched in Figure 3.

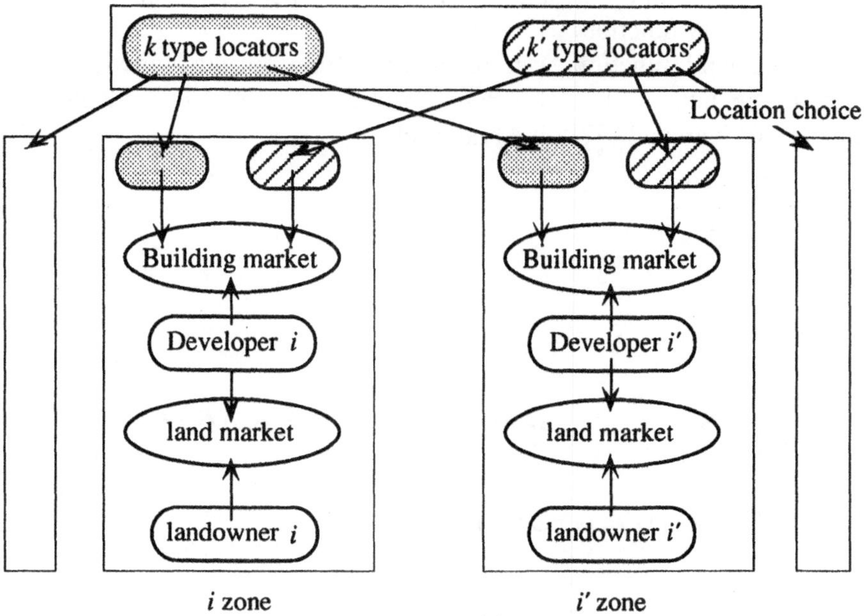

Fig. 3. Outline of the Model Structure

Definition of surplus functions

We, first, define surplus functions of economic agents. Although there exist consumption/production behaviours described as utility/profit maximisation, behind such surplus functions, we define surplus a priori without formulating such behaviours. However, this doesn't mean that our surplus functions are inconsistent with such microeconomics behaviours.

Definition of location surplus

Since a locator demands for floor space, we assume that k type locator's realised level of utility in i zone can be directly measured as consumer's surplus

$$V_{ki}(R_i, A_{ki}) = \int_{R_i}^{\infty} q_k(s)ds + A_{ki} \tag{1}$$

where V denotes level of location surplus. R is floor rent, A utility term dependent on location attributes, q individual floor demand, and s a dummy in integral. Nakamura, Hayashi, and Miyamoto (1981) and (1983), first proposed the concept of location surplus. However, Ueda(1992) redefined it to be consistent with optimisation behaviours in microeconomics, by using a quasi-liner utility function.

Floor demand functions can be derived from surplus functions, from Hotelling's lemma if the locator is a firm, and from Roy's identity (for example, see Varian(1992)), if a household.

$$\frac{\partial V_{ki}}{\partial R_i} = -q_k(R_i) \tag{2}$$

Here, we assume,

$$\frac{dq(R_i)}{dR_i} < 0 \quad \text{(3.a)}, \qquad \lim_{R_i \to R_i^{\max}} q_k(R_i) = 0 \quad \text{(3.b)}$$

where $R_i^{\max}$ is a upper boundary of R_i.

<u>Definition of developer's profit</u>

A representative developer in i zone has a profit π_i^D, which is a function of floor rent R_i, and land rent P_i.

$$\pi_i^D = \pi^D(R_i, P_i) \tag{4}$$

From Hotelling's lemma, floor supply Q_i^s is

$$Q_i^s = \frac{\partial \pi^D(R_i, P_i)}{\partial R_i} = Q^s(R_i, P_i) \tag{5}$$

and land demand L_i^d is

$$L_i^d = -\frac{\partial \pi^D(R_i, P_i)}{\partial P_i} = L^d(R_i, P_i) \tag{6}$$

Here we assume the following properties

$$\frac{\partial Q^s(R_i,P_i)}{\partial R_i} > 0 \qquad (7.a), \qquad \frac{\partial Q^s(R_i,P_i)}{\partial P_i} < 0 \qquad (7.b),$$

$$\lim_{R_i \to 0} Q^s(R_i,P_i) = 0 \qquad (7.c),$$

$$\frac{\partial L^d(R_i,P_i)}{\partial P_i} < 0 \qquad (7.d), \qquad \frac{\partial L^d(R_i,P_i)}{\partial R_i} > 0 \qquad (7.e),$$

$$\lim_{P_i \to 0} L^d(R_i,P_i) = \infty \qquad (7.f)$$

and

$$-\frac{\partial L^d(R_i,P_i)}{\partial R_i} = \frac{\partial \pi^D(R_i,P_i)}{\partial P_i \partial R_i} = \frac{\partial \pi^D(R_i,P_i)}{\partial R_i \partial P_i} = \frac{\partial Q^s(R_i,P_i)}{\partial P_i} \qquad (8)$$

<u>Definition of landowner's profit</u>

Profit of a representative landowner in i zone is,

$$\pi^L{}_i = \pi^L(P_i) \qquad (9)$$

Also from Hotelling's lemma, land supply, L^s is a function of land rent as,

$$L_i^s = \frac{\partial \pi^L(P_i)}{\partial P_i} = L^s(P_i) \qquad (10)$$

In addition, we assume that

$$\frac{\partial L^s(P_i)}{\partial P_i} > 0 \qquad (11.a), \qquad \lim_{P_i \to P_i^{\max}} L^s(P_i) = L_i^{\max} \qquad (11.b)$$

where $P_i^{\max}$ is a upper boundary defined as well as $R_i^{\max}$.

Location choice behaviour

From Miyagi (1986), location choice behaviour of a k type locator is formulated as the following mathematical programming, which can be interpreted as maximisation of expected maximum level of location surplus with a random variable of Gumbel's distribution.

$$S(V_k) = \max \sum_{i=1}^{I} \{a_{ki}V_{ki} - (\frac{1}{\theta})a_{ki}(\ln a_{ki} - 1)\} \qquad (12.a),$$

$$s.t. \quad \sum_{i=1}^{I} a_{ki} = 1 \qquad (12.b)$$

where S is a value function denoting expected maximum level of location surplus. $V_k = {}^t[V_{k1} \cdots V_{ki} \cdots V_{kI}]$ is a vector associated with location surplus of each zone. a_{ki} is a choice probability that a k type locator chooses i zone. θ is a parameter representing a degree of uncertainty in choice. Programming (12) gives, as its solution,

$$S(V_k) = (\frac{1}{\theta})\ln\{\sum_{i=1}^{I} \exp(\theta V_{ki})\} \qquad (13.a),$$

$$a_{ki} = \frac{\exp(\theta V_{ki})}{\sum_{i'=1}^{I} \exp(\theta V_{ki'})} \qquad (13.b)$$

Number of k type locators chooses i zone is,

$$N_{ki} = N_{kT} \cdot a_{ki} \qquad (14)$$

For the convenience in the later discussion, we define vectors,

$$N_k = {}^t[N_{k1}, \cdots N_{ki}, \cdots, N_{kI}] = N_{kT}\,{}^t[a_{k1}, \cdots a_{ki}, \cdots, a_{kI}]$$
for all $k = 1, \cdots, K$

and

$$N = [N_1, \cdots N_k, \cdots, N_K]$$
for all locators

where

$$
\begin{aligned}
N \in \Omega_N \\
&= \{N_1 | \sum_{i=1}^{I} N_{ki} = N_{kT}, N_{ki} \geq 0\} \times \cdots \\
&\quad \times \{N_k | \sum_{i=1}^{I} N_{ki} = N_{kT}, N_{ki} \geq 0\} \times \cdots \\
&\quad \times \{N_K | \sum_{i=1}^{I} N_{Ki} = N_{KT}, N_{Ki} \geq 0\}
\end{aligned}
$$

Furthermore, we should note, here, that logit model has properties,

$$\frac{\partial S(V_k)}{\partial V_{ki}} = a_{ki} \qquad (15.a), \qquad \frac{\partial a_{ki}}{\partial V_{ki}} = \theta a_{ki}(1 - a_{ki}) \qquad (15.b),$$

$$\frac{\partial a_{ki}}{\partial V_{kj}} = -\theta a_{ki} a_{kj} \qquad (15.c)$$

Equilibrium conditions

Market equilibrium

In a sense of Walrasian multimarkets equilibrium, all markets are cleared simultaneously. Hence, market equilibrium is defined as,

for building markets ,

$$\sum_{k=1}^{K} N_{ki} q_k(R_i) - Q^s(R_i, P_i) = 0, \qquad R_i \in [0, R_i^{\max}],$$
$$\text{for all } i = 1, \cdots, I \qquad (16)$$

and, for land markets,

$$L^d(R_i, P_i) - L^s(P_i) = 0, \qquad P_i \in [0, P_i^{\max}],$$
$$\text{for all } i = 1, \cdots, I \qquad (17)$$

Here, because of convenience for notation, we define

$$R={}^t[R_1,\cdots,R_I]\in\Omega_R=\left[0,R_1^{\max}\right]\times\cdots\times\left[0,R_i^{\max}\right]\times\cdots\times\left[0,R_I^{\max}\right]$$

$$P={}^t[P_1,\cdots,P_I]\in\Omega_P=\left[0,P_1^{\max}\right]\times\cdots\times\left[0,P_i^{\max}\right]\times\cdots\times\left[0,P_I^{\max}\right]$$

Thus, the system of equations (16) and (17) states the Walrasian multimarkets equilibrium, and means that aggregate demand is equal to aggregate supply in each market, or that a excess demand denoted by the LHS of (16) and (17) is zero.

Location equilibrium

In the state where levels of location surplus have not already be fractured by changes in allocation of locators, none of locators has an incentive to relocate. In that state, (14) with (13) still gives number of locators. By letting all V_{ki} represent such levels of location surplus, we can use (14) with (13), as conditions for location equilibrium.

Simultaneous equilibrium of market and location

Number of locators N_{ki}, ($k=1,\cdots,K$, and $i=1,\cdots,I$) affects R_i and P_i ($i=1,\cdots,I$) through (16) and (17), and inversely R_i and P_i do N_{ki} through (14) with (13). Since all endogenous variables $N=[N_1,\cdots N_k,\cdots,N_K]$, $R=[R_1,\cdots R_i,\cdots,R_I]$, and $P=[P_1,\cdots P_i,\cdots,P_I]$, are thus mutually dependent, they should be determined simultaneously. We call a system of equations to give conditions for such a solution simultaneous equilibrium. The conditions are,

$$V_{ki}(R_i,A_{ki})=\int_{R_i}^{\infty}q_k(s)ds+A_{ki}$$

for all $i=1,\cdots,I$ and $k=1,\cdots,K$ (18.a) [=(1)]

$$a_{ki}=\frac{\exp(\theta V_{ki})}{\sum_{i'=1}^{I}\exp(\theta V_{ki'})}$$

for all $i=1,\cdots,I$ and $k=1,\cdots,K$ (18.b) [=(13.b)]

$$N_{ki}=N_{kT}a_{ki} \quad \text{(18.c) [=(14)]}$$

$$\sum_{k=1}^{K}N_{ki}q_k(R_i)-Q^s(R_i,P_i)=0\,,\qquad R_i\in[0,R_i^{\max}],$$

for all $i = 1,\cdots,I$ (18.d) [=(16)]

$$L^d(R_i,P_i) - L^s(P_i) = 0\,, \quad P_i \in [0, P_i^{\max}]\,,$$
for $i = 1,\cdots,I$ (18.e) [=(17)]

In the urban system characterised by the above equations, number of locators N_{ki} interact with each other only though market equilibrium stated by (18.d) and (18.e). In more detail, this means that the system excludes *"externalities"* such as mutual merits of agglomeration in retail or business sectors, or, environmental damages of manufacturing sectors spillovered to residential locators in the neighbourhood. Hence, we call the equilibrium stated by the above system of equations a "*simultaneous equilibrium with no externalities* (SENE)". From the points of applicability of the model, the SENE model is still of use in the case that we may neglect externalities, for example, when we estimate only residential location by treating allocations of other types of locators as exogenous variables. However, to make our model more applicable, we should modify it so that we can deal with externalities as endogenous factors. Since magnitudes of externalities are location-dependent, we make utility term of locational attribute A_{ki} reflect externalities by setting it as a function,

$$A_{ki} = A_k(G_i, N) \qquad \text{for all } i = 1,\cdots,I \text{ and } k = 1,\cdots,K \tag{19}$$

where $G_i = {}^t[G_{1i},\cdots G_{ei},\cdots,G_{Ei}]$ denotes a vector associated with $e = 1,\cdots,E$ item of exogenous attributes of i zone, say, that G_{1i} denotes availability of sewage service, and that G_{2i} a volume of greenery, and so on. Thus, adding (19) to the conditions (18), we can define "*simultaneous equilibrium with externalities* (SEE)". Of course, we can regard SENE as a specified SEE by rewriting (19) as

$$A_{ki} = A_k(G_i) \qquad \text{for all } i = 1,\cdots,I \text{ and } k = 1,\cdots,K \tag{20}$$

As a summary of this subsection, we state the equilibrium in our model in two manners;

SEE) equilibrium conditions are, (18) and (19)

SENE) equilibrium conditions are, (18) and (20)

Interdependency of variables determined by the SENE and the SEE is shown in Figure 4.

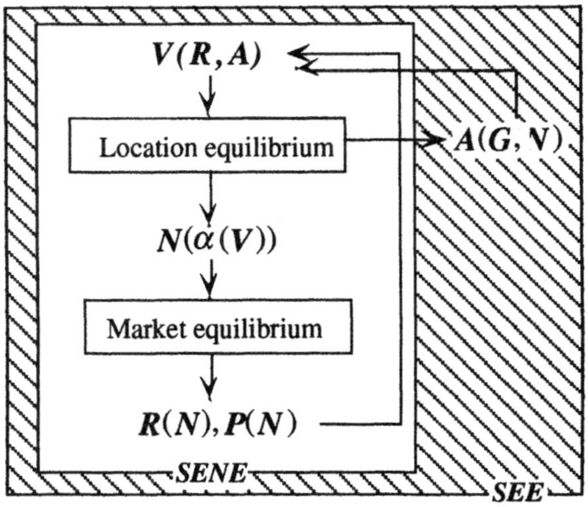

Fig. 4. Interdependency of Variables in Equilibrium

3.3 Properties of the Model

Programmability

We will discuss programmability of an equilibrium. Here, programmability (See, Smith and Bernstein(1993)), means that a certain type of programming formulation can represent an equilibrium model. It is, needless to say, important not only operationality or computability of the model, but also from the theoretical points of economics, in particular, interpretation of the relation of "invisible hand" with social welfare maximisation. Thus, we can sometimes investigate properties of an equilibrium model indirectly through those of the equivalent programming formulation. Most of studies on transport network equilibrium have shown such merits of programmability concept.

Programming formulation of the SENE model

It is obvious that the location equilibrium in SENE, stated by (18.a), (18.b), (18.c), and (20), is programmable because we have formulated the logit model itself as programming (12). From Varian(1984), the market equilibrium conditions stated by (16) and (17) can be converted into a programming formulation. Hence, from the fact that both the market equilibrium and location equilibrium in SENE are programmable, we state the SENE model as a programming formulation.

Although we cannot explain the detail of its derivation because of limitation of space, here we can propose the programming formulation equivalent to the SENE model. (See Ueda (1992) and Ueda et al (1993)).

$$-SS^{SENE} = \max_{[R,\ P]} \quad -\left[\sum_{k=1}^{K} N_{kT} S(V_k(R)) + \sum_{i=1}^{I} \{\pi^D(R_i, P_i) + \pi^L(P_i)\} \right] \tag{21}$$

$$s.t. \quad [R,\ P] \in \Omega_R \times \Omega_P$$

where A_{ki} is suppressed because of convenience for notation.

Recalling properties of logit model shown in (13.a), we have first order condition of the programming (21),

$$\begin{aligned} &\sum_{k=1}^{K} N_{kT} a_{ki}(V_k(R)) q_k(R_i) - Q^s(R_i, P_i) = 0 \\ &L^d(R_i, P_i) - L^s(P_i) = 0 \qquad \text{for all } i = 1, \cdots, I \end{aligned} \tag{22}$$

Practical use of programming formulation

The programmability of the SENE is so useful in the application of our model as that of user's equilibrium in traffic assignment models, especially in the model of Beckmann's type formulation. However, we should apply the SEE model so far as we aim at more general modelling that can deal with significant externalities in an urban system. Although it is unfortunate that the SEE model itself is not programmable as discussed in Ueda(1992) and Ueda et al(1993), we propose a practical algorithm for computing the equilibrium. Our idea is that we introduce the mathematical programming equivalent to the SENE model as one step into an algorithm for searching the SEE. This is in a line of the idea for the model that deals with the network equilibrium with asymmetric user's costs. We show the outline of the algorithm in Figure 5.

3.4 Examples of Application

Case study

We applied our model to the Hiroshima region to verify the exact computability. Since we are now still in the earlier stages of the development of the model, we can show only a few results of the application. However, they might be sufficient for evaluating the performance of our model.

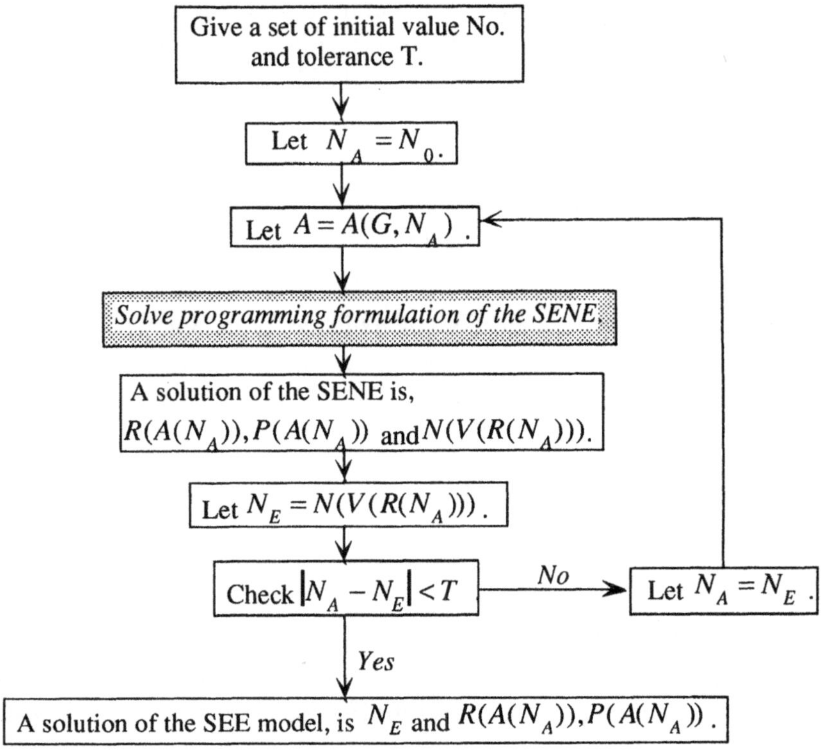

Fig. 5. Algorithm for Searching an SEE by Using the SENE Model

The reason why we choose Hiroshima for our case study area is that Hiroshima municipality office has a GIS (Geographical Information System) within which necessary data, in particular,flooruse data, for the application of our model are accumulated. Thus, our application is the first attempt, in Japan , for integrating a landuse model with GIS.

We simulated allocations of locators, landuse, and flooruse in 1990 as a test run, and then compared them with observed distributions in order to evaluate the performance of our model.

Specification of functions and estimation of their parameters

We specified the functions as shown in Table 1. Types of locators in the model are only residential and retail & business. Distribution of other types of locators such as manufacturing or public sector are fixed as observed. We show results of estimation of parameters in Table 1. Although some of parameters show low t-values, we can say that specification of functions and estimation of parameters were almost sufficiently successful for the early stages of our study.

Table 1. Specifications of Functions and Estimation of Parameters

(1-a) Location surplus

Location surplus of residential locator (k=1)

$$V_{1i} = \alpha_0 \int q_1(s)ds + \alpha_1 \ln T + (\alpha_2 / \alpha_4)\sum_j (N_{2j})\alpha^3 \ln(-\alpha_4 t_{ij}) + \alpha_5 z_i$$

T is average commuting time (min.)
t is travel time between i and j zone (min.)
z is average availability of sewage service
and N is in terms of persons.

$$q_1(R_i) = \beta_0 - \beta_1 R_i$$

q is measured by (m²/persons)
R is (Yen/m²/year)

α_s and β_s	parameter value (value of t)	Coefficient of determination R^2	
α_0	2.23×10^{-5} (2.23)		0.921
α_1	-2.26×10^{-1} (−3.90)		
α_2	1.53(6.29)		
α_3	5.57×10^{-1}(3.31)	0.538	
α_4	5.75×10^{-2} (9.57)		
α_5	2.14×10^{-1} (2.07)		
β_0	1.64×10^{2} (15.1)	0.946	
β_1	5.04×10^{-3} (8.88)		

Location surplus of Retail & Service business locator (k=2)

$$V_{2i} = \gamma_0 \int q_2(s)ds + (\gamma_1 / \gamma_2)\sum_j (N_{2j})\ln(-\gamma_2 t_{ij}) + (\gamma_3 / \gamma_4)\sum_j (N_{1j})\ln(-\gamma_4 t_{ij})$$

Second term is accessibility between R. & B. locators.
Third term is acessibility to residential locators, which is a proxy of hinterland size,
and N is in terms of persons.

$$q_2(R_i) = \delta_0 - \delta_1 R_i$$

q is measured by (m²/employees)
R is (Yen/m²/year)

γ_s and δ_s	parameter value (value of t)	Coefficient of determination R^2
γ_0	1.14×10^{-6} (2.19)	0.889
γ_1	1.39 (2.62)	0.889 (cont.)
γ_2	5.75×10^{-2} (***)	0.889 (cont.)
γ_3	3.40×10^{-5}(6.97)	0.889 (cont.)
γ_4	5.75×10^{-2} (***)	0.889 (cont.)
δ_0	1.95(6.22)	0.972
δ_1	7.77×10^{-6} (8.88)	0.972 (cont.)

γ_2, γ_4 was set by α_4 because of the lack of data.

(1-b) Developer's profit

Developer's Profit in residential zone (i=m)

$$\pi_i^D = \varepsilon_0 R_i^{1/(1-\varepsilon_1-\varepsilon_2)} \cdot P_i^{-\varepsilon_1(1-\varepsilon_1-\varepsilon_2)}$$

$$0 < \varepsilon_1 + \varepsilon_2 < 1 \qquad 0 < \varepsilon_1, \varepsilon_2$$

ε_s	parameter value (value of t)	Coefficient of determination R^2
ε_1	6.29×10^{-1} (4.98)	0.614
ε_2	2.61×10^{-1} (1.43)	

Developer's Profit in residential zone (i=2m)

$$\pi_i^D = \varsigma_0 R_i^{1/(1-\varsigma_1-\varsigma_2)} \cdot P_i^{-\varsigma_1(1-\varsigma_1-\varsigma_2)}$$

$$0 < \varsigma_1 + \varsigma_2 < 1 \qquad 0 < \varsigma_1, \varsigma_2$$

ς_s	parameter value (value of t)	Coefficient of determination R^2
ς_1	4.90×10^{-1} (5.89)	0.805
ς_2	4.74×10^{-1} (3.23)	

(1-c) Landowner's profit

Landowner's Profit in residential zone (i=m)

$$\pi_i^L = (1/\eta_0)\ln\{\exp \eta_0 P_i + \exp \eta_0 \eta_1\}(L_i^{\max} - L^{t-1}) + P_i L^{t-1}$$

L^{t-1} is land supply in the previous period exogenous in the present period in simulation.
P is measured by (Yen/m²/year)

η_s	parameter value (value of t)	Coefficient of determination R^2
η_0	1.90×10^{-4} (2.86)	0.621
η_1	1.22×10^{4} (4.62)	

Landowner's Profit in commercial zone (i=2m)

$$\pi_i^L = (1/\kappa_0)\ln\{\exp \kappa_0 P_i + \exp \kappa_0 \kappa_1\}(L_i^{\max} - L^{t-1}) + P_i L^{t-1}$$

κ_s	parameter value (value of t)	Coefficient of determination R^2
κ_0	5.23×10^{-5} (3.23)	0.715
κ_1	3.51×10^{4} (5.12)	

Results of test run simulation

In the test run simulation, we searched the SEE by the algorithm that we have already shown in Figure 5. For the total numbers of locators by each type, we adopted the values forecast by Hiroshima Municipality Office. We divide the area into 16 zones. A half of them are residential zone and the other half are commercial zone. A pair of $i = 2m - 1$ and $i = 2m$ ($m = 1, \cdots 8$) zones is included in an identical A-level zone in person trip survey.

Computing time for getting the equilibrium, using a workstation SUN SPARC1, was about 7 minutes in average for one set of initial values, and even in the longest, about 10 minutes. Thus, our model is computationally efficient in practice.

We compared results of the simulation with the observed data, as shown in Figure 6. Judging from coefficients of determination R^2 noted in those figures, we can conclude that our model has a good performance in simulating allocation of locators, landuse and flooruse.

4. Next Stages of the System Development

Although the main parts of the system have been already developed, we are still in the early stages of the development. There still remain tasks that we should tackle with in the next stages. As examples of them, the following tasks are mentioned.

i) The applicability of the whole system must be verified. This means that, since each parts of the integrated system and peripheral subsystems were applied into different case study areas, we have not verified whether or not all of them can be well integrated, that is, the total applicability of the system. Then, for this purpose, we should apply all subsystem in the same case study area to check the coordination between them.

Some tasks specific to the urban transport and activity location model are still remaining.

ii) Since urban transport and activity location model requires too much computing times and too much volume of data base in the case that we apply it to a large metropolitan area like Tokyo, we should modify the model for the purpose of quick computing and less data requirement. One of the ideas in this line is that many zones where impacts of urban policies are not significant are to be aggregated into a few representative large zones, while project areas should be still segmented into many small zones. This modification might be successful.

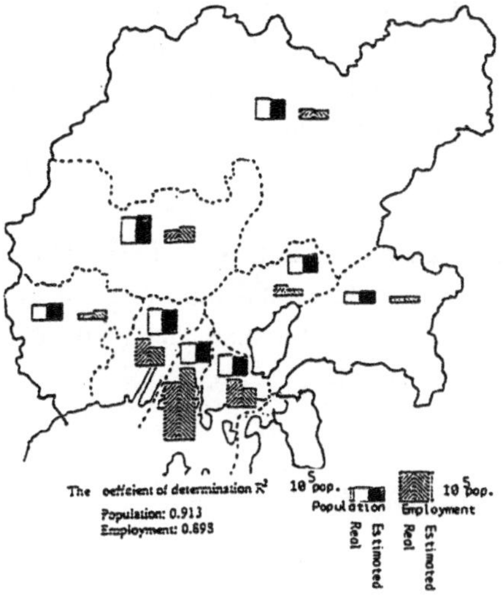

The result of application 1

(comparison of estimated population and employment with real in 1990)

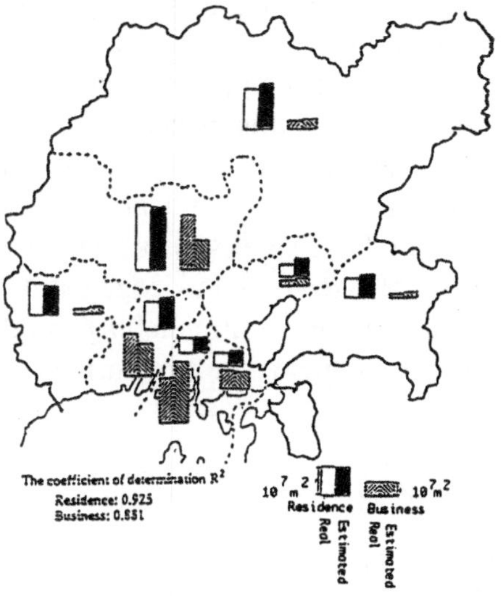

The result of application 2

(comparison of estimated floorspace with real in 1990)

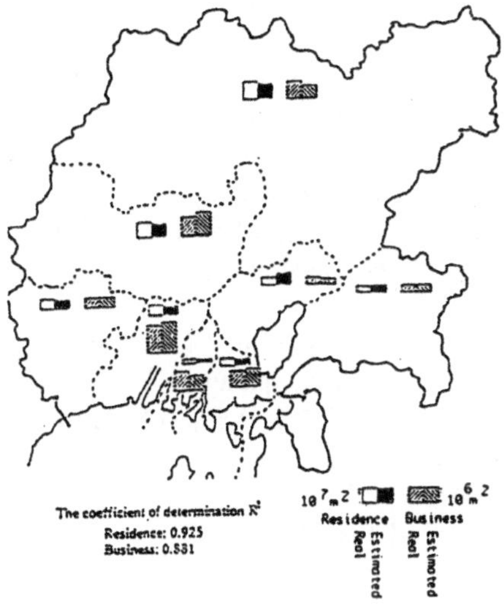

The result of application 3

(comparison of estimated land area with real in 1990)

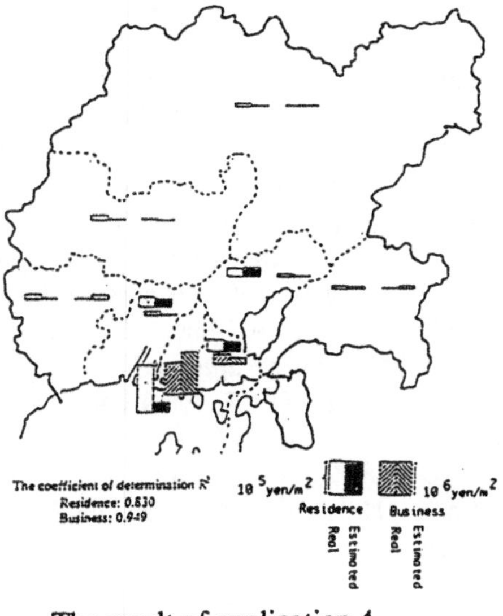

The result of application 4

(comparison of estimated land prices with real in 1990)

Fig. 6. Result of Application of the Model to Hiroshima

iii) For the forecasting of activity locations at a microscopic spatial coverage such as a 500 meters square mesh, which is sometimes required in practical planning, the modelling based on the equilibrium theories as explained in this paper might have poor persuasive power. This is because at such a microscopic level, activity locations are simultaneously dependent on many items of location attributes, which cannot be considered in simple formulations. Then, so-called expert system based on knowledge extracted from professional land evaluators, or neural network modelling, might be effective because they can deal with such many items of attributes simultaneously.

References

Anas, A. (1982) *Residential Location Market and Urban Transportation,* Academic Press

Miyagi,T. (1986) On the formation of a stochastic user equilibrium model consistent with the random utility theory, *Proc. of 4th World Conference on Transportation Research*, pp.1619-1635.

Nakamura, H. *et al.* (1981) A Land Use Model for Suburban Area, *Proc. of JSCE*, No.309, pp.103-122, JSCE (in Japanese)

Nakamura, H. *et al.* (1983) A Land Use Transport Analysis System for a Metropolitan Area, *Proc. of JSCE*, No.335, pp.141-152, JSCE (in Japanese)

Nakamura, H. and Ueda, T. (1993) Environmental Issues with Urban Growth (in *Urban Environment*, Nakamura, H. (eds.)) , Gyosei (in Japanese)

Nakamura, H. *et al.*(1993) A GIS integrated city planning system based on systems analysis, *Proc. of JSCE*, No.476, pp.67-76, JSCE (in Japanese)

Smith, T, and Bernstein, D. (1993) Programmable Transport Network Equilibria, (in *Structure and Change in the Space Economy*, Laksmann, T. and Nijkamp, P (eds.)), pp.91-130, Springer-Verlag

Ueda, T. (1992) A General Equilibrium Model with Redefined Location Surplus, *Infrastructure Planning* No.10, pp.183-190, JSCE (in Japanese)

Ueda, T. *et al.* (1993) Land Use Model based on the General Equilibrium of Land and Building Markets, *Porc. of International Conference of Land Problem and Urban Policy*, in Kyoto

Varian, H.R. (1984) *Microeconomic Analysis*, 2nd eds. Norton

Varian, H.R. (1992) *Microeconomic Analysis*, 3rd eds. Norton

CHAPTER 17

INCORPORATING ENVIRONMENTAL PLANNING DECISIONS IN TRANSPORT PLANNING: A MODELLING FRAMEWORK

Michael A.P. Taylor

1. Introduction

The transport planning process has evolved considerable sophistication involving modelling of existing land use, road network and travel patterns, and forward projection using these models to predict future travel demand and road network needs. Transport demand modelling of this type is in regular use at the regional, urban and local area scales of transport planning and the modelling procedures have now devolved to readily available and widely used packages on personal computers.

However, the development of these transport demand modelling procedures has not kept pace with changing community demands, which now focus significant attention on travel demand management (TDM) and require the inclusion of environmental factors in the planning, design and implementation of transport infrastructure. Traffic on roadways is a significant cause of environmental degradation in urban areas, contributing to air pollution and noise, as well as causing problems of congestion, safety and intrusion. Current practice for dealing with these environmental matters usually involves an environmental impact assessment for planned works projects. While these impact assessment procedures are important, and will remain so, particularly in providing ameliorative measures for the worst

Y. Hayashi and J. Roy (eds.), Transport, Land-Use and the Environment, 337–358.

environmental consequences of any project, their current weakness is that they come too late in the planning process - the important route location decisions have been made many years previously through travel demand modelling.

Recent research by Brown and Patterson (1990) introduced a novel approach to the assessment of the impacts of traffic noise, as an indication of the possibilities for a range of traffic-generated pollutants. They demonstrated that the noise impact of a planned road network could be explicitly included in travel demand modelling and network planning, i.e. when the network is still being developed, modelled and tested, rather than after the preferred route or potential alternative routes have been selected. In this regard the modelling work of Taylor and Anderson (1982, 1984, 1986, 1988) on estimating pollutant emissions from traffic streams is important, for this research focussed on the extension of traffic network models to provide information on emissions and to model the relationships between levels of traffic congestion, fuel consumption and pollutant emissions.

This paper considers the development of a computer-based method to assess the impacts of transport policies and adverse environmental impacts over a regional study area. It formulates a combined traffic assignment-trip distribution model for use at the strategic network level which is sensitive to alternative transport policy measures (e.g. in TDM), and which includes submodels for predicting fuel consumption and emissions. It indicates how these are being included in an environmental impact assessment package for application in traffic planning. This environmental impact assessment package consists of a set of PC-based computer models, linked through a common data structure, and making extensive use of interactive graphics displays.

Environmental and energy impacts represent two important community concerns influenced by the performance of road traffic systems. Fuel conservation and environmental degradation are of special importance (Jost, Ullrich and Waldeyer, 1987). Particular concerns have arisen with respect to:

(a) fuel consumption, and the means for conserving liquid fuels;
(b) air pollution, especially the emissions of gases and particulates from motor vehicles and current concerns about greenhouse gas emissions, and
(c) noise, vibration and visual intrusion.

2. Estimating the Environmental Impacts of Road Traffic

One procedural difficulty that has dogged planners and engineers has been how to assess the relative effects, merits and disadvantages of alternative transport infrastructure proposals at the planning stage. Survey methods for assessing levels of pollution are available for the study of existing conditions [e.g. Taylor and Young (1988); Maccarrone (1989)]. These methods cannot be applied to proposed developments, and alternative means for the appraisal of alternatives are required. This issue was addressed in the 1970s by Wigan (1976), who provided the following methodology for predicting the environmental impacts of road traffic:

(1) collate data on a link-by-link basis on road type, width, number of households, amount of activity by category of land use, etc;
(2) obtain traffic flow data, including traffic composition and travel time, speed and delay;
(3) develop a database that can provide the required link-by-link data to apply models of fuel consumption, emissions and pollutant dispersion;
(4) apply a framework that defines the conditions under which the consumption, emissions and dispersion models can be applied;
(5) generate indices of pollutant loads and environmental impacts (e.g. number of households subjected to a given noise level over a specified time interval);
(6) prepare tabular and graphical representations of this information as histograms, pollution load maps, etc, and
(7) indicate levels of individual and community annoyance under different pollution loadings.

Given the logical, 'common-sense' nature of this methodology, it may come as something of a surprise to realise that it has seldom, if ever been fully applied in practice! All too often transport planners and engineers have considered only the generation of pollution at its sources, not where that pollution will end up and who will be affected by it. The package described here follows Wigan's methodology through the construction of a combined model system, comprising a traffic network model (capable of producing a number of alternative travel patterns in response to differing transport policies), a family of emissions and fuel consumption models, a pollution dispersion model, and a land use impact model.

The basic scheme of the system is given in Figure 1. A traffic network model is used to produce (by simulation or forecasting) the levels of traffic flow and travel conditions on a study area network, under the given traffic management scheme. Models of vehicle fuel consumption and emissions under the modelled traffic conditions are then used to estimate the traffic system fuel usage and the levels and spatial distribution of pollution generation. This information, coupled with data on the meteorological conditions, may then be used as input to a pollution dispersion model, which estimates the spread of the pollution over the study area, so providing the modelled levels and spatial distribution of the pollution. The land use impact model superimposes the pollution levels on the land uses and populations in the study area to determine the likely sites and extent of environmental problems resulting from the traffic system.

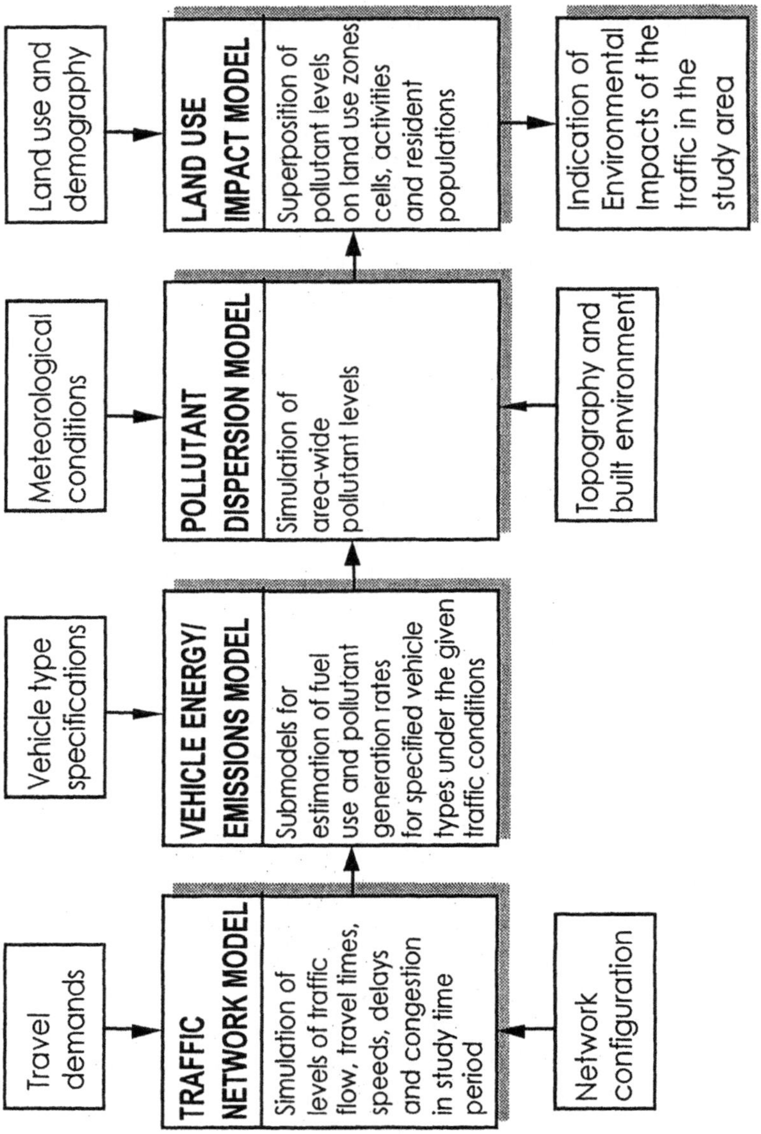

Fig. 1. Schematic Modelling System for Assessing Environmental Impacts

The application of the modelling system of Figure 1 to predictions of environmental impact and energy use depends on the accuracy of the traffic model in reproducing travel conditions on the network, and the validity of the vehicle performance models. The application to environmental impact analysis is based on the premise that, although the actual absolute levels of pollution may be affected by many other factors besides those included in the component models, the modelling system can reasonably detect relative differences in levels of pollution between alternative sets of traffic load distributions.

Further, the modelling approach means that a number of pollutants can be included together under the same sets of conditions, e.g. noise, gaseous emissions, and fine particulates. Thus alternative schemes may be ranked on a number of environmental quality objectives, and comparisons made between them.

3. Fuel Consumption and Emissions of Road Traffic

To gain an insight into the methods for assessing the severity of possible pollution problems, we must first consider the different fuels used in road transport, the range of pollutants generated by road traffic, the indications of environmental problems from these pollutants.

3.1 Road Transport Fuels

Virtually all of the road vehicle fleet is powered by petroleum-based liquid fuels. In most places, the principal fuel is petrol (gasoline), either as unleaded petrol (ULP) or as leaded ('super-grade') petrol, with some use of diesel fuel and liquid petroleum gas (LPG). Unleaded petrol (ULP) is the more modern fuel, and is of growing importance in countries like Australia as the proportion of the national vehicle fleet using ULP increases.

The use of super-grade (98 octane) petrol leads to emissions of the pollutants carbon monoxide (CO), volatile organic compounds (VOC), nitrogen oxides (NO_X) and particulate lead, besides emissions of water vapour and carbon dioxide. ULP also causes emissions of these pollutants, with the obvious exception of particulate lead. However, the more-modern catalytic convertor technology employed with ULP-powered vehicles means that lesser quantities of these gaseous pollutants will be emitted, per litre of fuel used. Diesel fuel is widely used for large vehicles, and occasionally by passenger cars. Diesel engines offer greater fuel efficiency (more kilometres travelled per litre) and significant reductions in emissions of carbon monoxide. On the other hand, they may produce more VOC, as well as sulphur oxides (which are largely absent from petrol engine emissions). Diesel fuel may also produce more carbon dioxide per litre of fuel used (Young, 1992). LPG offers a cheap alternative to petrol for some vehicles. Vehicles powered by LPG are marginally less polluting than equivalent vehicles using super-grade petrol. Taylor and Anderson (1982) provided summary comparisons of the polluting effects of these different fuels.

3.2 Pollutants from Road Traffic

Air pollution in urban areas typically consists of primary emissions such as carbon monoxide, volatile organic compounds (hydrocarbons), oxides of nitrogen and oxides of sulphur, and fine particulates (such as dust, soot and lead). In addition, carbon dioxide is also produced in quantity, although this has not been commonly regarded as a pollutant (as it is an important natural component of the atmosphere). Current concerns about emissions of greenhouse gases require that emissions of carbon dioxide should be included in any environmental impact assessment involving gaseous emissions. Urban road traffic is a significant source of such pollutants. Derived pollution such as photochemical smog results from the chemical reaction of some of the primary pollutants (e.g. the VOC and nitrogen oxides) under conducive atmospheric and meteorological conditions. Indications of the magnitude of environmental problems facing metropolitan areas were given in Taylor (1990).

Studies in the UK [e.g. Hothershall and Salter (1977)] and elsewhere [e.g. OECD (1980)] indicate that road traffic is the single most important source of noise pollution in urban areas. Hothershall and Salter found that traffic was the primary source of noise pollution at more that 60 per cent of the sites they investigated.

An important consideration for transport planning is the extent to which traffic systems operations and traffic congestion contribute to pollution loads and energy consumption. Thus there is a need to establish methods and relationships that link traffic movements and travel conditions to the environmental and energy variables. Following the ideas of Wigan (1976) and Brown and Patterson (1990), it is essential that attention be given to the effects of the traffic-generated pollution on community groups and land uses.

4. Predicting Environmental Impacts

The environmental impacts modelling system of Figure 1 provides a means for estimating the area-wide dispersion of pollution from a road network. Previous applications of this modelling framework have focussed on local area studies, e.g. Taylor and Anderson (1988). The framework may also be applied at the strategic network level, and an initial application of this kind was reported by Taylor and Anderson (1984). For application of the general system of Figure 1, the necessary information to be supplied or generated comprises the total flows and travel conditions (travel time, delays, queuing, congestion) on links in the network, the volume and composition of the traffic stream (in terms of vehicle and/or fuel type). Emission and fuel consumption rates may then be estimated by aggregating the contributions of the component traffic streams. The network is then treated as a set of line sources of each pollutant. The emissions from these sources may then be spread over the study region using the dispersion model, and the concentrations of pollution at different sites examined.

4.1 Congestion Models

A number of functional forms relating travel conditions to traffic flows at the link level are available [see Branston (1976) and Rose, Taylor and Tisato (1989) for some reviews of such functions]. One suitable function is the Davidson function, for which the most practical form is given by equation (1):

$$c = c_0\left\{1 + J\frac{\mu}{1-\mu}\right\} \qquad \mu < \rho$$

$$c = c_o\left\{1 + J\frac{\rho}{1-\rho}\right\} + \frac{J}{(1-\rho)^2}(\mu - \rho) \qquad \mu \geq \rho \tag{1}$$

where c is the link travel time, c_0 is the free-flow link travel time, μ is the volume-capacity ratio and J is an environmental parameter that reflects the road type and abutting land use development (and hence the level of friction within the traffic stream). Volume-capacity ratio is defined as the ratio of traffic volume (q) to link capacity (S). The linear extension of the curve for $\mu < \rho$ (where $\rho < 1$ is a pre-determined constant, usually in the range (0.85, 0.95)) provides a finite definition of the function for all finite volume-capacity ratios. It also allows for over-saturation of the link [see Taylor (1984)].

A new function, similar to Davidson's function but based on recent research on delays at traffic signals, has been proposed by Akcelik (1991) and is now finding application in transport network modelling.

The Davidson function provides one example of a relationship between travel time and volume that can be used to influence both the amount of traffic using a link and the emissions and fuel consumption on that link. How this may be done is the subject of the latter part of this paper. Alternative forms for congestion functions, such as Akcelik's function, could be used in similar fashion, as demonstrated by Berka and Boyce (1994).

4.2 Segmentation of Vehicle Flows

Changing fleet composition and the contributions of different vehicle types and trip classes to fuel usage and pollution are important in TDM, e.g. to see how such changes might affect pollution levels. [The differences in energy and environmental performance between pre-1986 and post-1986 Australian vehicles is one such issue. Trip class might include different categories of travellers, e.g. through traffic and local traffic, private, commercial and business travel, etc.] If q(e) is the total vehicle volume on link e then

$$q(e) = \sum_k q_k(e) \tag{2}$$

where $q_k(e)$ is the volume of trip class k vehicles on e. If p_{km} is the proportion of type m vehicles in trip class k then the flow $q_m(e)$ of type m vehicles is given by equation (3). It therefore follows that if $E_m(X)$ is the mean rate (per unit length) of emission (consumption) of pollutant (fuel) X by a type m vehicle then $TE_e(X)$, the total rate of emission (consumption) of X on link e is given by equation (4).

$$q_m(e) = \sum_k p_{km} q_k(e) \tag{3}$$

$$TE_e(X) = \sum_m E_m(X) p_m q(e) \tag{4}$$

In the common situation where trip class data are not readily available or cannot be accommodated in the computations, then an equivalent formulation can be used

$$TE_e(X) = \sum_m E_m(X) p_m q(e) \tag{5}$$

where p_m is the proportion of type m vehicles in the traffic stream.

Thus if models can be established to predict $E_m(X)$ for a range of traffic conditions then total pollution loads and fuel consumption can be estimated. These models will have the ability to suggest differences in energy and environmental impacts for changes in levels of traffic flow and congestion and for changes in vehicle fleet composition.

The basic form of such models is known, but only limited data (for a restricted number of vehicle types) has been available. Current research is enlarging the database of available vehicle types, and the first suitable data for unleaded petrol (ULP) cars is provided later in this paper. Segmentation of vehicles into size and/or fuel type classes in the manner suggested provides the means to derive reasonably accurate estimates of fuel consumption and emissions in transport network models. This poses some substantial new requirements for data and forecasting, and suitable contemporary data are becoming available (e.g. Woolley and Young, 1994) or can be found by research and investigation using methods such as those described by Bowyer, Akcelik and Biggs (1985).

5. Emission/Consumption Models for Traffic Streams

Four levels of fuel consumption and emissions modelling were proposed by Biggs and Akcelik (1986). Their models are:

(a) an *instantaneous model*, that indicates the rate of fuel usage or pollutant emission of an individual vehicle continuously over time;
(b) an *elemental model*, that relates fuel use or pollutant emission to traffic variables such as deceleration, acceleration, idling and cruising, etc. over a short road distance (e.g. the approach to an intersection);
(c) a *running speed model*, that gives emissions or fuel consumption for vehicles travelling over an extended length of road (perhaps representing a network link), and
(d) an *average speed model*, that indicates level of emissions or fuel consumption over an entire journey.

The instantaneous model is the basic (and most detailed) model. The other models are aggregations of this model, and require less and less information but are also increasingly less accurate. The running speed model is suitable for application in strategic networks, for it can be used at the network link level.

5.1 Instantaneous Model

This model is suitable for the detailed assessment of traffic management schemes for individual intersections or sections of road. It may be used for comparisons of the behaviour of individual vehicles under different traffic conditions. The variables in the model include instantaneous values such as speed v(t) and acceleration a(t) at time t. The instantaneous model gives the rate of emission/consumption (E/C) of X, including components for:

(a) the fuel used or emissions generated in maintaining engine operation, estimated by the idle rate (α);
(b) the work done by the vehicle engine to move the vehicle, and

(c) the product of energy and acceleration during periods of positive acceleration.

The energy consumed in moving the vehicle is further divided into drag, inertial and grade components. Part (c) allows for the inefficient use of fuel during periods of hard acceleration. The model is

$$\frac{dE(X)}{dt} = \alpha + \beta_1 R_T v + \left[\frac{\beta_2 M a^2 v}{1000}\right]_{a>0} \qquad \text{for RT} > 0$$

$$\frac{dE(X)}{dt} = \alpha \qquad \text{for RT} \le 0 \qquad (6)$$

where

v = speed (ms^{-1}),

a = instantaneous acceleration in ms^{-2},

R_T = total tractive force required to drive the vehicle, which is the sum of the drag, inertial and grade forces

M = vehicle mass in kg;

α = idling fuel consumption or pollutant emission rate;

β_1 = engine efficiency parameter (mL or g per kJ), relating E/C to energy provided by the engine,

β_2 = engine efficiency parameter (mL or g per ($kJ.ms^{-2}$)) relating E/C during positive acceleration to the product of inertia energy and acceleration.

R_T is given by

$$R_T = b_1 + b_2 v^2 + \frac{Ma}{1000} + MGg \times 10^{-5} \qquad (7)$$

where

g = gravitational acceleration in ms^{-2};

G = percentage gradient (negative downhill);

b_1 = drag force parameter relating *mainly* to rolling resistance, and

b_2 = drag force parameter relating *mainly* to aerodynamic resistance.

Both of the drag force parameters also reflect some component of internal engine drag. The model has been found to estimate the fuel consumption of individual vehicles to within five per cent. Its accuracy for emissions modelling remains to be established but a similar level could be expected. The five parameters α, β_1, β_2, b_1 and b_2 are specific to a particular vehicle, and the idling rate and energy efficiency parameters (α, β_1 and β_2) depend on the type of fuel or emission as well.

5.2 Running Speed Model

This model may be used for estimation of fuel consumption or emissions along a network link, and is thus the most suitable model for application in a transport network model. The data required to apply the model are travel time c_s (seconds), trip distance x_s (km), and stopped time c_i (seconds) over the route section. The vehicle is then assumed to travel at a constant running speed v_r (km/h), where

$$v_r = \frac{3600 x_s}{c_s - c_i} \tag{8}$$

while moving. The model predicts the mean rate of pollution emission or fuel consumption E_s (g or mL per km per vehicle) as

$$E_s = \max\left\{ f_r + \frac{\alpha t_i}{x_s}, \frac{\alpha t_i}{x_s} \right\} \tag{9}$$

where f_r is the fuel consumption or pollutant emission per unit distance (mL/km or g/km) excluding stopped time effects (i.e. while cruising at constant speed v_r), and is given by

$$f_r = \frac{3600\alpha}{v_r} + A + Bv_r^2 + k_{E1}\beta_1 + ME_{k+} + k_{E2}\beta_2 ME_{k+}^2 + gk_G\beta_1 M \frac{G}{100} \tag{10}$$

E_{k+} is the sum of *positive* kinetic energy changes per unit mass per unit distance along the road section (ms^{-2}), which may be estimated from

$$E_{k+} = \max\{0.35 - 0.0025 v_r, 0.5\} \tag{11}$$

as described by Bowyer, Akcelik and Biggs (1985). The calibration parameters k_{E1}, k_{E2} and k_G may be estimated from

$$k_{E1} = \max\left\{0.675 - \frac{1.22}{v_r}, 0.5\right\} \quad (12)$$

$$k_{E2} = 2.78 + 0.0178 v_r \quad (13)$$

$$k_G = 1 - 1.33 E_{k+} \quad \text{for } G < 0$$
$$= 0.9 \quad \text{for } G > 0 \quad (14)$$

A prediction of running speed is needed to complete this link-based model of emissions and consumption, and if this cannot be observed directly then [from Bowyer, Akcelik and Biggs (1985)] an estimate of the running speed v_r (km/h) may be made from equation (15), given knowledge of the overall average link travel speed v_s (km/h):

$$v_r = \max\left\{8.1 + 1.14 v_s - 0.00274 v_s^2, v_s\right\} \quad (15)$$

This model provides estimates of fuel consumption within 10-15 per cent of observed values for travel over road sections of at least 0.7 km. Road gradient plays a major role in determining the accuracy because of the non-compensatory effects of positive and negative gradients. Longer section lengths will give improved accuracy. The accuracy of this formula for emissions modelling remains to be determined.

5.3 A Fuel/Emissions Model for Transport Network Analysis

A combination of a congestion function such as the Davidson function shown by equation (1) and the E/C running speed model defined by equation (9) may then be used to estimate the E/C rate for a link of a given type operating at a given volume-capacity ratio. This combined model can then be included in a transport network model (e.g. for traffic assignment or for combined distribution and assignment). Tables 1 and 2 provide the necessary parameter values for the generation of specific E/C functions for the average pre-1986 Australian car for different road types in terms of one-way link volume/capacity ratios.

Figures 2-5 show functions for fuel consumption and emissions of carbon monoxide, carbon dioxide and VOC, for dual carriageway arterial roads in inner, middle and outer metropolitan locations for the pre-1986 car. A value of $\rho = 0.9$ was used to generate these relationships.

Table 1. Davidson Function Parameters (after Taylor (1984))

Link description		Davidson parameters		
Road type	Area	S (veh/h/lane)	J	c_0 (min/km)
Undivided, multilane	Inner	1344	0.475	1.15
Undivided, multilane	Middle	1765	0.468	0.85
Undivided, multilane	Outer	1741	0.486	0.80
Undivided, with LRT		1317	0.350	1.15
Divided	Inner	1607	0.374	1.06
Divided	Middle	1861	0.415	0.83
Divided	Outer	1911	0.419	0.79
Freeway		3055	0.138	0.70

Table 2. Emissions and Fuel Consumption Parameters for pre-1986 Australian Average Car (after Bowyer, Akcelik and Biggs (1985), Akcelik (1990))

Parameter	Fuel	VOC	Carbon monoxide	Nitrogen oxides	carbon dioxide
α (s^{-1})	0.444	0.0022	0.0139	0.0006	1.0212
β_1	0.090	0	0.015	0.001	0.207
β_2	0.0450	0.0040	0.0250	0.0002	0.1035
b_1	0.333	0.333	0.333	0.333	0.333
b2	0.00108	0.00108	0.00108	0.00108	0.00108
M (kg)	1200	1200	1200	1200	1200
Unit	mL/km	g/km	g/km	g/km	g/km

Table 3 shows fuel consumption and carbon dioxide emission parameters for a representative Australian ULP vehicle (a 4-cylinder Toyota Camry sedan). Figures 6 and 7 show the derived link functions for fuel consumption and carbon dioxide emissions for this vehicle.

Table 3. Carbon Dioxide and Fuel Parameters for the Toyota Camry

Parameter	α (s^{-1})	β_1	β_2	b_1	b_2	M (kg)
Fuel usage (mL/km)	0.294	0.068	0.041	0.455	0.00056	1250
Carbon Dioxide (g/km)	0.764	0.177	0.107	0.455	0.00056	1250

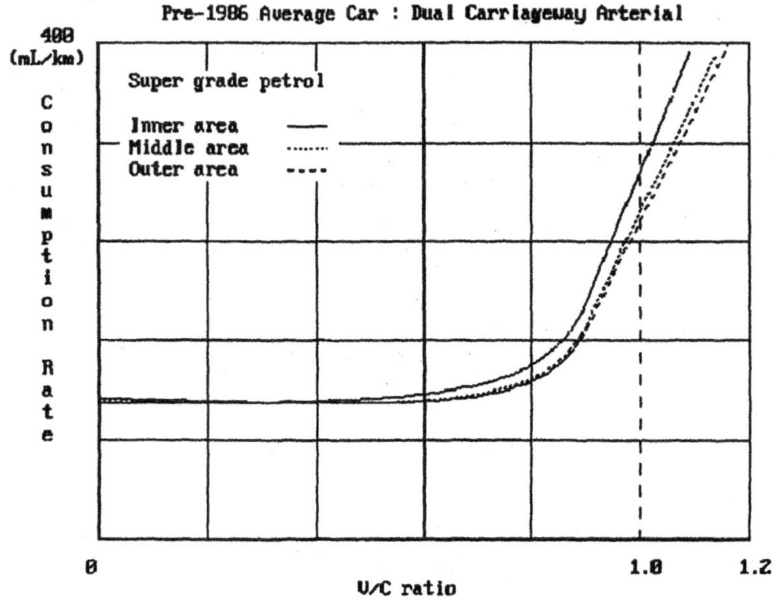

Fig. 2. Derived link function for fuel consumption (pre-1986 average car)

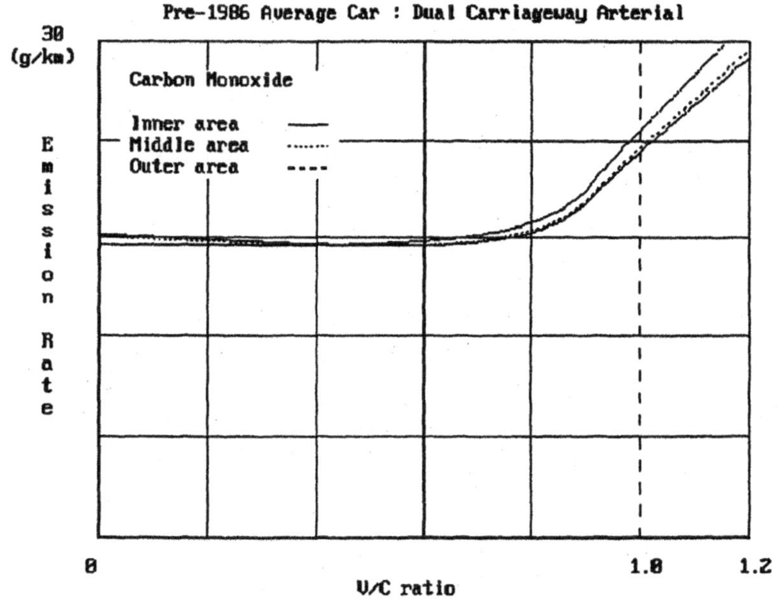

Fig. 3. Derived link function for emissions of carbon monoxide (pre-1986 average car)

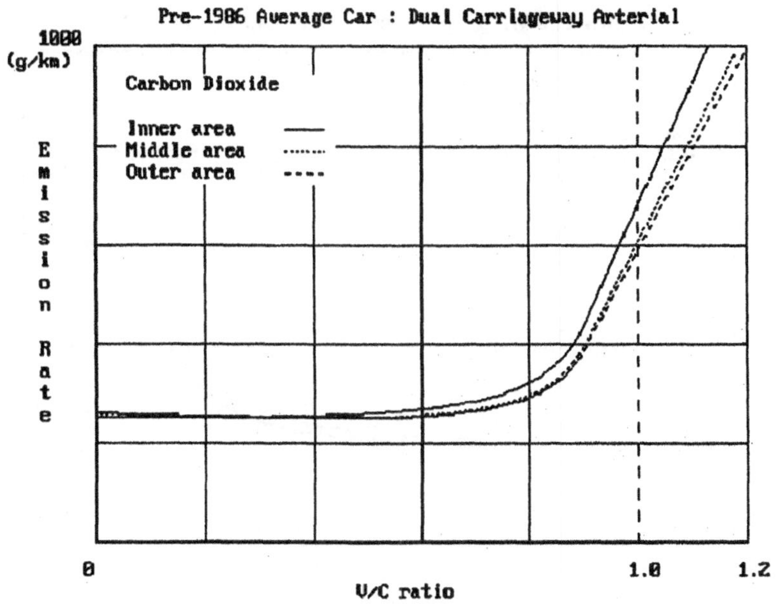

Fig. 4. Derived link function for emissions of carbon dioxide (pre-1986 average car)

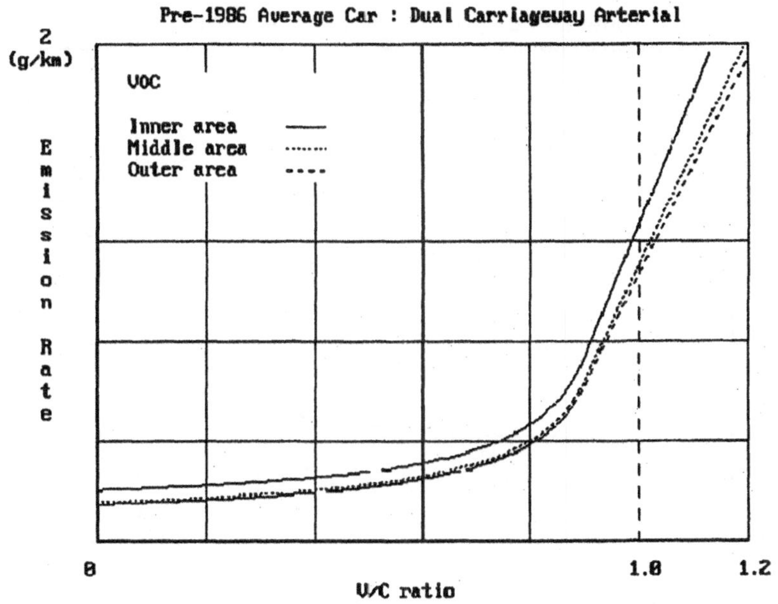

Fig. 5. Derived link function for emissions of VOC (pre-1986 average car)

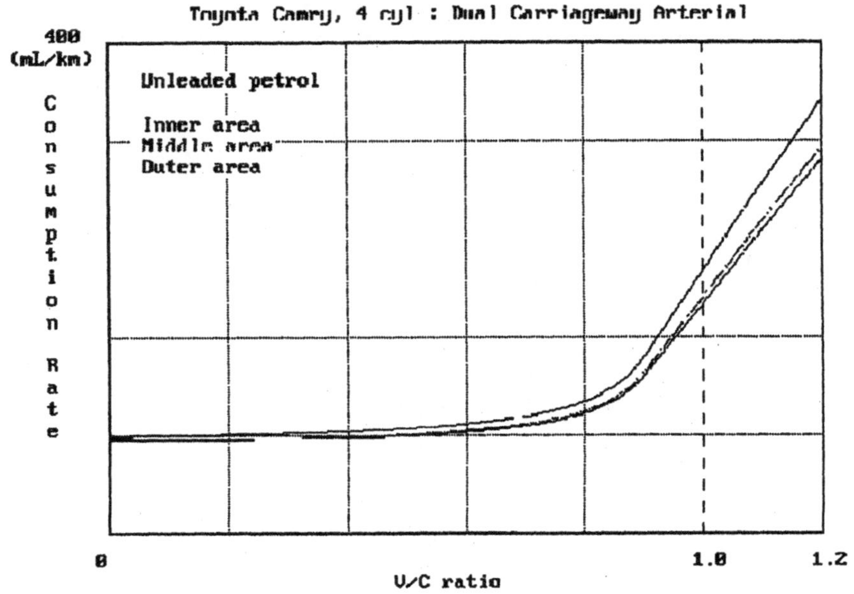

Fig. 6. Derived link function for fuel consumption by ULP Car (1991 Toyota Camry)

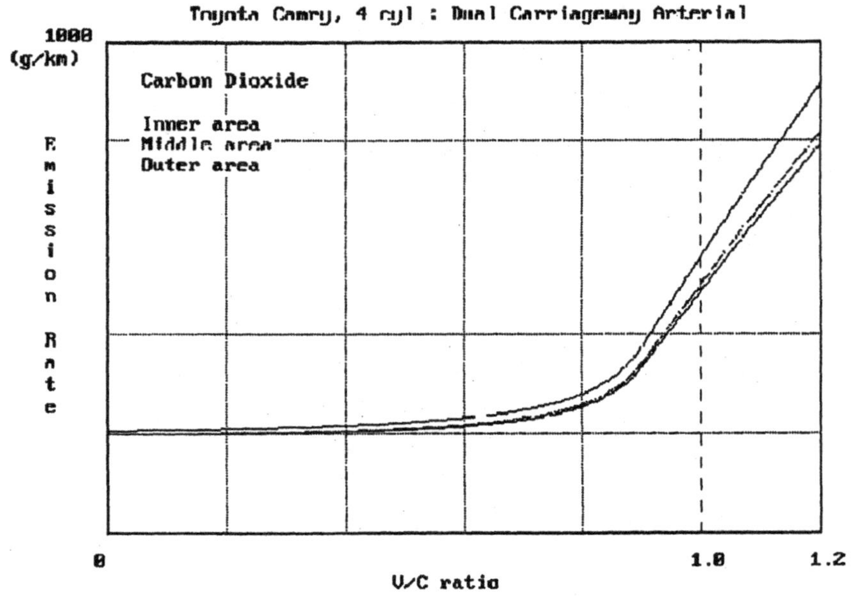

Fig. 7. Derived link function for carbon dioxide emissions by ULP Car (1991 Toyota Camry)

6. Applications in Transport Network Planning

Two conceptual models of traffic assignment that are responsive to volume/capacity ratios were given by Wardrop (1952). Wardrop's principles are:

(1) drivers could select routes that minimises their own individual travel times, on the basis that all other drivers are making their own individual decisions and that these decisions are made independently. Under the resulting flow patterns, all of the alternative routes used for a specified journey will have equal travel times, and these travel times will be less than those on any other possible route for that trip. The resulting flow pattern is stable, for no one driver can change route and gain any advantage by doing so. The resulting model is the 'individual travel time minimisation model'. This model provides a realistic simulation of present-day driver route choice behaviour. Alternatively,

(2) drivers could select routes so that the overall amount of travel (vehicle-hours of travel) in the network is minimised. This principle requires complete cooperation and sharing of information between drivers. It leads to a 'system travel time minimisation model' flow pattern with the minimum amount of total travel for the supplied (fixed) travel demand, but this flow pattern is unstable as individual drivers may find alternative routes that offer them quicker individual travel times.

These conceptual models may be stated in mathematical form as follows. The equilibrium assignment model for fixed (inelastic) travel demand is an expression of Wardrop's first principle (individual travel time minimisation). This model formulation provides a useful macroscopic simulation of travel on a metropolitan network. It may be written as the following non-linear optimisation problem, for which a convergent solution may be found (as indicated, for example, in Taylor (1984)):

$$Z = \min\left\{\sum_{e}\int_{0}^{q(E)} c_E(x)dx\right\} \tag{16}$$

subject to the continuity of flow constraints

$$T_{ij} = \sum_{r} X_{rij} \qquad \text{for all i, j} \tag{17}$$

$$q(e) = \sum_{ijr} \delta_{eijr} X_{rij} \qquad \text{for all i, j} \tag{18}$$

and the non-negativity constraints

$$q(e) \geq 0 \qquad \text{for all e} \tag{19}$$

and

$$X_{rij} \geq 0 \qquad \text{for all r, i, j} \tag{20}$$

where δ_{eijr} = 1 *if and only if* e is in path r from i to j,
= 0 otherwise

X_{rij} is the number of trips using path r between i and j, and the function $c_e(q)$ is the congestion function for link e.

The equivalent system-wide travel time minimisation problem may be written as a similar optimisation problem, with objective function

$$Z = \min\left\{\sum_e q(e)c_e(q(e))\right\} \tag{21}$$

with the same conservation of flow constraints.

Given the flow pattern corresponding to either of these traffic assignment models, the total fuel consumption and emissions generated can be estimated using the link E/C relationships described in the previous section.

The Wardrop principles may be treated as meeting different economic objectives for network travel, if travel time is taken as one possible alternative measure of travel cost. Jewell (1967) expanded this argument by suggesting a third principle for traffic assignment: that the ultimate pattern of flow in a network will satisfy some explicit economic objective, for instance minimum generalised travel cost or minimum fuel consumption (both either individual or system-wide). Thus direct substitution of the link E/C functions for the congestion function $c_e(q)$ would yield assignment models that could generate traffic patterns corresponding to minimum fuel use or minimum pollution generation for both user-optimal route choice (objective function equation (16)) and system-optimal route choice (objective function equation (21)). A generalised cost function including travel time, fuel consumption, pollutant emissions, money costs etc could also be proposed and solved for these objectives.

Substitution of a link fuel consumption or pollutant emission function in the objective functions described by equations (16) and (21) will yield user equilibrium or system minimisation network flow patterns corresponding to minimum fuel usage or pollution emission objectives. Further, the network flow pattern stemming from the imposition of road pricing (or a 'congestion tax') can be found by considering the marginal cost of travel on links. If the marginal travel cost on link e is c_{me} where

$$c_{me} = \frac{\partial C_{Te}}{\partial q_e} \tag{22}$$

and C_{Te} is the total travel cost on the link, given by

$$C_{Te} = c_e q(e) \tag{23}$$

then, for the Davidson function defined by equation (1), the marginal cost of travel is given by:

$$\begin{aligned} c_m &= c_0 \left\{ 1 + J \frac{\mu(2-\mu)}{(1-\mu)^2} \right\} && \mu < \rho \\ &= c_0 \left\{ 1 + J \frac{\rho}{1-\rho} + J \frac{(2\mu - \rho)}{(1-\rho)^2} \right\} && \mu \geq \rho \end{aligned} \tag{24}$$

An equilibrium assignment using the marginal link cost function of equation (24) for all network links will yield the system-wide travel time minimisation flow pattern that would arise from using the objective function of equation (21). An interesting case that deserves attention is that where 'congestion pricing' is only imposed on a sub-set of links within the network (e.g. as would occur when congestion pricing was used in a central business district but not in the surrounding network).

6.1 Consideration of Elastic Travel Demand

In the case that travel demand (as represented by the trip matrix T_{ij}) is regarded as elastic, i.e. the trip distribution (destination choice) will vary depending on the congestion levels in the network, then an alternative model formulation is in order. The combined distribution-assignment model proposed by Evans (1976) and explored by Boyce, LeBlanc and Chon (1988) provides an equivalent formulation to the equilibrium assignment model, and may be solved by a similar mathematical programming approach. On the assumption that the trip distribution can be explained by the entropy-maximising model

$$T_{ij} = r_i P_i s_j A_j \exp(-\lambda c_{ij}) \tag{25}$$

where P_i is the trip production of i, A_j is the trip attraction of j, c_{ij} is the travel cost between i and j and λ, r_i and s_j are calibration constants, then the elastic-demand user-equilibrium traffic assignment model is

$$Z = \min\left\{\sum_{e} \int_{0}^{q(e)} c_e(x)dx + \frac{1}{\lambda}\sum_{j} \sum_{j} T_{ij}\left[\ln(T_{ij}) - 1\right]\right\} \tag{26}$$

subject to the constraint equations (17)-(20). This model may be treated in identical fashion to the equilibrium assignment model for fixed travel demand. It has considerable promise as a transport network model for use in analysis of TDM programs. With the addition of fuel and emissions relationships of the form discussed in this paper, it offers a useful means to examine the ways in which variations in vehicle fleet composition, travel demand patterns and congestion levels will influence energy consumption and pollution emissions from urban transport systems.

An application of the enviornmental and energy impacts model described above to the combined trip distribution, modal split and trip assignment model described by Boyce, Lupa and Zhang (1994) is also feasible and will be the subject of future research.

7. Conclusions

Fixed demand or elastic-demand transport network models of the type defined described in this paper, using selected definitions of travel costs and alternative congestion functions, can be usefully employed within the general environmental impact assessment framework of Figure 1, given suitable functions that relate emissions and energy consumption of link traffic streams to link volumes. The capability then exists to use a set of network flow models that can compare the transport, energy and environmental impacts of alternative transport and environmental policies and vehicle technologies. Current TSC research is developing a comprehensive family of link-based emissions and energy functions for a range of vehicle types (including post-1986 passenger cars, trucks and buses). Considerations of alternative definitions of travel costs to include a set of congestion, economic, energy and environmental variables are also being made. The outcome of this research will be a strategic transport network model that may be applied to questions involving considerations of TDM. As such, it may point the way to some new policy-sensitive and relevant transport models.

Acknowledgements

The research and model development described in this paper form part of a research project on 'Environmental impacts of road traffic' supported by the Australian Research Council, the South Australian State Energy Research Advisory Council and the Australian Road Research Board.

References

Akcelik, R (1990) Cost and emission data *SIDRA Update Note May 1990*, Australian Road Research Board, Melbourne (mimeo).

Akcelik, R (1991). Travel time functions for transport planning purposes: Davidson's function, its time-dependent form and an alternative travel time function. *Australian Road Research 21* (3), 49-59.

Anderson, M (1985) POLDIF: dispersion of air and noise pollution from urban road networks. *DBR Computer Manual CM85/2*, CSIRO Division of Building Research, Melbourne.

Berka, S and Boyce, D E (1994). Implementation and solution of a large asymmetric network equilibrium model. Paper presented at *Tristan II, the Triennial Symposium on Transportation Analysis*, Capri, Italy.

Biggs, D C (1988). ARFCOM - models for estimating light to heavy vehicle fuel consumption *Research Report ARR152*. Australian Road Research Board, Melbourne.

Biggs, D C and Akcelik, R (1986). Estimation of car fuel consumption in urban traffic. *Proceedings of 13th Conference of the Australian Road Research Board 13* (7), 124-132.

Boyce, D E, LeBlanc, L J and Chon, K S (1988). Network equilibrium models of urban location and travel choices: retrospective survey. *Journal of Regional Science 28*, 159-183.

Boyce, D E, Lupa, M and Zhang, Y (1994). Introducing 'feedback' into the four-step travel forecasting procedure vs the equilibrium solution of a combined model. *Transportation Research Record* (in press).

Bowyer, D P, Akcelik, R and Biggs, D C (1985). Guide to fuel consumption analyses for urban traffic management. *Special Report No 32*, Australian Road Research Board, Melbourne.

Branston, D M (1976). Link capacity functions: a review. *Transportation Research 10*, 233-236.

Brown, A L and Patterson, N S (1990). Noise assessment when it matters: environmental evaluation integrated with road network planning. *Proceedings of 15th Conference of the Australian Road Research Board 15* (7), 61-78.

Evans, S P (1976). Derivation and analysis of some models for combining trip distribution and assignment. *Transportation Research 10* (1), 37-57.

Horowitz, A J (1989). Tests of an ad hoc algorithm of elastic-demand equilibrium traffic assignment. *Transportation Research B 23B* (4), 309-313.

Hothershall, D C and Salter, R J (1977). *Transport and the Environment.* Crosby Lockwood Staples, London.

Jewell, W S (1967). Models for traffic assignment. *Transportation Research 1* (1), 31-46.

Jost, P, Ullrich, J and Waldeyer, H (1987). Étude a grande échelle des émissions de polluants sur autoroute: Abagasgrossversuch. *RTS, Revue de l'INRETS 14*, 5-12.

Maccarrone, S (1989). Air pollutant levels near roads. *Australian Road Research 19* (3), 234-242.

OECD (1980). *Conference on Noise Abatement Policies.* Organisation for Economic Co-operation and Development, Paris.

Rose, G, Taylor, M A P and Tisato, P (1989). Estimating travel time functions for urban roads: options and issues. *Transportation Planning and Technology 14* (1), 63-82.

Taylor, M A P (1984). A note on using Davidson's function in equilibrium assignment. *Transportation Research B 18B* (3), 181-199.

Taylor, M A P and Anderson, M (1982). Modelling pollution and energy use in urban road networks. *Proceedings of 11th Conference of the Australian Road Research Board 11* (6), 1-17.

Taylor, M A P and Anderson, M (1984). Energy and emissions in a metropolitan road transport system. *Proceedings of 12th Conference of the Australian Road Research Board 12* (6), 63-78.

Taylor, M A P and Anderson, M (1986). Assessing environmental quality. *In* Newton, P W and Taylor, M A P (eds). *Microcomputers for Local Government Planning and Management.* Hargreen Publishing Co, Melbourne, 192-200.

Taylor, M A P and Anderson, M (1988). Modelling the environmental impacts of urban road traffic with MULATM-POLDIF: a PC-based system with interactive graphics. *Environment and Planning B 15*, 192-200.

Taylor, M A P and Young, W (1988). *Traffic Analysis: New Technology and New Solutions.* Hargreen Publishing Co, Melbourne.

Wardrop, J G (1952). Some theoretical aspects of road traffic research. *Proceedings of the Institution of Civil Engineers, Part II 1* (2), 325-378.

Wigan, M R (1976). The estimation of environmental impacts for transport policy assessment. *Environment and Planning B 8*, 125-147.

Woolley, J E and Young, T M (1994). Environment: the third dimension of the land-use transport interaction. *Papers of the Australasian Transport Research Forum 19*, 223-239.

Young, T M (1992). Environmental and energy impacts of road traffic: Quarterly report no. 2. *UMATS Working Paper 92/1*, Unit for the Modelling and Analysis of Transport Systems, University of South Australia.

CHAPTER 18

USING COMBINED NETWORK EQUILIBRIUM MODELS FOR ENVIRONMENTAL ASSESSMENTS OF LAND USE-TRANSPORTATION SCENARIOS

Lars Lundqvist

1. Introduction

During recent years simple combined models of modal split and route choice have been applied to both small scale networks and large scale networks in the Stockholm region. This occurred 10-15 years after the development of theoretically sound algorithms for route choice equilibria (traffic equilibria) about 1975 and the formulation of combined models (market equilibria) in the late 70s and early 80s. A natural next step is to take into account also the feed-back effects between trip distribution and combined modal split/route choice equilibria. Even further, the interdependencies between land-use patterns and transportation market equilibria need to be analyzed.

In this paper we will report on the development of modern combined network equilibrium models and their applications to environmental assessments in the Stockholm region. Possibilities of linking advanced models to user friendly software will be discussed as well as extensions of the models from work trips to other trip types drawing upon recently estimated travel demand systems based on the Stockholm travel survey 1986/87.

Most of the model development mentioned above has been related to

Y. Hayashi and J. Roy (eds.), Transport, Land-Use and the Environment, 359–381.

evaluations of the 1991 Regional Plan for the Stockholm region, to projections based on the "Dennis" package of transportation investments and financing in the Stockholm region and to forecasts of travel patterns in future scenarios of land-use and transportation of the Mälar Region. We will briefly introduce each of these studies.

Since the middle of the 80s four main goals of regional planning in Stockholm have been stated:

- create good environmental conditions;
- encourage the development of industry and commerce;
- improve the balance between the different parts of the region; and
- utilize the urban areas efficiently.

Although these goals emphasize both efficiency, distribution and environmental concerns, coordinating growth with a good environment can be viewed as the main issue raised in the recent planning documents.

Handling emissions from road traffic is given high priority. In the English summary of the 1990 Regional Plan Outline, a strategy for treating road traffic emissions was outlined, including land reservations for railway and strategic road investments, public transit oriented location of housing and work-places and expansion of tele communications. The strategy also emphasized encouragment of high speed rail instead of air traffic, relocation of ports to external locations, introduction of inner city road tolls and modification of taxation rules in order to reduce car use. The development of low emission engines should be stimulated and a declaration of the Stockholm region as an environmental protection area was discussed. A very similar strategy for infrastructure development was included in the final Regional Plan Proposal 1991.

The 1990 Regional Plan Outline indicated an aim to improve the subregional balance between housing and work-places mainly through directing the future housing investments to areas with oversupply of work-places, i.e. mostly central areas. The location pattern of work-places can be characterized as a moderate decentralization. However, the overall evaluation of the proposed regional structure revealed an increase in average density: the slow decentralization could not compensate the growth of activities. The 1991 Proposal is somewhat more decentralized in terms of housing and somewhat less decentralized in terms of work-places than the 1990 Outline, the main components of which are illustrated in Figure 1. It should be noted that the central part of the region contains 46 per cent of the population and 53 per cent of the total emloyment in the late 80s. According to the 1990 Plan Outline 40-45 per cent of the new housing and 25-30 per cent of the new employment should be allocated to the central part of the region. In the 1991 Plan Proposal these shares are about 33 per cent (housing: lower bound; employment: upper bound).

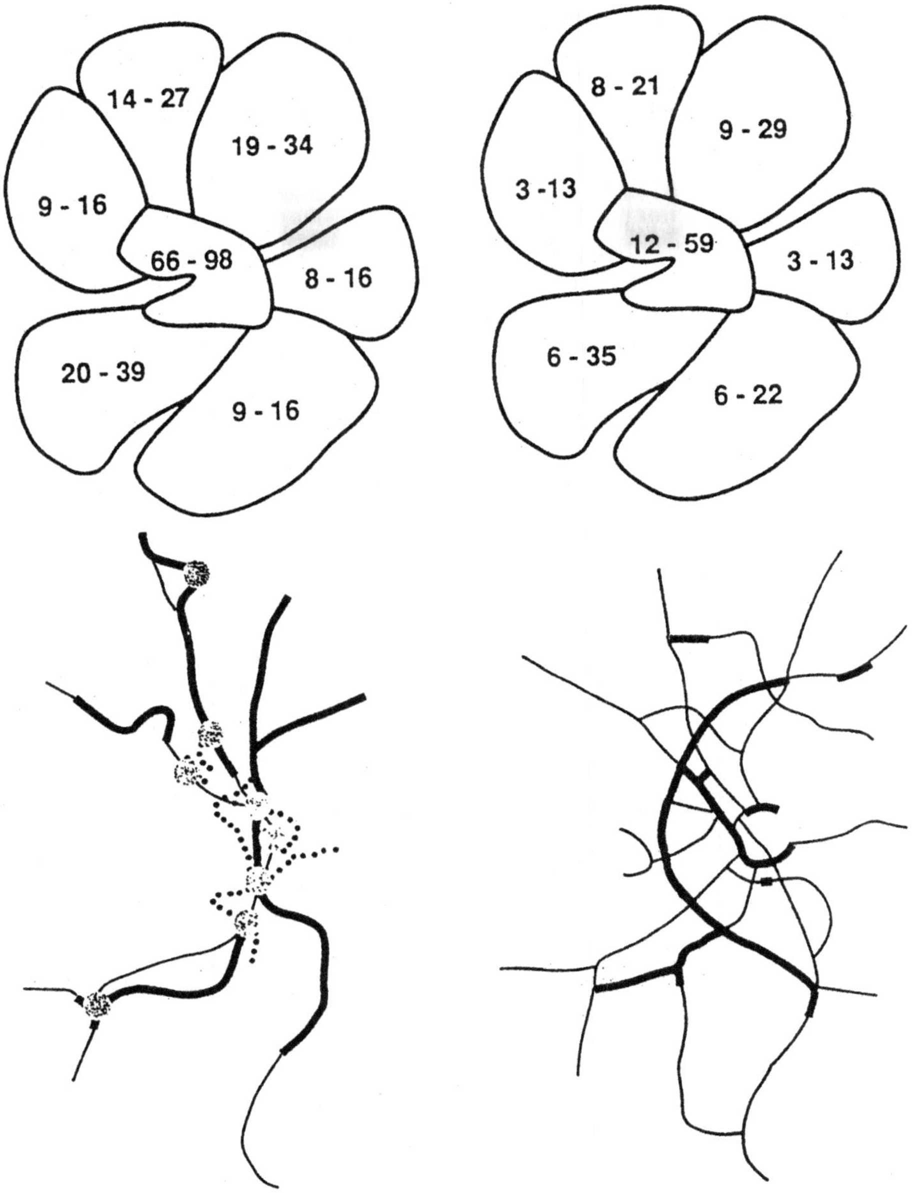

Fig. 1. Main components of the 1990 Regional Plan Outline. Top left: Housing production 1990-2020 at moderate and rapid economic growth (thousands). Top right: Increase of employment 1990-2020 at moderate and rapid economic growth (thousands). Lower left: Main rail investments 1990-2020 and main transfer nodes. Lower right: Main road investments 1990-2020.
Source: Stockholm County Council (1989).

During 1990-1992 a package agreement on transportation investments 1992-2006 and their financing was negotiated in the Stockholm region: the "Dennis" package, named after the leader of the negotiations Bengt Dennis, former head of the Bank of Sweden. The result of a preliminary agreement was included in the 1991 Regional Plan Proposal. The final agreement contains road investments for 18.2 billion Sw. cr. (mainly completion of an inner city ring road and a peripheral semicircular route to the west of Stockholm) and transit investments for 15.8 billion Sw. cr. (mainly increased rail capacity in central Stockholm, upgrading and extensions of the underground system and commuter train systems and a new light rail system connecting the inner suburbs), see Figure 2. 1.9 billion Sw. cr. are reserved for other road related investments (including a core grid of bus services in the CBD area and park-and-ride facilities). The road investments shall be financed by vehicle fees from an inner

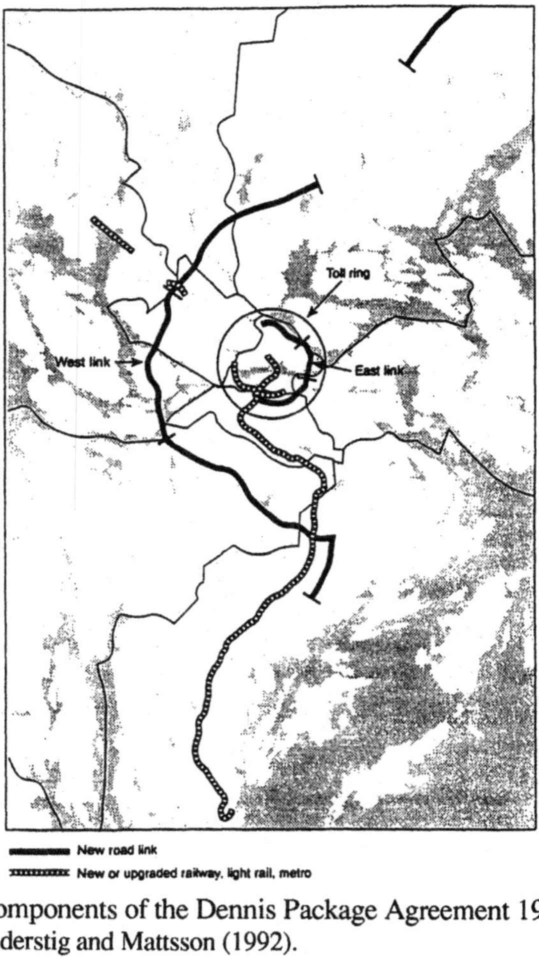

Fig. 2. Main Components of the Dennis Package Agreement 1992. From Anderstig and Mattsson (1992).

city toll ring located just outside the inner city ring road (15 Sw. cr. per inbound passage) and from the western link (5 Sw. cr. in each direction). The remaining funds are supplied by national government sources and the Stockholm County Council. The explicitly stated goals for the negotiation process were:

- to improve the environmental situation;
- to increase accessibility; and
- to promote the development of the Stockholm region.

The extended Stockholm region, the Mälar Region with its recently established Council, is made up by four counties and 35 municipalities. Its purpose is to promote a more competitive region in the European context through stonger integration and more efficient use of resources. The Council has commissioned future studies on the development of settlement systems, transportation patterns, environmental conditions, industries and services and educational systems. In addition to the Dennis package, nationally decided rail investments in the Mälar Region amounting to 12 billion Sw. cr. are scheduled for the next ten years. They include an upgraded railway ring around Lake Mälaren and a railway to Arlanda airport, see Figure 3. Also the road infrastructure of the Mälar Region will be further improved by nationally decided investments. As already mentioned, the Mälar Region Council has adopted studies of the settlement system, the infrastructure networks and the sustainability of environmental developments. Network equilibrium models of residential and employment location have been developed as part of these efforts.

Fig. 3. Railways in the Mälar Region with Commuter Train Operations (and some connecting railways).
Source: Folder from Tåg i Mälardalen AB.

2. Combined Network Equilibrium Models

2.1 The First Attempts: EMME/2 and CMA

In 1990 the first attempts to introduce combined models of mode and route choices were made in Stockholm. A model developed as part of a suite of research tools by professor David Boyce (see e.g. Boyce and Lundqvist (1987)) was implemented on personal computer and tested on a very aggregated Stockholm road network. This model (CMA - Combined Modal split and Assignment) was evaluated by comparing its results with the outcomes of the "variable demand" module of the transportation planning system EMME/2. The new variable demand module of EMME, permitting combined treatment of modal split and route choice was released in 1989 and installed on a mini-computer in the Office of Regional Planning and Urban Transportation.

For the purpose of testing and research the 1980 road network of the Stockholm county was represented by an aggregate network with 413 nodes and 966 links (with data on length, number of lanes and type of volume/delay function). Trips were "produced" in and "attracted" to 46 origin/destination zones. The transit mode was represented by an exogenous matrix of interzonal generalized costs and the total peak hour trip demand was exogenously specified in an interzonal OD-matrix. For this type of problems CMA and the variable demand module of EMME can be regarded as two implementations of basically the same algorithm: the partial linearization technique of Evans (1976) which generalizes the Frank and Wolfe algorithm for equilibrium assignment of Leblanc *et al.* (1975).

As expected, CMA and EMME/2 applied to the aggregate Stockholm network gave very similar results, see Svalgård (1990). The average auto travel times differed by 0.1 min comparing the CMA result after 90 iterations (with a step size in the line search less than 10^{-5}) with the EMME result after 120 iterations (with the step size still larger than 10^{-5}). The average difference in the number of interzonal trips by mode is 0.16.

2.2 A Prototype Formulation of a Combined Mode and Route Choice Model

As a basis for the following discussion of combined network equilibrium models for environmental assessments we will formulate a simple prototype model of combined mode and route choice, following Boyce and Lundqvist (1987). New features and ongoing developments will be related to this prototype formulation in the remaining part of section 2. Environmental assessments based on network equilibrium models will be discussed in sections 4 and 5.

This prototype formulation determines modal shares for transit and auto between each pair of zones in the morning peak hour and, simultaneously, the equilibrium route choice of car trips in a network with congested road links. It assumes an exogenously given transit network with prespecified travel times and travel costs between each pair of zones. The total number of trips between each pair of zones is also exogenously given.

Let us introduce the following notation:

g_{ij}^{m} = number of peak hour trips from zone i to zone j by mode m;

h_{ijk}^{au} = number of cars going from zone i to zone j by route k in the peak hour;

R = car occupancy factor;

$\gamma_1, \gamma_2, \gamma_3$ = generalised cost coefficients for travel time, travel cost and parking cost;

$s_a(v_a)$ = travel time on link a as a function of the link flow v_a;

K = car operation cost per unit distance;

d_a = length of link a;

p_j = parking cost in zone j;

t_{ij} = total public transit travel time from zone i to zone j (including waiting and transfers);

f_{ij} = public transit travel cost from zone i to zone j;

μ = sensitivity to generalised cost in the choice of travel mode;

G_{ij} = total number of trips in the peak hour between zone i and zone j;

δ_{ijka} = indicator equal to 1 if link a is part of route k between zone i and zone j, 0 otherwise;

$\bar{c}_{ij}^{m}$ = generalised travel cost per person for travel by mode m from zone i to zone j;

c_{ij}^{au} = equilibrium generalised travel time and operation cost by car from zone i to zone j (Lagrange multipliers, see below);

$\hat{c}_{ijk}^{au}$ = generalised travel time and operation cost by car along route k between zone i and zone j; and

P_{ij} = the set of routes connecting zone i with zone j.

The equivalent optimization problem for the combined mode and route choice network equilibrium can now be formulated as a convex, nonlinear programming model:

$$\min_{g_{ij}^{m},\, h_{ijk}^{au}} R\gamma_1 \sum_a \left[\int_0^{v_a} s_a(v).dv\right] + K\gamma_2 \sum_a d_a.v_a + \frac{\gamma_3}{R} \sum_{ij} p_j.g_{ij}^{au}$$
$$+\gamma_1 \sum_{ij} g_{ij}^{tr} \cdot t_{ij} + \gamma_2 \sum_{ij} g_{ij}^{tr} \cdot f_{ij} + \frac{1}{\mu} \cdot \sum_{ijm} g_{ij}^{m} \cdot \log\ g_{ij}^{m} \tag{1}$$

subject to the following constraints (with the Lagrange multipliers in parentheses):

$$\sum_m g_{ij}^{m} = G_{ij} \qquad (\beta_{ij}) \tag{2}$$

$$\sum_{k \in P_{ij}} h_{ijk}^{au} = \frac{1}{R}\ g_{ij}^{au} \qquad (c_{ij}^{au}) \tag{3}$$

$$v_a = \sum_{ijk} \delta_{ijka} \cdot h_{ijk}^{au} \tag{4}$$

$$h_{ijk}^{au} \geq 0\ ,\ g_{ij}^{m} \geq 0 \qquad (\theta_{ijk}^{au}\ ,\ \Phi_{ij}^{m}) \tag{5}$$

By introducing notation for generalized costs ((6)-(7)) and route costs (8), the formulation of the optimum conditions below is simplified:

$$\bar{c}_{ij}^{tr} := \gamma_1 \cdot t_{ij} + \gamma_2 \cdot f_{ij} \tag{6}$$

$$\bar{c}_{ij}^{au} := \frac{1}{R}(c_{ij}^{au} + \gamma_3 p_j) \tag{7}$$

$$\hat{c}_{ijk}^{au} := R\gamma_1 \sum_a s_a(v_a)\delta_{ijk}^{a} + K\gamma_2 \sum_a d_a \delta_{ijk}^{a} \quad \forall k \in P_{ij} \tag{8}$$

The optimum conditions can after some calculations be stated as follows ((9)-(11)):

$$\hat{c}_{ijk}^{au} = c_{ij}^{au} \quad \forall k \in P_{ij}\ ,\ h_{ijk}^{au} > 0 \tag{9}$$

$$\hat{c}^{au}_{ijk} \geq c^{au}_{ij} \quad \forall k \in P_{ij} \ , \ h^{au}_{ijk} = 0 \tag{10}$$

$$g^{m}_{ij} = G_{ij} \frac{\exp(-\mu \bar{c}^{m}_{ij})}{\sum_{m} \exp(-\mu \bar{c}^{m}_{ij})} \tag{11}$$

These conditions merely require that the Wardrop user equilibrium is obtained (for any origin and destination, the costs on all used routes are the same (9), and the costs on unused routes are not lower (10)) and that the modal split follows a logit model based on generalized per capita travel costs (including parking costs), (11). If another constraint is introduced in addition to (2)-(5), requiring the total number of transit trips to equal the observed volume, the generalized transit cost will contain a constant term and hence reproduce the conventional form of a bimodal logit model.

2.3 EMME/2

Beginning in 1990 the use of the EMME "variable demand" module for combined mode and route choice projections in large scale networks started. Within the Office of Regional Planning and Urban Transportation a Stockholm road network with 850 origin/destination zones, about 2200 nodes and about 5600 road links has been used for traffic projections with regard to transportation and land use policies. When evaluating proposals for the Regional Plan 1990, manual iterations between a modal split model and EMME auto assignments were carried out (transit assignments in EMME were used to produce an interzonal transit generalized cost matrix).

In late 1990 the modal split model was implemented in the EMME "variable demand" module and the combined model was used for projecting the effects of preliminary alternatives in the Dennis negotiations on transportation investments and user charges. One year later also the Stockholm Streets and Traffic Administation started to use large scale "variable demand" auto assignment when evaluating various toll schemes to be included in the package agreement. The use of combined mode and route choice models has made traffic forecasting easier and more consistent by eliminating the need for manual iterations. Such efforts are still required, however, when the impact of transportation changes on the origin/destination travel pattern or land-use transportation interactions need to be taken into account. We will return to ongoing developments of integrated approaches below (section 2.5).

As touched upon, the first large scale combined model used in the Stockholm planning context was a bimodal mode and route choice model similar to the prototype formulation stated in the previous section. The next improvement was to also include a third mode: walk/cycle. Since about one year the EMME/2 package permits multiclass equilibrium assignment. In the final assessments of the Dennis package computed this year, the multiclass assignment option was used for simultaneous analysis of toll sensitive private car traffic and toll insensitive distribution and business car trips. Before these two classes had been assigned sequentially.

One further development related to EMME/2 should be mentioned. A heuristic OD matrix adjustment module is available, see Spiess (1990). It has been used to adjust the OD matrix in order to reproduce traffic counts (within on average 5 per cent) on a subset of the network containing about 500 links. A similar concern for an exact fit between the observed modal split and the modal split according to the model can be met with incremental logit demand functions easily implemented in the EMME/2 variable demand module (see EMME/2 News, April 1993).

2.4 CMAWIN

The CMA research code for the combined mode and route choice problem was used in education after the first evaluation against EMME/2 reported above. Interactive changes of key model parameters were made possible together with presentation of some key results on the screen.

A Windows-based PC software (CMAWIN) has recently been developed, aiming for a more elaborate interactive program (Lundqvist, 1993). The decision support system provides facilities for checking and changing the input data and for displaying the results. It is built around a road map of the urban area with "view", "add" and "delete" functions for zones, nodes and links in addition to the zoom function. Input data can be changed directly in various windows on the screen and by adding and deleting nodes and links. The decision support system is programmed in C++ by Martin Lundqvist. It calls the FORTRAN research code for solving CMA and reports the results in terms of a summary table and a wide range of histograms for travel times and trip volumes by mode. Additional options for displaying results are being added and full comparisons of results from various scenarios will also be possible in the near future. The inclusion of modules for impact assessments (e.g. economic evaluation and environmental effects) form high priority tasks on the agenda for future developments.

The CMAWIN model deviates from the prototype formulation (see section 2.2) in three respects: it contains a transit modal constant, γ_4, (corresponding to a constraint on the total number of transit trips) , it contains a toll fee ($K \cdot d_a := K \cdot d_a + toll_a$) and the transit fare structure is simplified to a flat fare ($f_{ij}=F$). The latter simplification is of course easily lifted. The model has been calibrated and run on the 1980 data base (see section 2.1). For the purposes of demonstrations and course exercises the model runs have mostly been based on travel times only ($\gamma_1=1$; $\gamma_2=\gamma_3=0$).

2.5 Ongoing Developments: CODMA and SYNERGETIC

Two parallel ongoing projects aim at a further integration of destination, mode and route choices for a number of trip purposes in order to arrive at a more general transportation market equilibrium. In this kind of modelling approach, various trip purposes compete for network capacity and the origin-destination pattern of trips is endogenously adjusted. This means that the effects of new transport links can be analysed, not only in terms of mode and route choices, but also in terms of changes in trip length. It still remains to include relocation and trip generation effects to cover the full spectrum of potential impacts.

CODMA

The first of these projects is part of a thesis work by Torgil Abrahamsson. The project started in 1990 by developing a *simultaneous* trip distribution, mode and route choice model within the combined network equilibrium tradition, see Abrahamsson and Lundqvist (1990). This model (CODMA) can be obtained as a generalization of the prototype model in section 2.2 by substituting the trip matrix constraint (2) for the typical origin and destination constraints of a doubly constrained gravity model (12-13):

$$\sum_{jm} g_{ij}^{m} = O_i \tag{12}$$

$$\sum_{im} g_{ij}^{m} = D_j \tag{13}$$

The resulting model has the well known form:

$g_{ij}^{m} = A_i O_i B_j D_j \exp(-\mu \bar{c}_{ij}^{m})$, where A_i and B_j are "balancing factors" related to the constraints (12)-(13) respectively.

The Wardrop user equilibrium conditions and the logit type modal split function are also reproduced by the optimum conditions. The model is solved by a slight generalization of the partial linearization algorithm (see section 2.1) for combined mode and route choice problems. Improvements of the basic algorithm (bi-proportional balancing; direction finding) have been studied and implemented, see Abrahamsson (1991). Abrahamsson and Lundqvist (1990) suggested a maximum likelihood (ML) approach for the estimation of model parameters based on the endogenously determined travel costs. With only two parameters (μ and γ_4 (transit modal constant); we assume that γ_1=1 and γ_2=γ_3=0), the estimation can easily be performed through trial-and-error runs with CODMA until the appropriate ML equations are satisfied.

More recently *nested* trip distribution and modal split models have been developed within a similar combined network equilibrium framework, Abrahamsson and Lundqvist (1993). Using exactly the same constraints as the simultaneous model (e.g. (3)-(5),(12)-(13)) and splitting the entropy term of the objective (the last term of (1)) into two terms with different weighting parameters (cost sensitivities) reflecting origin-destination choice and conditional mode choice, the traditional nested gravity-logit model can be derived. Here, the trip distribution generalized cost is a log-sum composition of modal genera-lized costs (mode choice is conditional on destination choice). With a different way of splitting of the entropy term into two terms reflecting mode choice and conditional destination choice, the reverse nested logit-gravity model can be derived.

The full information maximum likelihood equations for estimation of model parameters become more complicated in the case of nested demand models: the equations corresponding to lower level parameters obtain a weighted structure (instead of the model generated upper level trip pattern, a weighted sum of the observed and model generated trip pattern is used). Estimation of parameters with endogenous travel costs is easily performed in the case of the traditional nested model. In the case of the reverse nested model the weighted ML equation corresponding to the estimation of the balancing factor B_j is in conflict with the destination constraint (13). This conflict could be resolved by iteratively solving the estimation problem (with the weighted constraints corresponding to lower level parameters) and the combined network equilibrium problem (with the correct unweighted balancing constraints: (13)). A second possibility could be to view the balancing factors as endogenous variables (prices), which are not to be estimated. The remaining parameters may then be estimated using endogenous travel costs and the unweighted balancing constraint (13). Alternatively, assuming good correspondence between observed and model generated modal shares, the model can be run in the estimation phase with the weighted balancing constraints replacing (13). These three estimation procedures for the reverse model will be compared in future research.

The combined nested destination-mode choice and assignment models have been subject to initial estimations and test runs on future scenarios, see Abrahamsson and Lundqvist (1993), using data bases of the size mentioned in section 2.1. The models need to be further developed in terms of generalized cost structures: introduction of operation costs, fare structures and toll schemes. The doubly constrained nested combined model structures are best suited for commuter trips. They should be supplemented by singly constrained models for private trips and business trips in order to reflect the competition for capacities in the road network (compare SYNERGETIC).

SYNERGETIC

A model system for travel projections in the Mälar Region was recently developed as part of a study of the market for regional commuter trains (see Figure 3), Lundqvist and Mattsson (1992). Separate nested demand models were estimated for work trips as well as for directed business trips and other personal trips. Three modes were considered: car, public transport and walk/cycle. The destination choice model was specified as doubly constrained in the case of work trips and singly (production) constrained for the two other trip purposes. The parameters of the model system were estimated by employing the full information maximum likelihood technique. For business trips and other private trips only Stockholm data were available. The preferred model was of the reverse nested logit/gravity type for all trip purposes. However due to simplified implementation, the simultaneous model structure was selected for work trips. The demand models were estimated using exogenously computed travel times and travel costs from base year transit and auto assignments. In the case of future scenarios, the demand system was iterated with EMME/2 network equilibrium assignments. The estimations were performed on a zonal subdivision of 214 zones while the projections were performed on a disaggregate subdivision of about 1200 zones.

Based on the experiences from the Mälar Region study, a new travel forecasting system (SYNERGETIC) is currently being developed for the Stockholm region by the Institute for Regional Analysis (INREGIA), see Karlsson and Svalgård (1993). In its full version, its ambition is to cover both peak hour and low traffic periods and to integrate transportation and land-use analysis in a user-friendly environment. So far peak hour travel demand models have been estimated for four trip purposes: work trips, business trips, school trips and other personal trips. In the case of work trips and other personal trips separate models were estimated for individuals with or without car availability. A household based model for car availability of individuals is being developed. As in the Mälar Region study business trips and other personal trips are modeled by singly (production) constrained models. Work trips and school trips are modeled by doubly constrained models. Three modes are used for trips with car availability: car, transit and walk/cycle, while for trips without car availability the mode choice is restricted to transit and walk/cycle.

Full information maximum likelihood estimations for each of the six trip categories have indicated preference for the reverse nested logit/gravity model type in all cases except school trips. However, in order to ease implementation the work trip models were constrained to the simultaneous choice structure (compare the Mälar Region study). The estimations were carried out on a 263 zone data base from the 1986/87 Stockholm Travel Survey. The forecasting system is going to be executed on an 850 zonal subdivision of the Stockholm County. The travel demand model system is presently being programmed within the EMME/2 system. This could be done in the iterative way (used in the Mälar Region study) through iterations between the travel demand models and fixed demand assignments. It could also be done in a more integrated combined way by iterating between the doubly constrained demand models and a multiclass variable demand auto assignment, which includes the singly constrained demand models. The work trip models will later be incorporated in the IMREL land-use/travel demand model system, see Anderstig and Mattsson (1992), to be based on the peak hour level of service data from the travel forecasting system.

3. Environmental Goals and Assessments

The emphasis on environmental issues in recent transportation and land-use planning documents was illustrated in section 1. The present intensive debate about environ-mental consequences of the Dennis package provides a further indication of the importance attached to these issues. In this section the decisions on environmental goals will be summarized and the current environmental situation in Stockholm will be outlined. With a focus on traffic related environmental impacts, the methodology of impact assessments will be discussed.

The nationally decided goals on reductions of sulphur, nitrogen oxides and carbon dioxide emissions have been adopted by municipal and county level authorities in the Stockholm region together with guidelines for certain concentrations, total deposits and traffic noise:

- the 1980 level of emissions of *sulphur* should be reduced by 65 per cent to 1995 and by 80 per cent to the year 2000;
- the 1980 level of emissions of *nitrogen oxides* should be reduced by 30 per cent to 1995 with a further ambition to achieve a reduction by 50 per cent to the year 2000;
- the level of emissions of *carbon dioxide* should not increase beyond the 1988 level;
- the *concentrations* of carbon monoxide, nitrogen dioxide, sulphur dioxide, soot and particles should not exceed specified threshold values (averaged over specified time periods; in some cases expressed as 98 percentiles) in the year 2000;
- critical levels (depending on soil type) of *total deposits* of nitrogen and sulphur should not be exceeded; and
- specified guidelines for maximum *traffic noise* in various types of floor spaces and open spaces should be fulfilled.

From global environmental considerations long term reductions of carbon dioxide emissions by 50 per cent during the next 50 years are recommended by the National Environmental Protection Agency. Similarly, strong reductions of other green house gases and substances stimulating the formation of (ground) ozone are recommended.

Traffic emissions play an increasing role in the environmental debate. This is due to the fact that point source emissions from industry and from power generation and space heating plants within the region have been (or are being) reduced drastically. The sulphur dioxide emissions from combustion have been reduced by more than 50 per cent in the Stockholm region during the 80s. In this situation dispersed emissions from transportation and consumption activities become more important. Figure 4 shows the total emissions in tons 1990 and their distribution over sources. The transportation sector is the totally dominating source of nitrogen oxides emissions, with road traffic accounting for about 50 per cent. The total emission of nitrogen oxides has been marginally reduced in the late 80s and the reduction is expected to continue due to the growing share of catalytic converters in the stock of cars. About 50 per cent of the carbon dioxide emissions emanates from the transportation sector with road traffic contributing by about 25 per cent.

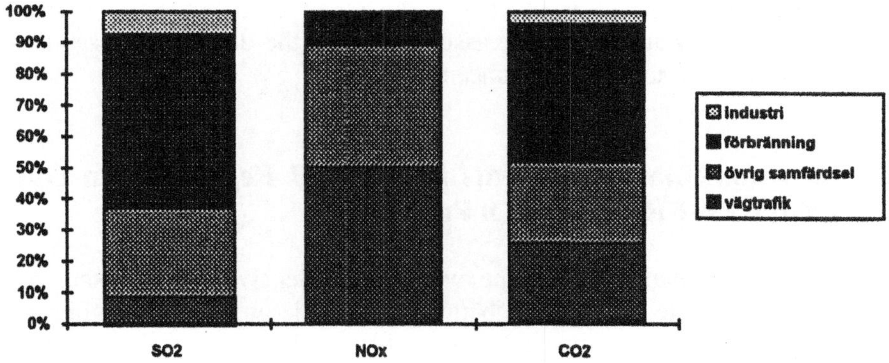

Fig. 4. Sectoral Shares Of Total Emissions (sulphur dioxide, nitrogen oxides and carbon dioxide) in the Stockholm Region 1990. Contributions (from bottom) from Road Traffic, other Transportation, Combustion and Industry.
Source: Stockholm County Board (1993).

Hence, transportation represents large shares of important pollutants produced within the region. Emissions from other regions and other nations constitute an increasing part of the total deposition in the Stockholm region: 80 per cent of the total national nitrogen deposits originates from international sources. 50 per cent of depositions in Stockholm City is estimated to originate from external sources. On the other hand only one third of Swedish emissions are deposited within the nation. The Stockholm region is estimated to export more than twice the amount of imported emissions. It should be concluded that transportation and road traffic are significant sectors in terms of their production of regionally controllable emissions.

In comparison to many other cities in continental, eastern and southern Europe, the environmental situation in Stockholm can probably be rated as satisfactory: the water permits swimming and fishing, the high share of transit users and the strict control of point sources of emissions have positive impacts on air quality. However, according to the national and regional ambitions stated above, further improvements are required to secure health standards and ecological sustainability.

Network equilibrium models have been used to forecast the environmental impacts of the transportation sector in various future land-use transportation scenarios. The levels of emissions and noise are strongly related to the vehicle miles of travel and vehicle types but are also depending on velocities, congestion etc. Combined models of mode and route choice implemented in EMME/2 have been adopted as the main analytical tool, while at the same time major efforts are spent on developing more integrated approaches (see the previous section). The emissions projected from vehicle flows can be transformed to roadside concentrations by applying some assumptions on the distribution of emissions and on the background level of concentrations.

4. Environmental Assessments in the 1990 Regional Plan Outline and the 1991 Regional Plan Proposal

The 1990 Regional Plan Outline was fairly extensive in the discussion of strategies to handle the conflicts between economic growth and environmental sustainability, as already touched upon in the introduction. To a large extent this discussion dealt with the transportation systems. The actual environmental assessments were less explicit. In the main text, much attention was paid to how projections of various emissions were influenced by alternative transportation policy measures. The development of emissions of nitrogen oxides as reported in Figure 5 was published in the 1991 Proposal.

The main message is that the reduced emission rates due to catalytic converters is counterbalanced by the expected growth in vehicle miles of travel. The goal of 30 per cent reduction compared to the 1980 level can only be met after the year 2000 in the case of strong traffic policy measures (rail investments, area licensing scheme in the inner city, public transit oriented settlement location). The expected growth in vehicle miles of travel according to the 1990 Regional Plan Outline was 8 per cent 1988-2000 and 33 per cent 1988-2020 (morning peak hour). For the inner city area, the total vehicle miles of travel was forecasted to decrease by 35 per cent 1988-2000 and by 19 per cent 1988-2020 due to introduction of the proposed area licensing scheme. These projections were achieved by manual iterations between fixed demand equilibrium assignments in EMME/2 and a mode choice model with car and transit modes only.

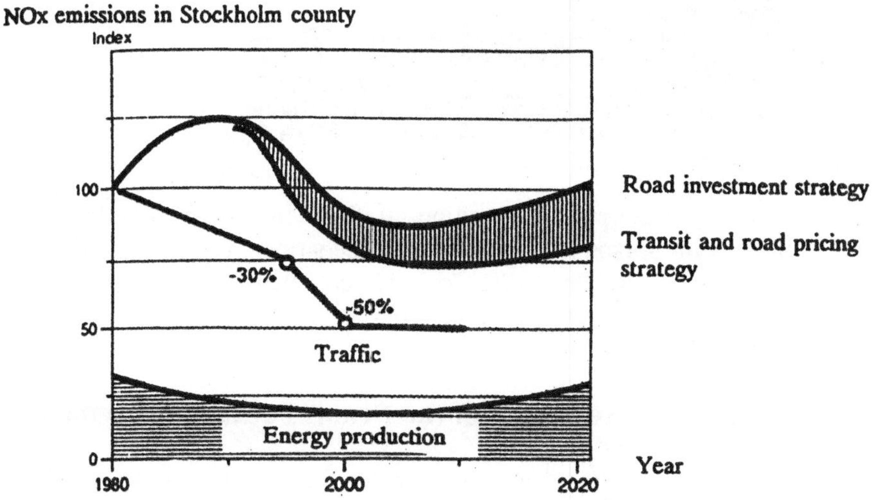

Fig. 5. Emissions of Nitrogen Oxides in Stockholm County.
Source: Stockholm County Council (1991).

The goal concerning carbon dioxide was judged to be even more difficult to achieve, since these emissions are related to the total use of fossil fuels and hence, in the case of transportation and with a given technology of cars, these emissions are closely linked to the total vehicle miles of travel. Still, the official carbon dioxide goal is considered to be very modest. Only new vehicle types running on electricity or bioenergy fuels can drastically reduce the emissions of carbon dioxide.

The goal concerning sulphur can be met through improved combustion processes and exhaust control measures. Road traffic generates a very small share of sulphur depositions and the major part is imported. The imported level by itself leads to acidification above the critical level in soil types with low buffering capacity, see Figure 6.

The preliminary agreement on the Dennis package was incorporated into the 1991 Regional Plan Proposal. A more detailed discussion of traffic forecasts and environ-mental impacts based on the finally decided Dennis agreement follows in the next section.

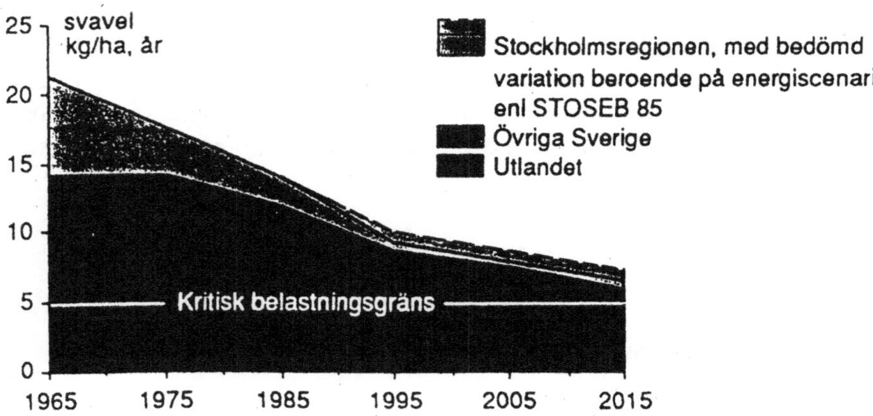

Fig. 6. Sources of Sulphur Deposition in Southern Stockholm (from top: Stockholm region (various energy scenarios), rest of Sweden, and international) Related to the Critical Load Level.
Source: Stockholm County Council (1989).

5. Environmental Assessments of the Dennis Package Agreement

Although the regional political agreement on the Dennis package was reached in 1992, only recently the government issued financial guarantees for the first set of projects, which are still subject to approval in local environmental impact assessments. In the meantime the public debate on the package agreement is intensive. Demonstrations against road investments are taking place. As part of the environmental assessments an investigation was recently conducted by the Stockholm County Board and the Stockholm County Office of Regional Planning and Urban Transportation. In the report "Dennis and the environment" traffic-environmental interactions today and in the future are analyzed. Traffic projections using multiclass combined mode and route choice network equilibrium formulations in EMME/2 form the core of the analysis. Emissions, concen-trations and noise are calculated on the basis of vehicle flows and velocities from the model and assumptions concerning the composition of vehicle types. The projections have been worked out jointly by the Stockholm City Planning Office, The Institute of Regional Analysis and the National Road Agency (Region Stockholm).

The total traffic volume in the morning peak hour is estimated to grow by 8-16 per cent between 1990 and 2005 depending on economic growth rate. In the slow growth case car and transit trips show similar growth rates (8 and 6 per cent respectively) while in the high economic growth alternative the number of car trips increases much faster than the number of transit trips (21 and 11 per cent respectively). The development of the total vehicle miles of travel in various parts of the region is shown in Figure 7. It is clearly shown that the road tolls are projected to decrease the inner city traffic by about 25 per cent 1990-2005. In the county as a whole the total vehicle miles of travel increases by 16-29 per cent.

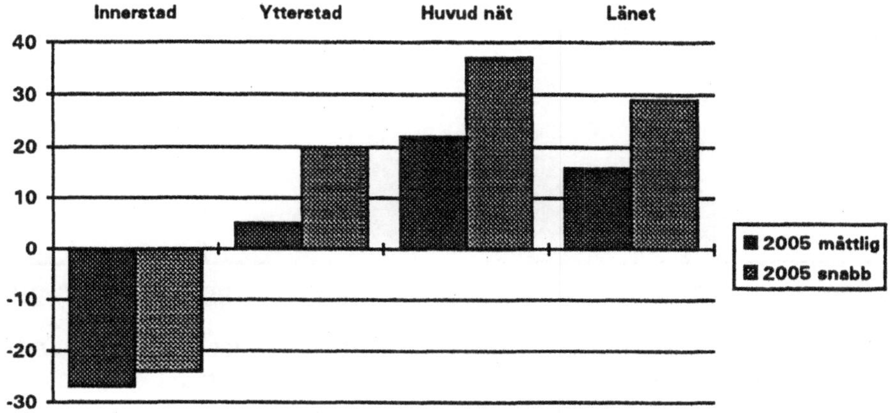

Fig. 7. Percentage Change of the Vehicle Miles of Travel During the Morning Peak Hour 1990-2005 in the Inner City, the Inner Suburbs (including the inner ring road), the Main Roads and the County Total. High and Slow Economic Growth Scenarios.
Source: Stockholm County Board (1993).

The projected development of emissions based on these traffic forecasts can be summarized as follows:

Nitrogen oxides: The emissions are projected to decrease by on average 65 per cent in the inner city and by 45 per cent in the county 1990-2005. The goal of 30 per cent reductions of the county total 1980-1995 can not be met. It will be reached a few years later, see Figure 8. Also the recommended reduction by 50 per cent 1980-2000 can not be met until after the year 2005. The long term prospects are very much depending on the future develoment of vehicles and traffic growth, compare Figure 5.

Carbon dioxide: The total emissions in the county are projected to decrease by 10 per cent 1990-2005 in the slow growth rate scenario and to increase by 5 per cent in the high growth rate scenario. The goal of zero growth 1988-2000 can be met in the slow economic growth scenario but not in the high growth case.

Carbon monoxide: Drastic reductions due to catalytic converters imply that these emissions will not be problematic.

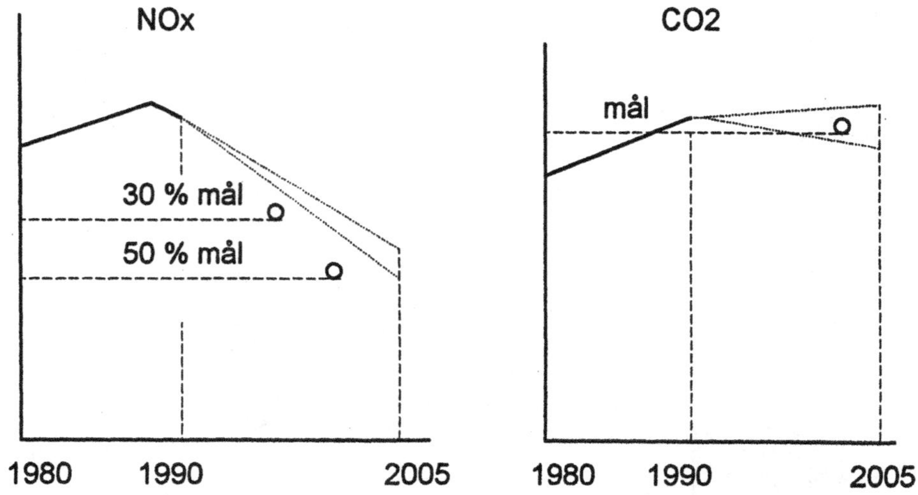

Fig. 8. Emissions of Nitrogen Oxides and Carbon Dioxide from Road Traffic in the Stockholm County Compared to the Stated Goals (circles). Dotted lines represent the development according to high and low growth scenarios.
Source: Stockholm County Board (1993).

Hydro carbons: Emissions are projected to decrease by on average 80 per cent in the inner city and by 60 per cent in the county.

The average concentrations of nitrogen oxides at the inner city street level are projected to decrease by 50-60 per cent 1990-2005 to concentrations clearly below the recommended maximum values. At the roof level a decrease by about 40 per cent is projected. Near the ring road and in the ring road tunnels much higher concentrations are projected (up to 10 times the inner city roof level value).

The dry deposition of nitrogen oxides is projected to decrease by 20-50 per cent in the inner city. The wet deposition is to a large extent depending on external sources, see section 3.

The noise disturbances from road traffic are projected to decrease by on average 11 per cent in the inner city and to increase by on average 9 per cent in the county (kilometer roads with equivalent noise level above 65 dBA, 24 hours). The number of individuals in locations with noise disturbances is estimated to decrease by 15-20000.

Instead of comparing the results for 2005 with 1990 it may be interesting to compare the 2005 results with a projected reference scenario. Such a scenario may be difficult to define. One such attempt has been made in the report "Dennis and the environment". The reference scenario only includes projects that have been planned earlier, no road tolls and none of the new big investment projects. In comparison with this reference scenario the Dennis package seems even more favourable for the year 2005: the inner city vehicle miles of travel are reduced by 23-33% and the county vehicle miles of travel are reduced by 5-10%. The Dennis package performs better in relation to the reference scenario at high economic growth. Figure 9 shows average (over growth cases) volumes of emissions in the reference and Dennis scenarios 2005 and the corresponding 1990 levels (index=100) for the inner city and for Stockholm county. It can be seen that the Dennis package leads to an improved environmental situation as compared to the reference scenario.

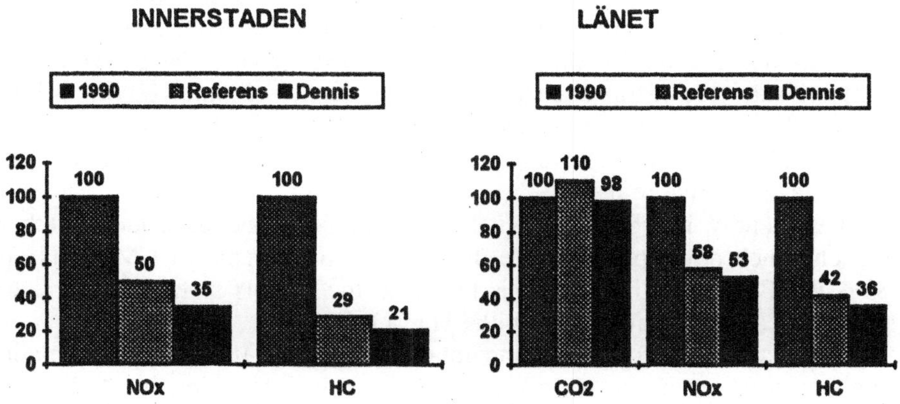

Fig. 9. Relative Emission Levels 1990 (Index=100) and in the Reference and Dennis Scenarios 2005 (average values for two growth alternatives) for the Inner City (left) and for Stockholm County (right).
Source: Stockholm County Board (1993).

The ongoing controverse on the Dennis package is mainly related to the reliability of scenario assumptions and traffic projections. The forecasts are performed using state-of- the-art combined mode and route choice network equilibrium models. Effects on the trip distribution and on relocation of activities have not been taken into account in the results reported above.

These effects have been studied separately using an integrated model of residential and employment location (IMREL), see Anderstig and Mattsson (1992). The combined effects on trip lengths from land-use, trip distribution and modal split adjustments in 2005 due to the Dennis package have been estimated to an increase by about 8 per cent for all modes and by 3 per cent for the car mode. This should be compared to the 10 per cent increase in average car trip length resulting from the land-use changes of the Regional Plan. An additional decentralizing land-use effect of the Dennis package emerges in favour of in particular the north-eastern direction for residences and the northern and north-western direction for workplaces.

6. Future Developments

Although state-of-the-art methodology in terms of a multiclass combined mode and route choice network equilibrium model and a well researched land-use transportation model have been used for land-use and traffic projections related to the Dennis package, there is still a potential for further improvements. With even more developed tools available, a wider range of land-use transportation policies can be studied in environmental impact assessments.

The combined network equilibrium model should be developed to include trip distribution, mode choice and route choice. It should also be developed to cover non-commuting trip types. Alternative approaches with these ambitions were discussed above in the context of CODMA and SYNERGETIC. The controversial issue of trip generation as a consequence of improved accessibility also needs to be carefully studied and integrated within the models.

The submodels for estimating emissions, concentrations, depositions and noise may be further refined.

User-friendly interfaces are vital for simplifying the utilization of these models. Such developments were discussed in the context of CMAWIN and SYNERGETIC. Another option is to imbed the models in transport oriented geographical information systems like TransCAD.

There is a need for better integration of the land-use and transportation analysis. Attempts in this direction will be made by integrating the IMREL and SYNERGETIC model systems discussed above. An alternative approach (TRANSACT), generalizing the combined network equilibrium model to also include land-use patterns and densities, is being developed as part of an international comparative study, see Lundqvist *et al.* (1992).

The increasing weight attached to quality of life aspects in many societies implies that well-founded methods for analyzing the land-use/transportation/ environment interactions become more and more demanded. The way of coping with these interactions may turn out to be an important factor in the competition between regions in the future. The relations between intraregional policy measures and interregional attractivity would then constitute an additional level adding to the complexity of the analysis.

Acknowledgements

Permissions to reproduce the following figures are gratefully acknowledged:
Figure 1 and 5-6: Stockholm County Council Office of Regional Planning and Urban Transportation.
Figure 2 is adapted from Anderstig and Mattsson (1992).
Figure 4 and 7-9: Stockholm County Board.

References

Abrahamsson, T. and Lundqvist, L. (1990) A combined trip distribution, modal split and assignment model applied to Stockholm, Paper presented at the 30th European RSA Congress, Istanbul

Abrahamsson, T. and Lundqvist, L. (1993) Implementation and application of nested combined models for trip distribution, modal split and route choice - a comparative study of model formulations in the Stockholm region, Paper presented at the 40th North American Meetings of the RSAI, Houston

Anderstig, C. and Mattsson, L.-G. (1992) Policy applications of an integrated land-use transport model in the Stockholm region, Paper presented at the 6th World Conference on Transport Research, Lyon

Boyce, D.E. and Lundqvist, L. (1987) Network equilibrium models of urban location and travel choices: alternative formulations for the Stockholm region, *Papers of the Regional Science Association 61*, 93-104

Evans, S.P. (1976) Derivation and analysis of some models for combining trip distribution and assignment, *Transportation Research 10*, 37-57

Karlsson, S. and Svalgård, S. (1993) SYNERGETIC - a new travel demand and land-use model for the Stockholm region, integrated with EMME/2, Paper presented at the 21st PTRC annual meeting, Manchester

Leblanc, L.J., Morlok, E.K. and Pierskalla, W.P. (1975) An efficient approach to solving the road network equilibrium traffic assignment problem, *Transportation Research 9*, 309-318

Lundqvist, L. (1993) A simple DSS for urban traffic planning, Paper presented at the workshop on Advances in Methodology and Software in Decision Support Systems, IIASA

Lundqvist, L., Kim, T.J. and Rho, J.H. (1992) On the application of an integrated land-use transportation model to Stockholm, Paper presented at the IV World Congress of the RSAI, Palma de Mallorca

Lundqvist, L. and Mattsson, L.-G. (1992) Modelling travel demand in an extended metropolitan region, Paper presented at the 32nd European RSA Congress, Brussels

Spiess, H. (1990) A gradient approach for the O-D matrix adjustment problem, Centre de Recherche sur les Transports, Université de Montréal

Stockholm County Council (1989) Regional Plan 1990 for Stockholm County - proposal for consultation (in Swedish), The Stockholm County Council Office of Regional Planning and Urban Transportation

Stockholm County Council (1991) The Regional Plan 1991 for Stockholm County 1990-2020 - final proposal (in Swedish), The Stockholm County Council Office of Regional Planning and Urban Transportation.

Stockholm County Board (1993) Dennis and the environment (in Swedish), Report 1993:6, Stockholm County Board.

Svalgård, S. (1990), Combined models for mode and route choices - a comparative analysis of models for network equilibrium (in Swedish), Memo 1990:3, Stockholm County Council of Regional Planning an Urban Transportation

CHAPTER 19

A SEQUENTIAL LAND USE/TRANSPORTATION MODEL WITH EXTERNALITIES: LINKING THE DYNAMICS OF REGIONAL ECONOMIC GROWTH AND URBAN SPATIAL STRUCTURE

Jong Gook Seo, James E. Moore II and Peter Gordon

1. Introduction

1.1 Research Objective

One of the least satisfactory features of regional analysis is the gulf between the studies of regional economic change and the study of regional spatial structure. Recent regional economic analysis concerns empirical and theoretical developments in growth theory, econometric modelling, and input-output techniques; but are rarely concerned with spatial structure. Similarly, studies of spatial structure are generally undertaken in a static context seldom related to the process of regional economic change. We contend the sectoral composition of a region's economy exerts an important influence on the spatial structure of the region.

Interdependence between activities is an important factor in the growth of regions. Interactions between agents makes the location decision of one agent dependent on the location decisions of other agents. Input-output relationships are important determinants of clustering both within and between activities. Interdependency is further influenced by standard

Y. Hayashi and J. Roy (eds.), Transport, Land-Use and the Environment, 383–411.

structural transformations in the composition of demand, trade, production, and factor use in a developing economy.

Neoclassical approaches such as Fisher (1935) and Clark's (1957) development stages theory, Kuznets' (1957, 1966) modern economic growth theory, and Lewis's (1954) dual economy theory suggest that structural change is essentially a byproduct of economic growth. Based on these three underlying theories, Chenery (1960) develops general models of structural change that link changes in the composition of consumer demand to rising per capita income.

We contend that the process of metropolitan economic growth drives transformations in the spatial structure of the activity system. Our study depicts the dynamics of land use patterns, integrating Chenery's regional economic development processes into an activity location model. Structural transformations are revealed by nonproportional growth across sectors. Economic development produces changes in input-output relationships that are translated into updated transhipments between activities.

Our research model is a simulation that accounts for interactions between

(1) *a priori* profitabilities,
(2) transport costs defined by a congestive transportation network,
(3) externalities
(4) relocation costs, and
(5) technological change.

These factors tractably explain the evolution of an urban economy, and the effect of this evolution on urban structure.

1.2 Trends in Regional Spatial Structure

Contemporary metropolitan areas are characterised by decentralised patterns of employment. As a large metropolitan region grows, a point is reached at which economic activity and population begin to relocate from the metropolitan centre to subcentres situated within the metropolitan region. The emergence, growth, decline, and obsolescence of individual urban subcentres is part of a dynamic process resulting from simple economic behaviour. Some authors characterise this evolving form as a counterurbanisation that implies erosion of a single centred metropolis, and as a process of population de concentration characterised by decreasing densities and increasing local homogeneity (Berry 1976). Others describe it as a dispersion of activities producing a random sprawl of tract housing, shopping malls, and industrial parks, each locating without any specific relation to particular focal points (Blumenffeld 1964). Many see this emerging form as a mixed blessing, increasing the consumption of land and other finite resources (Clark1954). Regardless of the perspective taken, the study of how such subcentres develop and their impacts on land values, population distribution, and travel patterns is central to the investigation of land use and transportation interaction.

Empirical studies abound. In 1980, 57 percent of all office space in the US was located in urban centres and 43 percent in suburbs; by 1986, 60 percent was in the suburbs, compared to 40 percent in the centres (Pisarski 1987). Originating in America's sunbelt (Cervero 1986), the suburban office boom has become nationwide, occurring even in older industrial areas. In greater Philadelphia and St. Louis, for example, suburban employment grew by 8 and 17 percent respectively between 1982 and 1986, contrasted by a loss in central city jobs over the same period (Orski 1986; Urban Land Institute 1987). Erickson (1983) describes the evolution of suburban economic activities in the US through 1960 as a process of "dispersal/differentiation," followed by a subsequent phase of "infilling/multinucleation." Hartshorn and Muller (1986) describe the changing of land uses in American suburbs in terms off our stages, including bedroom communities (pre-1960), independence (1960-70), catalytic growth (1970-80), and high rise / high technology (post-1980).

Many expect the trend toward decentralisation to accelerate in the coming years as America's economy continues to shift from a manufacturing base to an emphasis on service and information processing activities. Suburban areas offer cheaper land, reduced externalities, proximity to regional airports, smart buildings laced with fibre optic cables and advanced telecommunications equipment, pools of second wage earners, and country like amenities (Dowall 1987, Urban Land Institute 1986 and 1987).

Relevant specifications of urban space must represent condition sunder which policentrism might emerge, discussing where the centres maybe located (Richardson 1988). Blackley and Follain (1987) argue that accessibility to amenities other than workplaces need to be accounted for in locational equilibria. This implies that spatial externalities should be represented in location decisions.

1.3 Regional Economic Analysis

The relationship between a region's economy and its spatial structure should be viewed in dynamic terms. If sectors of an expanding regional economy are subject to technical change, there may be significant modifications in their locational characteristics. A similar outcome is expected if existing sectors are replaced with new sectors that have substantially different locational requirements.

Neoclassical economic theory treats economic growth as the result of the long term effects of capital formation, labour force expansion, and technological change under conditions of competitive equilibrium. "Shifts in demand and the movement of resources from one sector to another are considered relatively unimportant because the marginal returns in the use of these factors (labour and capital) should in principle be equal in all uses (Chenery, et al 1986)." Hence, there are no incentives for drastic structural change. Structural change is essentially a by product of growth, which in turn is produced by increases in factor supplies and productivity.

However, economic growth may also be treated as a result of transformations in the structure of production. These transformations are required to meet changing demand, and to make more productive use of technology. Conditions of imperfect foresight, incomplete competition, and limits to factor mobility imply that these structural changes are most likely to occur under conditions of disequilibrium (Chenery, et al 1986).

The relationship between growth and structural change can be divided into two parts. These are the effects of growth in per capita income on structure (the demand side); and the effects of changing structure and productivity increase on growth (the supply side). Fisher (1935) and Clark (1957) articulate a sequential path of economic development through which all societies progress. Under this development stages theory of growth, societies are assumed to experience changes in the dominant occupation of the labour forces. This process is explained by changes in comparative costs and changes in income elasticities of demand. Kuznets (1957, 1966) and many others place the study of industrial growth in a broader perspective. Instead of focusing so narrowly on the allocation of resources, Kuznets describes the increasing industrial outputs as part of the general transformation he identifies as "modern economic growth." Starting with the national accounts recorded by a number of countries, Kuznets measures the changes in the composition of consumption, production, trade, and other aggregate measures as income rises; demonstrating similarities between the growth patterns of the 1950s. He measures comparable development patterns both among countries and over time.

Building on these growth process theories, Chenery and others (Chenery 1960; Chenery, Shishido, and Watanabe 1962; Chenery and Taylor 1968; Taylor 1969; and Chenery and Syrquin 1975) develop general models of structural change. These models usually assume similar changes in the composition of consumer demand with rising per capita income. These multisectoral, cross country models address the demand side in terms of three sets of relationships. These are

(1) income elasticities of demand for each commodity,
(2) input-output relations that are a function of per capita income, and
(3) export demands and import proportions that reflect the factor endowments and policies of different groups of countries.

Under these conditions, the basic causes of structural change are technological progress, population growth, rising level of income, consequent changes in trading conditions, and the supply of external capital.

Chenery employs regression equations as reduced form representations of a detailed general equilibrium system. In this simplified model, the level of per capita income and population are the only explanatory variables, implying that all of these factors can be collapsed into income and market size effects. The model yields solutions for levels of consumption, production, and trade by sector as a function of the level of per capita income.

2. A Sequential Urban Land Use Model

The sequential urban land use model developed here consists of two major components,

(1) a discrete programming model of the market for urban land and transportation, and
(2) an interactivity flow system that accounts for structural transformations resulting from economic development.

Contemporary suburbs are interdependent, collectively comprising the metropolitan economy. This metropolitan economy is, in turn, part of a larger system of economies, engaging in trade with its hinterland, other metropolitan economies, and the rest of the world. At the same time, the metropolitan region is an economy with an evolving differentiation between suburbs, each of which exhibits specification in term of its activity characteristics.

Recognising interdependencies is the principal means of integrating the regional economic development process into an activity location model. The model is initialised by an exogenous economic structure and spatial pattern. Given an initial spatial pattern, the characteristics of establishments change. These economic changes result in structural transformation and nonproportional growth across sectors. Given income elasticities for each sector and an existing set of input-output relationships, the structural transformation model endogenises production levels and traffic intensities. Exogenous values in the discrete programming model describing segment revenues, externalities, and relocation costs might also be influenced by production levels. Given these updates, the discrete programming model identifies a new land use pattern. In the next time period, more economic structural changes are realised and the structural transformation model once again produces new traffic intensities.

2.1 The Market for Urban Land and Transportation

Urban spatial structure is the outcome of a process that allocates activities to sites. The process is principally one of transactions between owners of real estate and those who wish to rent or purchase space for their homes and businesses. These transactions are accomplished by the general rule of the market. We assume the urban area is divided into many discrete sites. These sites have different attributes. Each site belongs to an owner who is free to sell or lease his property. At the beginning of each transaction period, every establishment evaluates the merits of every site, and decides what price it would be willing to pay for access to each site.

The passage of time brings changes in the number and types of establishments bidding for access to locations. Existing establishments also change in terms of their characteristics. Households change in size, manufacturers acquire new production methods, and retailers shift product lines. Some sites change hands and some establishments move to new locations. As long as some establishments are moving, the pattern of accessibility and contiguity changes for other establishments. Even if site characteristics are fixed, these various changes accumulate overtime to cause significant shifts in the matrix of site bids.

In most contemporary regional I-O tables, the structural coefficients represent inter industry trade flows. Recent developments in combining input-output and transportation planning models have made it possible to construct comprehensive urban and regional activity models. A metropolitan area industry activity model divides the local economy into identifiable sectors along two dimensions, product (or industry) and geography. Transactions representing interactivity linkages are identified across industries and locations.

Gordon and Moore II (1989) and Moore II and Gordon (1990) formulate a sequential programming model that simulates the spatial evolution of modern cities. Locators are assumed to make decisions from a *ceteris paribus* perspective (Moore II and Gordon 1990). By solving a series of linear assignment problems that track urban land use over time, their model presents a sequence of urban location decisions resulting from locators' efforts to maximise net revenues by mitigating congestion costs and other externalities (Moore II and Gordon 1990). Network congestion and other effects are endogenous in each period, but traffic intensities between all activities i and j are exogenous. In the current study, interactivity flow systems are conditioned on economic development patterns that include changes in the composition of demand, trade, production, and factor use as functions of per capita income.

The arrival, departure, and ongoing bidding of activities constitute the principal mechanisms for spatial rearrangement. Unsuccessful bidders are consigned to a null site, or queue. Activities bid nothing for access to the queue, and there is no constraint on the number of activities that can locate there. To represent this process in a more complete way, Moore II and Gordon also introduce a nonbidding or null activity called "vacancy" that bids nothing for physical sites and can be assigned to any number of sites. When nonvacancy activities offer (sufficiently) positive bids for sites, existing vacancies are displaced.

Index activities from 1 to I and physical sites from 1 to M. Append an I+1st row accounting for vacancies, and an M+1st column corresponding to the null location, or queue. The augmented matrix that results is **A**, an initial $[(I+1)\times(M+1)]$ matrix of seminet revenues. At time 0, $\mathbf{A}(0)=[a_{im}(0)]$ is the profitability of plants i at m, ignoring externalities and transportation. That is, **A**(0) is the value to plant i of the attributes of site m independent of the locations of other plants.

The principal advantage of the solution procedure is that complex information about congestion and other externalities is assumed to flow from recent experience, allowing the sequential use of linear programs to emulate the decisions of locators. A flowchart describing this approach appears in Figure 1.

Step 6 is key, updating each activity's location bid. Given $\mathbf{A}(0)=[a_{im}(0)]$, the $[(I+1)\times(M+1)]$ matrix of seminet revenues (identified in Step 0); $\chi(t)=[\chi_{jn}(t)]$ the $\{[(I+1)\cdot(M+1)]\times 1\}$ vector of optimal location assignments from the previous time period (identified in Step 2); $\mathbf{F}=[f_{ij}]$, an exogenous matrix of traffic intensities between all activities i and j (identified in Step 3); $\mathbf{C}^*(t)=[c^*_{mn}(t)]$, the $(M\times M)$ matrix of user equilibrium link costs (identified in Step 4); and $\mathbf{E}^*(t)=[e^*_{imj}(t)]$, the $[I\times(I\cdot M)]$ matrix of potential spatial externalities imposed by each activity j at (fixed) location n on each activity i at (variable) location m (identified in Step 5); the bid for each locator i prepares for each site m is updated based on each locator's semi net revenues and anticipated experiences at all locations. That is, compute the $[(I+1)\times(M+1)]$ matrix of location bids

$$\mathbf{A}(t+1)=[a_{im}(t+1)]=[a_{im}(0)\cdot v_i(t)/v_i(0)]$$
$$+\Sigma_{j=1->I+1}\{\Sigma_{n=1->M}[c^*_{mn}(t)\cdot f_{ij}\cdot\chi_{jn}(t)+e^*_{imj}(t)]\}$$
$$(\text{for all } i=1->I+1, \text{for all } m=1->M+1) \qquad (1)$$

where $v_i(t)$ is the value added by activity i in time period t as identified in the structural transformation model. Given fixed site characteristics, activities will still change their production levels as a result of changes in the cost of primary materials. Consequently, $a_{im}(0)$ is updated in each time period relative changes in the value added levels associated with each activity. The values $a_{im}(0) = 0$ if i is vacancy and/or m is the null site.

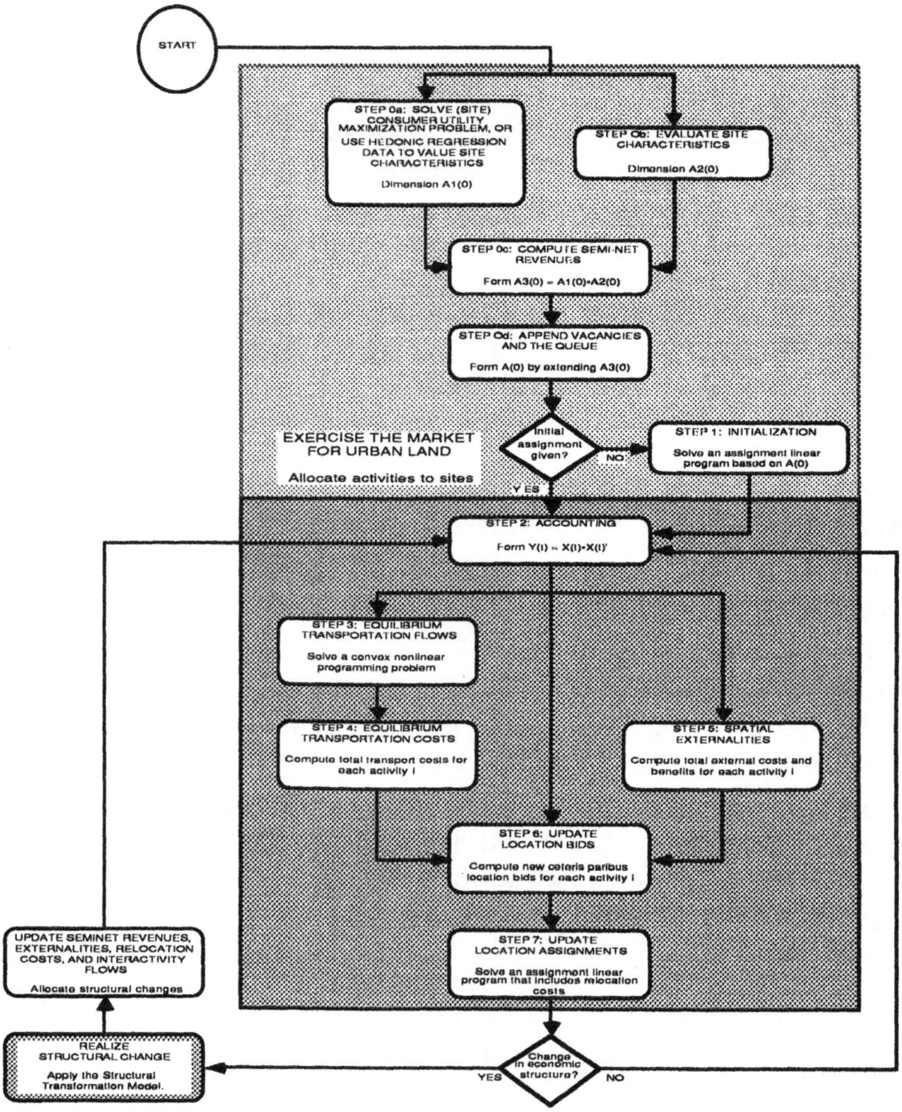

Fig. 1. Algorithmic Representation of the Sequential Urban Land Use Model: An Extended Version of the Moore II and Gordon Model (1990)

Step 7 is a market clearing operation. Given the matrix $\mathbf{A}(t+1) = [a_{im}(t+1)]$, and an exogenous $[(I+1) \times 1]$ vector $R = [R_i]$ consisting of activity specific relocation costs; solve the following linear program.

Maximise

$$\Sigma_{i=1->I+1}\Sigma_{m=1->M+1}\{a_{im}(t+1) - R_i \cdot [1 - \chi_{im}(t)]\} \cdot \chi_{im}(t+1) \qquad (2)$$

subject to

$$\Sigma_{i=1->I+1}\chi_{im}(t+1) = 1$$
(for sites $m = 1\text{-} > M$; i.e., for all sites m, except the null site) (3)

$$\Sigma_{m=1->M+1}\chi_{im}(t+1) = 1$$
(for activities $i = 1-> I+1$; i.e., for all activities i in I, except vacancy) (4)

$$\chi_{im}(t+1) \geq 0$$
(for all activities $i = 1-> I+1$, for all sites $m = 1-> M+1$) (5)

where $\chi_{im}(t)$ is exogenous to time period t+1 (Moore II and Gordon 1990).

Steps 8 and 9 impose structural transformations associated with economic development on the location assignment model.

2.2 Summary of the Structural Transformation Model

There are two dominant approaches to multisectoral analysis in modelling structural transformation. Leontief (1951; Leontief, et al 1953) was the first to use the input-output (I-O) approach to study structural changes in the American economy. Leontief's research analyses the effects of changes in input coefficients between 1919 and 1939 on the structure of production and labour use. External trade and domestic demands are held constant. Chenery, Shishido, and Watanabe (1962) use a similar procedure to trace transformations in the structure of Japanese production between 1914 and 1954 in response to changes in demand, trade, and technology.

Computable general equilibrium (CGE) models are more sophisticated alternative originally derived from the I-O perspective. Johanson (1960) linearised Walras' general equilibrium model for Johanson's empirical application to the Norwegian economy. Johanson employed Leontief's input-output system to describe interindustry relations, but he included demand and production functions that depend on relative prices. Such CGE models are developed further in a number of recent studies (Taylor and Black 1974; Dervis 1975; J. de Melo 1977, M.de Melo 1979; de Melo and Robinson 1980; Celasun 1986) that focus on issues of international trade, growth, economic structure, and income distribution in developing and developed countries. These models are sometimes called price endogenous models because all prices must adjust until the decisions made in the productive sphere of the economy are consistent with the final demand decisions made by households and other autonomous decision makers.

Significant differences between the two approaches arise in cases where relative prices change substantially. In the long term, the most important price is the rising cost of labour, which leads to the substitution of capital for labour and to changes in comparative advantage. CGE approaches can distinguish between capital-labour substitution and technological change. I-O approaches aggregate these effects.

As noted above, the structural transformation of a developing economy may be defined as the set of changes in the composition of demand, trade, production, and factor use that takes place as per capita income increases. Although the CGE approach is clearly preferable inmost contexts on theoretical grounds, the important advantages of the CGE approach are offset by the limited data on prices and capital stocks. Also, the relevant production functions are cumbersome, and must be approximated from observations scattered across both space and time. Lastly, there are problems in proving the existence of an equilibrium solution (Diewert and Wales 1985). Consequently, we use the input-output approach to explain the locational changes resulting from structural transformation. Data permitting, a CGE model can substituted directly.

The research model derives activity growth functions from a general equilibrium model that allows for changes in the composition of demand and in factor proportions. The general equilibrium models of Walras (1954), Leontief (1951), and Dorfman, Samuelson, and Solow (1958) customarily omit elements that would lead to persistent differences in growth rates. These elements include limited natural resources, changing factor supplies, nonhomogeneous consumption functions, economies of scale, and international trade. Accounting for imports, exports and intersectoral requirements defines four determinants of the level of production. These include three components of demand and one alternative source of supply. The accounting identity for this system is

$$X_p = D_p + W_p + E_p - M_p \tag{6}$$

where X_p is domestic production of commodity p, D_p is domestic final use of p, W_p is use of p by other producers, E_p is the export of p, and M_p is the import of p.

Intermediate demand W_p for a commodity p depends on output levels from the sectors using p, on the substitutability of other inputs for p, and on the variation in the relative prices of inputs. Based on previous work involving international comparisons (Houthakker 1957; Chenery and Watanabe 1958; Taylor 1969; Chenery and Syrquin 1980 and 1986), price effects are suppressed on the assumption that per capita income incorporates the effects of all these explanatory variables. Thus the function for intermediate use of commodity p is

$$W_p = \Sigma_k \alpha_{pk} \cdot X_k = \Sigma_k \alpha_{pk} \cdot (Q_k + M_k) \qquad (7)$$

where the α_{pk} are input-output coefficients, X_k is the total output of commodity k, and Q_k is the sum to total intermediate purchases and value added in the production of commodity k.

Structural change is often defined by sectoral shifts, which may include changes in any component of demand or value added by production. Alternatively, changes in structure can also be measured as sector specific deviations from proportional growth across sectors. Under the assumption of proportional growth, equation (14) can be expressed for time t

$$X_p\wedge(t) = X_p\wedge[Y(t)] = \lambda(t) \cdot X_p[Y(0)] \qquad (8)$$

where $X_p\wedge(t)$ indicates total production of commodity p proportional toper capital income at time t, and

$$\lambda(t) = Y(t) / Y(0) \qquad (9)$$

is the proportionate increase in income between periods time 0 and time t.

In general, these proportional benchmarks will not be realised. Deviations from proportional growth can be expressed as follows

$$\Delta X_p(t) = X_p(t) - X_p\wedge(t) = \Delta W_p + \Delta D_p - \Delta M_p + \Delta E_p \qquad (10)$$

Thus, deviation from proportional growth in each sector can be traced back to deviations from proportional growth in intermediate demand, final demand, imports, and exports. Equation (10) implies several alternative decompositions of structural change that depend on import substitution, and the nature of changes in interactivity structure.

The explanatory variables for the determinants of sector growth depend on the degree of openness of the economy, its trade pattern, and its rate of growth. The United Nations (1963) tested eight proxy variables for these factors in estimating growth patterns for individual industrial sectors. This and other studies of economic development patterns has led to the identification and measurement of a number of structural changes associated with rising income. As a result, income level has been used as an overall index of development as well as a measure of output. We employ income per capita as an explanatory variable on the assumption that per capita income incorporates effects of all other explanatory variables.

To investigate structural changes implied by sectoral deviations from proportional growth, we need to measure the income elasticities of domestic production X, domestic final demand D, exports E, and imports M for each sector p. Regression analysis provides a convenient vehicle. Regression has been widely used to compare and explain the uniform patterns of industrial growth measured by Chenery (1960), Kuznets (1966), Chenery and Taylor (1968), and Chenery and Syrquin (1975, 1980).

At the national level, economic development takes place in an environment in which trading opportunities and technology are constantly changing. The growth functions derived from cross sectional analysis describe the adaptation of countries at different levels of income to conditions of technology and trade existing at one point in time. Ideally, these states indicate the path that a typical country would follow if its income increased so rapidly that conditions of trade and technology were relatively constant (Kuznets 1957, Chenery 1960).

Estimated income elasticities depend on the type of function fitted. The double logarithmic function is preferred for most international comparisons (Houthakker 1957; Chenery and Watanabe 1958; Taylor 1969; and Chenery and Syrquin 1980 and 1986). Chenery (1960) and United Nations (1963) show that the logarithmic form fits the available data much better than a linear function for most sectors. Houthakker's (1957) findings support this assumption in the case of household consumption. We use linear logarithmic regression equations in which the value of each determinant of sector growth depends on per capita income. For example, the function for final domestic use per capita is

$$\log(D_p) = \log(\beta_{p0}) + \beta_{p1} \cdot \log(Y) \tag{11}$$

where β_{p0} is the initial state of final use of commodity p implied by data series, β_{p1} is an income elasticity for the consumption of commodity p, and Y is per capita income.

Consider the hypothetical regional economy summarised in Table 1. Based on updated estimates of domestic production $X[Y(t)]$, domestic final demand D[Y(t)], exports E[Y(t)], and imports M[Y(t)] for each sector p; we will apply equations (6) and (7) to compute intermediate use W(t). The various phenomena associated with economic development can lead to technological changes within any and all sectors, and there are several ways these changes might be represented in the matrix of technical coefficients.

Table 1. A Hypothetical Regional Economic System

		Processing								Purchasing			
	Inputs \ Outputs	Sector 1	Sector 2	Sector 3	Sector 4	Sector 5	Sector 6	Households	Total Intermediate Use W_p	W_p / Total Output	Final Demand D_p	Exports E_p	Total Output X_p
Processing	Sector 1	1,000	1,500	100	200	500	600	1,400	5,300	0.838	600	500	6,400
	Sector 2	500	400	700	100	300	800	1,700	4,500	0.763	800	600	5,900
	Sector 3	700	200	800	100	500	300	500	3,100	0.775	600	300	4,000
	Sector 4	1,100	100	200	800	600	400	400	3,600	0.923	300	0	3,900
	Sector 5	400	0	100	1,400	300	200	900	3,300	0.825	500	200	4,000
	Sector 6	200	600	700	600	200	600	800	3,700	0.804	500	400	4,600
	Household	1,600	1,800	700	500	700	900	100	6,300	0.875	900	0	7,200
	L_k Total Intermediate Purchases	5,500	4,600	3,300	3,700	3,100	3,800	5,800	29,800		4,200	2,000	36,000
Payments	L_k / Total Outlays	0.859	0.780	0.825	0.949	0.775	0.826	0.806		0.828			
	v_k Value Added	700	1,200	400	200	600	600	1,200	4,900				
	v_k / Total Outlays	0.109	0.203	0.100	0.051	0.150	0.1340	0.167	0.136				
	Q_k $L_k + v_k$	6,200	5,800	3,700	200	3,700	4,400	7,000	34,700				
	M_k Imports	200	100	300	0	300	200	200	1,300				
	X_k Total Outlays	6,400	5,900	4,000	3,900	4,000	4,600	7,200	36,000				36,000

Viewed from this perspective, the fundamental problem is generic. Given new distributions for the row and column marginals of a matrix, the objective is to make best use of the information content in the original matrix to update the matrix entries in a way that satisfies the conditions imposed by the new marginals. We rely on the biproportional adjustment method (Hewings 1977, 1982) used to update input-output, migration, and trip interchange tables. Biproportional adjustment minimises the I-divergence, i.e., the information gain, of the posterior array relative to the a priori array. Other approaches to the same problem include linear and quadratic programming, and variational inequalities (Nagurney 1993). These approaches differ in terms of how distances between the a priori and posterior matrices are defined, and in terms of the algorithms used to address the constrained optimisation problems that result.

Our use of biproportional adjustment does not provide an endorsement of one penalty function versus another. We do not presume to know if technological changes imply the creation of new technologies, or substitutions between existing technologies. Further, we do not know which adjustment procedure maps best to this mixed process of innovation and choice. We elect biproportional adjustment because the theoretical and computation aspects of the procedure are well understood, because the positivity of the initial array ensures the positivity of the unique solution to the problem, and because it operates directly on technical coefficients rather than on flows.

Ideally, Leontief sectors are aggregations of activities producing a single product by similar techniques. Given the variety of products produced by typical plants, realising a close approximation of this concept is impossible. In empirical inter industry studies, a productive sector corresponds to a grouping of processes and products that may differ in some respects. Still, an aggregate sector of production activities may be satisfactory for a Leontief model even if the activities involved do not have uniform inputs of primary factors.

Table 1 describes flows between sectors, yet the discrete programming model identifies locators at the level of activities. Consequently, sector flows updated by the structural transformation model will have to be disaggregated into activity flows before the land use model can be applied. The rules used to disaggregate a sector into constituent activities can be traced back to the rules for consolidating the sectors of a detailed input-output table. The rules of consolidation involve simple summation of flows in a particular base period. Let X_{ij} denote the flow from activity i to j; let D_i denote the final demand for activity i; and let X_i denote the total output of activity i.

$$X_i = \Sigma_{j=1->J} X_{ij} + D_i \,. \tag{12}$$

Generalising to any period, let the input coefficient α_{ij} denote the quantity of input from activity i that is needed to produce one unit of output j.

$$\alpha_{ij} = X_{ij} / X_j \,. \tag{13}$$

The flows between sectors p and k consist of flows between several constituent activities i in sector p and j in sector k. At the sectoral level,

$$\alpha_{pk} = X_{pk} / X_k \,. \tag{14}$$

Interactivity flows can be estimated from intersectoral flows by reversing the procedures implied by standard consolidation rules. If the activities defining a given sector have similar input-output characteristics, intersectoral flows can be disaggregated into an interactivity flows even if the activities vary with respect to the use of primary inputs. In terms of the abbreviated notation associated with Table 1, compute

$$f_{ij} = f_{pk} \cdot (X_{i\,in\,p} / X_p) \cdot (Q_{j\,in\,k} / Q_k) \tag{15}$$

where

$$X_p = \Sigma_{\text{all activities i in sector p}}\, X_i \text{, and} \tag{16}$$

$$Q_k = \Sigma_{\text{all activities j in sector k}}\, Q_j. \tag{17}$$

More generally,

$$f_{ij}[Y(t)] = f_{pk} Y[(t)] \cdot \{X_{i\,in\,p}[Y(0)] / X_p[Y(0)]\} \cdot \{Q_{j\,in\,k}[Y(0)] / Q_k[Y(0)]\}, \tag{18}$$

where $Y(0)$ denotes per capita income in the base year.

Because urban land use configurations are characterised by capital intensive land uses, input substitutions between land and capital are of special importance in an urban context. In this exercise, activities are classified based on the intensiveness of the land input. High, medium, and low land intensive activities correspond to low, medium, and high density land uses respectively.

2.3 Algorithmic Elements of the Structural Transformation Model

In each time period, interactivity flows are derived from income levels. A flowchart describing this approach appears in Figure 2.

Step 0: Estimation of Income Elasticities for Determinants of Intermediate Purchases

Under the I-O approach, imports are classified by purchasing sector, but our discussion of import substitution classifies imports in terms of output accounts. Thus we require transactions to be tracked insufficient detail to permit one set of import values to be converted to the other. Given cross regional or time series observations on X_p, D_p, E_p, and N_k, estimate income elasticities by specifying logarithmic regression equations in which the explanatory variable is per capita income. That is, estimate

$$\log\{D_p[Y(t)]\} = \log(\beta_{p0}) + \beta_{p1} \cdot \log[Y(t)], \tag{19}$$

$$\log\{X_p[Y(t)]\} = \log(\phi_{p0}) + \phi_{p1} \cdot \log[Y(t)], \tag{20}$$

$$\log\{E_p[Y(t)]\} = \log(\eta_{p0}) + \eta_{p1} \cdot \log[Y(t)], \text{ and} \tag{21}$$

$$\log\{N_K[Y(t)]\} = \log(\gamma_{p0}) + \gamma_{p1} \cdot \log[Y(t)]. \tag{22}$$

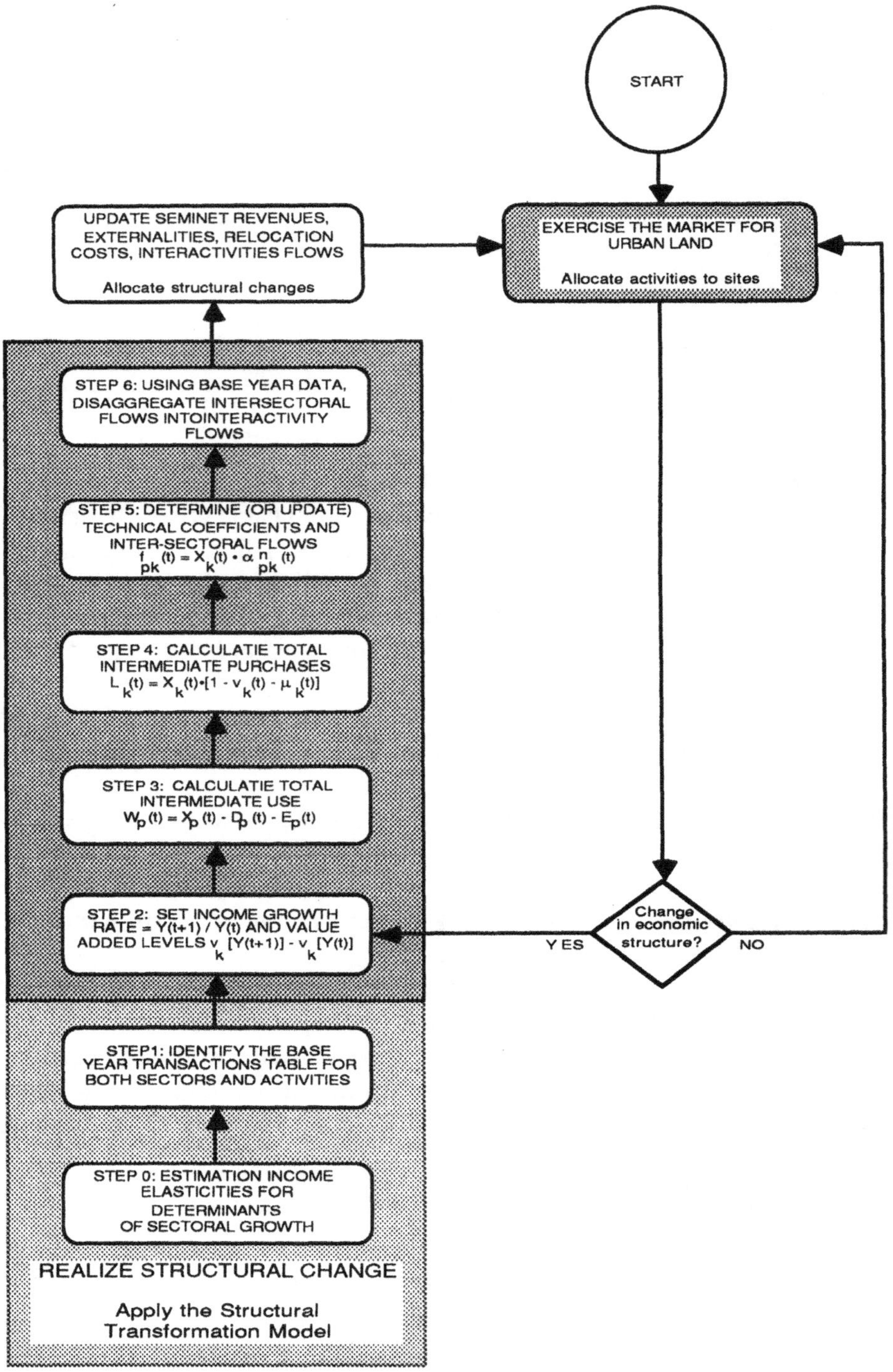

Fig. 2. Algorithmic Representation of Structural Transformation Model

where X_p, D_p and E_p are the domestic production, domestic final use, and exports associated with each sector p, respectively; N_k is the imports associated with each sector k, and Y(t) is the level of per capital income at time t. The subscript 0 denotes each data series' initial state. The coefficients β_{b1}, ϕ_{p1}, γ_{p1}, and η_{p1} are a vector of income elasticities for determinants X_p, D_p, E_p, and N_k, respectively.

Step 1: Identify Base Period Transactions Table for All Sectors and Activities

Establish a transactions table for the region. Let "p" denote a producing sector. Let "k" denote a purchasing sector. Identify base year values for determinants $D_p[Y(0)]$, $X_p[Y(0)]$, $N_k[Y(0)]$, and $E_p[Y(0)]$; technical coefficients $\alpha_{pk}[Y(0)]$, and the sum of intermediate purchases and value added in sector k $Q_k[Y(0)] = \{L_k[Y(0)] + N_k[Y(0)]\}$. Based on each activity's land use characteristics, decompose each sector p's total output $X_p[Y(0)]$ and outlays $Q_k[Y(0)]$ into corresponding activity totals $X_i[Y(0)]$ and $Q_j[Y(0)]$.

Step 2: Define Income Growth Rates and Value Added Ratios for Period t

Given the per capita income levels $Y(t)$ for each time period t and an initial state $Y(0)$, define an exogenous income growth rate $\lambda(t) = Y(t) / Y(0)$. In this exercise, we assume a 10 percent increase in per capita income per period. Define an exogenous value added vector $v_k(t) = v_k[Y(t)]$ for each sector k, and value added ratios $\omega_k(t) = v_k(t) / v_k(0)$.

Step 3: Calculate Intermediate Uses for Period t

Given the income elasticities β_{b1}, ϕ_{p1}, γ_{p1}, and η_{p1} determined in Step 0; the initial values for determinants $D_p[Y(0)]$, $X_p[Y(0)]$, $N_k[Y(0)]$, and $E_p[Y(0)]$ identified in Step 1; and the exogenous income growth rate $\lambda(t)$ and the value added ratios $\omega_k(t)$ defined in Step 2; determine the future values of determinants $X_p(t) = X_p[Y(t)]$, $D_p(t) = D_p[Y(t)]$, $N_k(t) = N_k[Y(t)]$, and $E_p(t) = E_p[Y(t)]$; and the future values of intermediate use $W_p(t) = W_p[Y(t)]$. That is, compute

$$D_p(t) = D_p(0) \cdot \lambda(t)^{\beta_p}, \tag{23}$$

$$X_p(t) = [v_p(0) / v_p(t)] \cdot X_p(0) \cdot \lambda(t)^{\phi_p}, \tag{24}$$

$$N_k(t) = N_k(0) \cdot \lambda(t)^{\gamma_p}, \tag{25}$$

$$E_p(t) = E_p(0) \cdot \lambda(t)^{\eta_p}, \tag{26}$$

$$W_p(t) = X_p(t) - D_p(t) - E_p(t), \text{ and} \tag{27}$$

$$\mu_k(t) = N_k(t) / X_k(t) \tag{28}$$

Step 4: Calculate Intermediate Purchases for Period t

Given the exogenous value added ratio $\omega_k(t)$ identified in Step 2, and the vector of import ratios $\mu_k(t)$ identified in Step 3, determine the values of intermediate purchases $L_k(t)$. That is, compute

$$L_k(t) = X_k(t) \cdot [1 - \omega_k(t) - \mu_k(t)]. \tag{29}$$

$$Q_k(t) = L_k(t) + v_k(t) \tag{30}$$

Step 5: Determine Technical Coefficients and Intersectoral Flows for Period t

Given the technical coefficients $\alpha_{pk}(0)$ identified in Step 1 or the coefficients $\alpha_{pk}(t-1)$ identified for the preceding time period, the intermediate uses $W_p(t)$ identified in Step 3, and the vector of intermediate purchases $L_k(t)$ identified in Step 4, update intersectoral flows $f_{pk}(t)$ via biproportional adjustment.

- **Step 5a:** Compute $\Sigma_{k=1->K} X_k(t) \cdot \alpha_{pk}(t-1) = W_p^1$. (31)

  Update $\alpha_{pk}^1(t) = \alpha_{pk}(t-1) \cdot [W_p(t) / W_p^1]$. (32)

- **Step 5b:** Compute $\Sigma_{p=1->P} X_k(t) \cdot \alpha_{pk}^i(t) = L_k^i$.

  Update $\alpha_{pk}^{i+1}(t) = \alpha_{pk}^i(t) \cdot [L_k(t) / L_k^i]$. (33)

- **Step 5c:** Compute $\Sigma_{k=1->K} X_k(t) \cdot \alpha_{pk}^{i+1}(t) = W_p^{i+1}$.

  Update $\alpha_{pk}^{i+2}(t) = \alpha_{pk}^{i+1}(t) \cdot [W_p(t) / W_p^{i+1}]$. (34)

- **Step 5d:** Repeat steps 5b and 5c until $W_p(t) / W_p^{i+1}$ and $L_k(t) / L_k^i$ approach 1.

  Update $\alpha_{pk}(t) = \alpha_{pk}^n(t)$ where n is the terminal value of i.

- **Step 5e:** Compute intersectoral flows $f_{pk}(t) = X_k(t) \cdot \alpha_{pk}^n(t)$. (35)

Step 6: Disaggregate Intersectoral Flows into Interactivity Flows for Period t

Given total activity outputs $X_i[Y(0)] = X_i(0)$ and activity outlays $Q_j[Y(0)] = Q_j(0)$ identified in Step 1, and intersectoral flows $f_{pk}(t)$ identified in Step 5, determine interactivity flows $f_{ij}(t)$. That is, compute

$$f_{ij}(t) = f_{pk}(t) \cdot [X_{i\ \mathrm{in}\ p}(0) / X_p(0)] \cdot [Q_{j\ \mathrm{in}\ k}(0) / Q_k(0)] \tag{36}$$

where

$$X_p(0) = \Sigma_{\mathrm{all}\ i\ \mathrm{in}\ p} X_i(0) \text{ and} \tag{37}$$

$$X_k(0) = \Sigma_{j\ \mathrm{in}\ k} X_j(0)\,. \tag{38}$$

Table 2 summarises the inputs and outputs of the structural transformation algorithm in terms of the numerical data in Table 1.

Table 2. Exercising the Structural Transformation Algorithm on the Entries in Table 1.

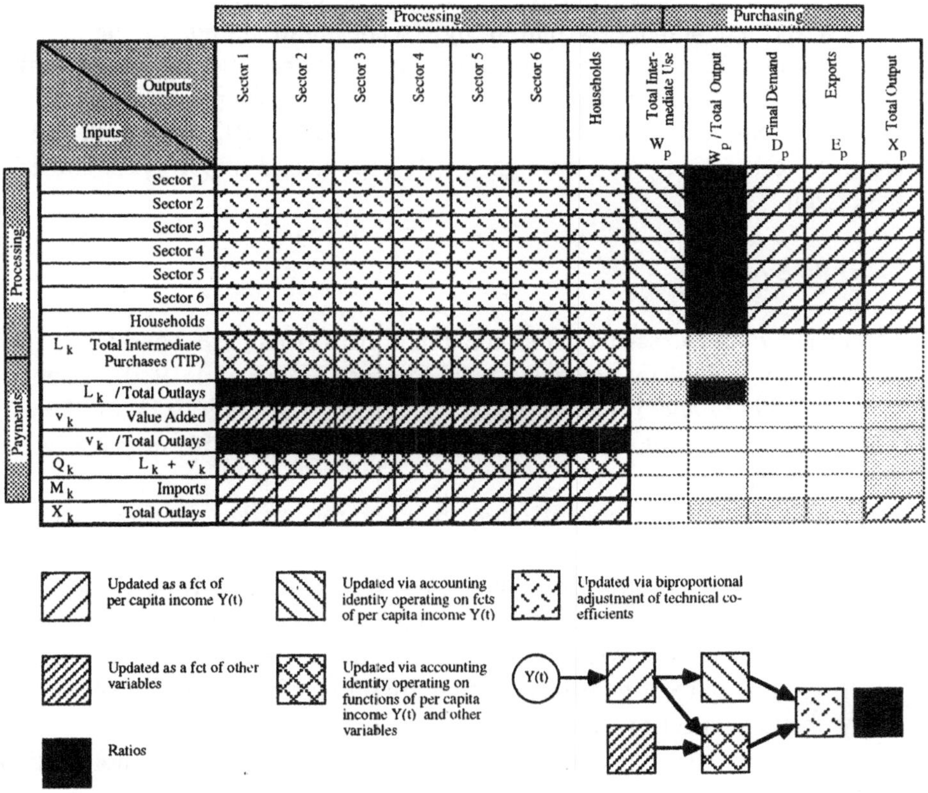

3. Numerical Applications

The simulation demonstrates the integrated model by investigating locational changes under conditions of economic growth. The results provide insight into the changes in the regional geography of advancing economies. Moreover, if this evolutionary approach can depict reality reasonably well, then this research may also help us to develop and test some useful empirical hypotheses.

3.1 Spatial Representation of a Hypothetical Metropolitan Region

The hypothetical area is a metropolis based region consisting of an urban area and a periphery. The focus of the study is an urban area consisting 21 hexagonal land use zones (Moore II and Wiggins 1988). The periphery is defined to be a dimensionless null site. Each physical site is initially occupied by an activity. The urban transportation system consists of aggregate links between zones. The network consists of congestive links connecting to nearest neighbours. The system is summarised in Figure 3.

If a located activity is outbid by vacancy and retires to the queue, interactivity shipments involving this activity are assumed to be imported through the null site. Otherwise, the absence of a key production activity would present an infeasibility (Moore II and Gordon1990). The null site is assumed to be a periphery through which imports and import enter and exit the region.

3.2 Data Synthesis

Parameter values describing the economic growth patterns bearing on this research are drawn from the work of Chenery (1960 and 1980), Kuznets (1966), Chenery and Taylor (1968), Chenery and Syrquin (1975 and 1980). These studies provide income elasticities explaining uniform patterns of economic growth.

Exercising the assignment component of the model requires matrices describing seminet revenues, transportation link costs and capacities, external effects, relocation costs, and intersite distances. Precursor exercises rely on synthetic data, and the literature provides little empirical information relevant to the assignment component of this research. A more realistic description of an existing urban configuration is preferred. However, we rely on synthetic data for two reasons. While an empirical exercise would permit us to forecast the trajectory of a real metropolis, it would not further elaborate the function of the model. Also, we want this work to remain as comparable as possible to precursor efforts. Our synthetic data set is available upon request.

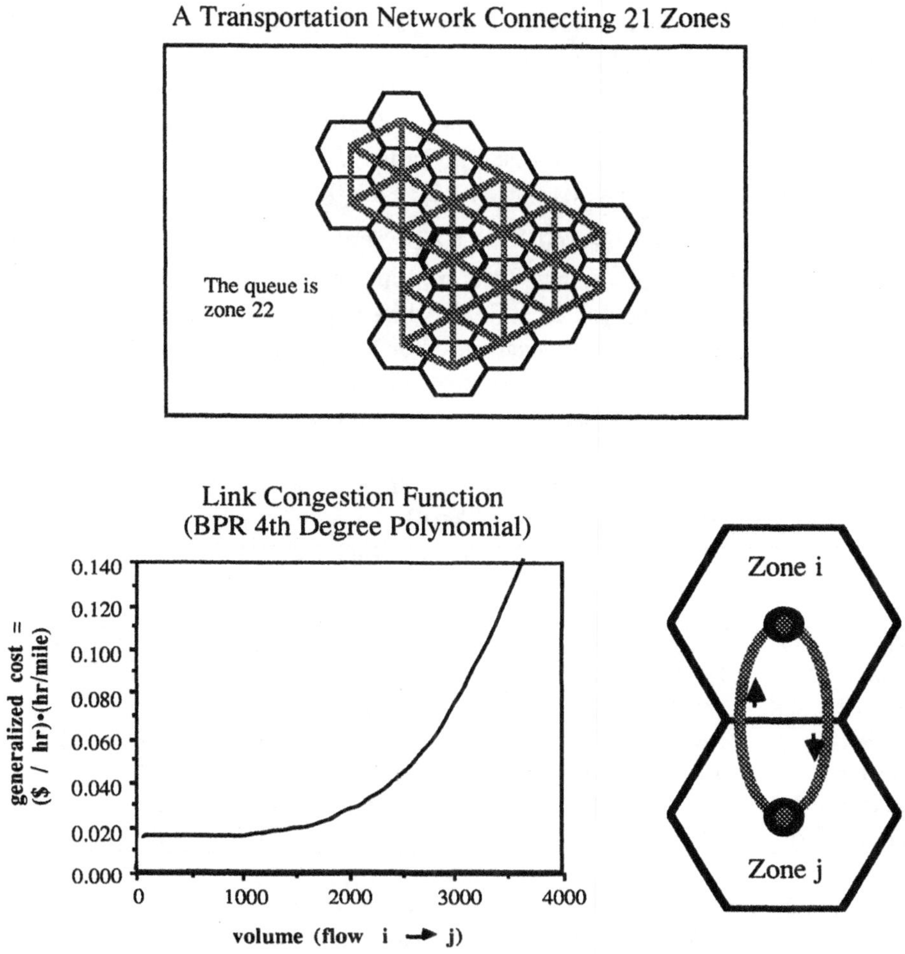

Fig. 3. A 21 Zone System and Associated Transportation Network.

3.3 Results

Development is often characterised by decline in the relative size of the manufacturing sector, almost always accompanied by a rising share of the service sector. Clark (1957) and Fisher (1939) argue that developing economies can be expected to move away from primary production activities toward service production. Because high income elasticities are associated with many service activities, it is argued that this sector only becomes large after the basic necessities are provided by the primary sector, and most demands for manufacturing goods are satisfied.

Such patterns imply nonproportional growth across sectors relative to increases in per capita income. The simulated ratio of production growth rate to income growth rate is summarised in Table 3. The production growth rate of sector p is $\Delta X_p(t) = [X_p(t) - X_p(t-1)] / X_p(0)$. The income growth rate is $\Delta Y(t) = [Y(t) - Y(t-1)] / Y(0)$ = 10 percent. The simulation produces significant differences across sectors in terms of deviations from proportional growth. Sector 2, a final primary production activity such as service, is the fastest growing sector. Sector 5, a primary production activity such as agriculture, is the slowest growing sector. Other sectors, such as manufacturing, fall in between these two extremes.

Table 3. The Ratio of Production Growth Rate to Income Growth Rate: $[\Delta X_p(t) / X_p(0)] / [\Delta Y(t) / Y(0)]$

	TIME PERIOD						
SECTOR	t = 1	t = 2	t = 3	t = 4	t = 5	t = 6	t = 7
Sector 1	1.508	1.542	1.574	1.605	1.636	1.665	1.695
Sector 2	2.128	2.227	2.324	2.421	2.516	2.610	2.703
Sector 3	1.863	1.939	2.016	2.094	2.172	2.251	2.331
Sector 4	1.821	1.889	1.957	2.024	2.092	2.159	2.226
Sector 5	0.716	0.704	0.693	0.682	0.672	0.662	0.652
Sector 6	1.386	1.413	1.440	1.467	1.494	1.520	1.547
Households	1.069	1.073	1.076	1.079	1.081	1.084	1.087

Another phenomenon is the substitution of manufactured goods for primary inputs. The combination of rising purchases by other sectors, together with the substitution of manufactured for primary commodities, produces rapid growth in the intermediate demand for manufactured goods. The corresponding increase in manufacturing output above that implied by proportional growth accounts for the greater part of structural change associated with development. In the simulation, the average shares of intermediate use in total domestic demand increase from 0.828 to 0.846. These increasing average shares of intermediate use imply an increasingly complex economic system. In addition, technological changes are implied by nonproportional growth in domestic demand, final demand, imports, and exports. These technological changes are summarised in Table 4.

A key development phenomenon is increasing use of intermediate industrial products. Lack of interdependence and linkage is perhaps the most typical characteristic of undeveloped economies. Increased use of intermediate inputs is characterises an increasingly complex economic system. As economies develop, their productive structures become more

Table 4. Changes in Technical Coefficients

TIME PERIOD t = 0

PRODUCING SECTOR	PURCHASING SECTOR Sector 1	Sector 2	Sector 3	Sector 4	Sector 5	Sector 6	House-holds
Sector 1	0.156	0.254	0.025	0.051	0.125	0.130	0.194
Sector 2	0.078	0.068	0.175	0.026	0.075	0.174	0.236
Sector 3	0.109	0.034	0.200	0.026	0.125	0.065	0.069
Sector 4	0.172	0.017	0.050	0.205	0.150	0.087	0.056
Sector 5	0.063	0.000	0.025	0.359	0.075	0.043	0.125
Sector 6	0.031	0.102	0.175	0.154	0.050	0.130	0.111
Households	0.250	0.305	0.175	0.128	0.175	0.196	0.014

TIME PERIOD t = 1

PRODUCING SECTOR	PURCHASING SECTOR Sector 1	Sector 2	Sector 3	Sector 4	Sector 5	Sector 6	House-holds
Sector 1	0.160	0.261	0.025	0.054	0.127	0.133	0.195
Sector 2	0.085	0.074	0.189	0.029	0.081	0.189	0.253
Sector 3	0.115	0.036	0.208	0.028	0.130	0.068	0.072
Sector 4	0.178	0.018	0.051	0.220	0.154	0.090	0.057
Sector 5	0.056	0.000	0.022	0.336	0.067	0.039	0.111
Sector 6	0.031	0.102	0.173	0.158	0.049	0.129	0.109
Households	0.237	0.291	0.165	0.126	0.165	0.185	0.013

TIME PERIOD t = 4

PRODUCING SECTOR	PURCHASING SECTOR Sector 1	Sector 2	Sector 3	Sector 4	Sector 5	Sector 6	House-holds
Sector 1	0.165	0.275	0.026	0.062	0.129	0.137	0.194
Sector 2	0.104	0.092	0.226	0.039	0.097	0.229	0.296
Sector 3	0.129	0.041	0.229	0.034	0.144	0.076	0.078
Sector 4	0.192	0.019	0.054	0.260	0.163	0.096	0.059
Sector 5	0.042	0.000	0.016	0.273	0.049	0.029	0.079
Sector 6	0.030	0.100	0.164	0.169	0.047	0.125	0.101
Households	0.205	0.255	0.139	0.120	0.140	0.159	0.011

TIME PERIOD t = 7

PRODUCING SECTOR	PURCHASING SECTOR Sector 1	Sector 2	Sector 3	Sector 4	Sector 5	Sector 6	House-holds
Sector 1	0.167	0.281	0.025	0.068	0.129	0.137	0.191
Sector 2	0.119	0.107	0.254	0.048	0.110	0.261	0.330
Sector 3	0.142	0.045	0.247	0.041	0.156	0.083	0.083
Sector 4	0.201	0.021	0.056	0.296	0.169	0.100	0.060
Sector 5	0.031	0.000	0.012	0.216	0.035	0.021	0.056
Sector 6	0.029	0.098	0.155	0.177	0.045	0.120	0.095
Households	0.181	0.228	0.120	0.114	0.122	0.139	0.009

roundabout in the sense that a higher proportion of output is sold to other producers than to final users. As with final demand, this phenomenon means a shift in output mix toward manufacturing and other sectors that use more intermediate inputs.

The simulation is initialised by an exogenous match between site sand activities. The subsequent sequence of patterns is summarised in Figure 4. Initially, six low density activities are outbid by vacancy. In each period, seminet revenues, externalities, and activity specific relocation costs are modified in light of value added levels. Locators update their bids in response to the environmental externalities and congestion costs experienced in their current locations. Based on these updated bids, the assignment linear program generates new location patterns. Activities are displaced to the queue when they cannot generate a positive bid that is highest for any site. Not all locators relocate simultaneously. Some activities do not change their locations, whereas some relocate and persist for only a few time periods. Some activities exchange their locations. Vacancies are created frequently. After period 7, the simulation results stabilise. Relocations are still possible, but occur much less frequently. This result presumes no special site improvements or new investments in infrastructure.

3.4 Sensitivity Analysis: Transportation Costs

The research model provides an integrated treatment of regional economic change and spatial structure. Seminet revenues, externalities, and relocation costs are defined by production / location decisions. Input-output relationships change as a result of economic development processes, and these changes are translated into transshipments between activities. Transportation costs, including congestion externalities, play an important role relative to other system elements. If the unit transportation costs are too high, then changes in economic structure will dominate the activities' location decisions because each activity's total transportation costs are rendered very sensitive to the changes in traffic intensities. Thus, it is important to determine whether unit transportation costs dominate other location factors.

We investigate four new scenarios in which free flow link costs are increased by 5, 10, 20, and 30 percent, respectively. The 5 and 10 percent increases in link costs induce no changes in the trajectory of location patterns. A 20 percent increase induces a few changes, the most conspicuous of which is a decrease in the number of vacancies. A 30 percent increase in link costs produces significant differences in locational patterns, and a more rapid convergence to stability.

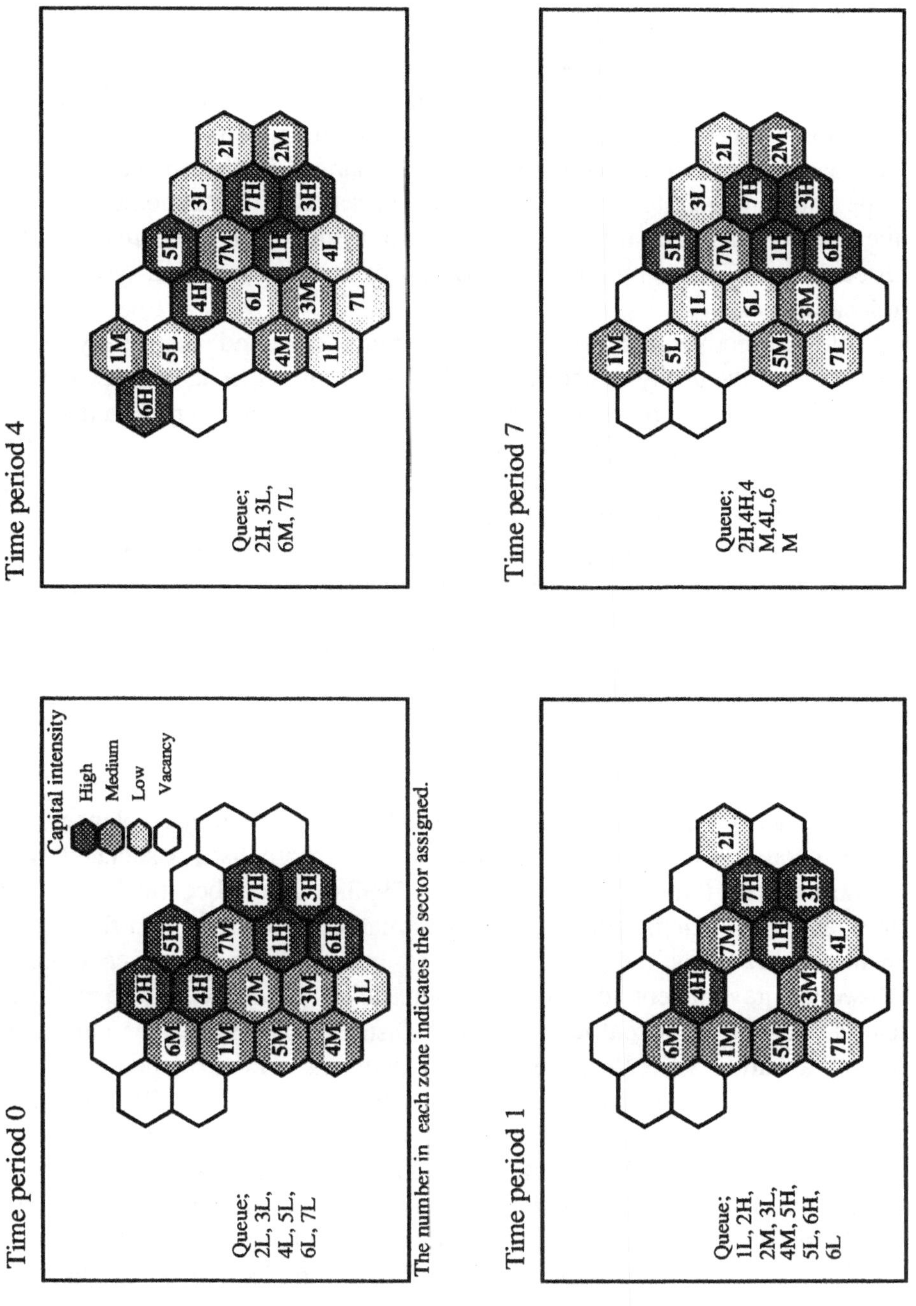

Fig. 4. Simulation Results for Periods 0, 1, 4, and 7

3.5 Conclusions

In contemporary metropolitan areas, decentralisation of activities is a dynamic process resulting from the interplay of simple economic behaviours. The model explained here demonstrates the locational behaviour of activities in a system subject to economic growth. Changes in the spatial structure appear to be related to the locational characteristics of the economic activity; the characteristics of the economic environment; external economies and diseconomies, including the congestability of the transportation system; and relocation costs.

The collective results of these simulations and sensitivity tests demonstrate the utility of decentralisation as a coping mechanism. Economic activities progressively relocate to decentralised locations to maintain access to each other. Gordon, Kumar, and Richardson (1988) contend that the relocation of activities within cities is guided by the desire to avoid congestion. Location of activities, intensity of land uses, means of production, origins, and destinations are all affected by the provision and pricing of transportation facilities, but an increase in transportation costs does not necessarily translate into centralisation. Indeed, the simulation suggests that decentralisation offers more advantage.

4. Policy Implications

The evolution of a policentric spatial structure, either planned or spontaneous, is a reasonable response to the externalises associated with monocentricity (Gordon and Richardson 1993a). The subcentre location of jobs and populations alleviates the external diseconomies of urban scale without sacrificing the benefits of area wide agglomeration economies. However, government intervention often slows these spatial and political shifts toward policentric patterns. Instead of pursuing ambitious decentralisation strategies, metropolitan planners tend to respond to local increases in urban growth, pollution, and traffic congestion. Consequently, resources are invested in infrastructure that exceeds prospective demand. Planners should promote the more efficient policentric structures critical to successful metropolitan growth, and avoid expensive interventions that might inhibit spatial restructuring (Gordon and Richardson 1993a).

The key issues are how, when, and where to intervene. Gordon and Richardson suggest an appropriate scope for planning and regulatory approaches. The first step is to identify when public intervention is justified. The second step is to evaluate the conditions under which market based strategies are less practical than regulation. In most circumstances, planners should draw policy guidance from market approaches. Market based measures use economic principles to alter consumption or production decisions. These include the institution of market exchange mechanisms, or the establishment of prices that reflect true costs. For example, congestion pricing is a pricing system that corrects the market failure inherent in the passenger transport system. Tradeable emissions rights perform similarly in the case of production. Firms that must pay a market price for the right to pollute will not pollute unless the revenues available from production exceed the social cost of emissions.

Because of (perceived) uncertainty concerning the benefits associated with market based approaches, public authorities favour regulation. This reflects a lack of information. As information brokers, planners have a role in forecasting future phenomena. Multiperiod forecasts describing the benefits and optimal budgets of pricing strategies and investments are particularly important. Integrated models of the sort proposed here organise, process, and improve information concerning the anticipated impacts of policies and public investment decisions. By keeping the economic role of externalities explicit, regulatory strategies can be compared to other approaches aimed at internalising externalities.

References

Berry, B. J. L. (1976) The Counter Urbanization Process: Urban America Since 1970. In B.J. L.Berry (ed.), *Urbanization and Counterurbanization. Sage*, Beverly Hills

Blackley, Dixie M and Follain, James R. (1987) Tests of Locational Equilibrium in the Standard Urban Model. *Land Economics 63*, 46-61

Blumenfeld, H (1964) The Urban Pattern. *Annals of the American Academy of Political and Social Science 352*, 74-83

Celasun, M. (1986) A General Equilibrium Model of the Turkish Economy, SIMLOG-1. *METU Studies in Development 13*, 29-94

Cervero, R. (1986) *Suburban Gridlock.* Rutgers University, New Brunswick

Chenery, H. B. (1980) Changes in Trade Shares and Economic Growth: Interactions Between Industrialization and Exports. *American Economic Association 70*, 281-287

______________. (1979), *Structural Change and Development Policy.* Oxford University Press, London

______________. (1960) Patterns of Industrial Growth. *American Economic Review 50*, 624-54

Chenery, H. B. and Clark, P. G. (1959) *Interindustry Economics.* Wiley, New York

Chenery, H. B., Shishido, S. and Watanabe, T. (1962) The Pattern of Japanese Growth. *Econometrica 30*, 98-139

Chenery, H. B. and Syrquin, M. (1986) Typical Patterns of Transformation. In H. B. Chenery, S., Robinson, S., and M. Syquin, M. (eds.), *Industrialization and Growth: A Comparative Study.* Oxford University Press, London

_____________. (1980) A Comparative Analysis of Industrial Growth. In R. C. O. Matthews (ed.), *Economic Growth and Resources*. St. Martin's Press Inc., New York

_____________. (1975) *Patterns of Development 1950-1970*. Oxford University Press, London.

Chenery, H. B. and Taylor, L. (1968) Development Patterns: Among Countries and Over Time. *The Review of Economics and Statistics 50*, 391-416

Chenery, H. B. and Watanabe, T. (1958) International Comparisons of the Structure of Production. *Econometrica 26*, 487-521

Chenery, H. B. et al. (1986) Alternative Routes to Development. In H.B. Chenery, S. Robinson, and M. Syquin, (eds.), *Industrialization and Growth: A Comparative Study*. Oxford University Press, London

Clark, F. P. (1954) Office Buildings in the Suburbs. *Urban Land 13*, 8-14

Clark, C. (1957) *The Conditions of Economic Progress*. Macmillan and Company, London

Dervis, K. (1975) Substitution, Employment, and Intertemporal Equilibrium in a Non-Linear Multi-sector Planning Model for Turkey. *European Economic Review 6*, 77-96

Diewert, W. E. and Wales, T. J. (1985) Flexible Functional Forms and Global Curvature Conditions. University of British Columbia, Vancouver

Dorfman, R., Samuelson, P. A., and Solow, R. M. (1958) *Linear Programming and Economic Analysis*. McGrawHill Book Company, Inc., New York

Erickson, R. A. (1983) The Evolution of the Suburban Space Economy. *Urban Geography 4*, 95-121

Fisher, A. G. B. (1935) *The Clash of Progress and Security*. Macmillan and Co., Ltd., London

Fisher, A. G. B. (1939) Production: Primary, Secondary, Tertiary. *Economic Record 15*

Gordon, P., Kumar, A., and Richardson, H. W. (1989) The Spatial Mismatch Hypothesis: Some New Evidence. *Urban Studies 26*, 315-326

Gordon, P. and Moore II, J. E. (1989) Endogenizing the Rise and Fall of Urban Subcentres via Discrete Programming Models. *Environment and Planning A 21*, 1195-1203

Gordon, P. and Richardson, H. W. (1993) Market Planning: Oxymoron or Common Sense? *Journal of the American Planning Association 59*, 347-352

Hartshorn, T. A. and Muller, P. O. (1986) Suburban Business Centres: Employment Implications. U.S Department of Commerce, Washington, DC

Hewings, G.J.D. (1977) Evaluating the Possibilities of Exchanging Regional Input-Output Accounts. *Environment and Planning A 9*, 927-944

Hewings, G.J.D. (1982) Regional and Interregional Interdependencies: Alternative Accounting Systems. *Environment and Planning A, 14*, 1587-1600

Houthakker, H. S. (1957) An International Comparison of Household Expenditure Patterns: Commemorating the Centenary Engel's Law. *Econometrica 25*, 532-551

Johansen, L. (1960) *A Multi-sectoral Study of Economic Growth*, North-Holland Publishing Company, Amsterdam

Kuznets, S. (1966) *Modern Economic Growth: Rate, Structure, and Spread.* Yale University Press, New Haven

______________. (1957) Quantitative Aspects of the Economic Growth of Nations II. Industrial Distribution of National Product and Labor Force. *Economic Development and Cultural Change 5*, suppl

Leontief, W. (1951) *The Structure of the American Economy, 1919-1939.* Oxford University Press, New York

Leontief, W., et. al. (1953) *Studies in the Structure of the American Economy*, Oxford University Press, New York

Lewis, W. A. (1954) Development with Unlimited Supplies of Labour. *Manchester School of Economic and Social Studies 20*, 189-192

Melo, J. de, (1977) Distortions in the Factor Market: Some General Equilibrium Approach *The Review of Economics and Statistics 54*, 398-405

Melo, J. de and Robinson, S. (1980) The Impact of Trade Policies on Income Distribution in a Planning Model for Columbia. *Journal of Policy Modeling 2*, 81-100

Melo, M. de, (1979) Agricultural Policies and Development: A Socioeconomic Investigation Applied to Sri Lanka. *Journal of Policy Modeling 1*, 201-216

Moore II, J. E. and Gordon, P. (1990) A Sequential Programming Model of Urban Land Development. *Socio-Economic Planning Science 24*, 199-216

Moore II, J. E. and Wiggins, L. L. (1990) A Dynamic Mills Heritage Model with Replaceable Capital. *Papers of the Regional Science Association 68*, 23-41

Nagurney, A. (1993) *Network Economics: A Variational Inequality Approach.* Kluwer Academic Publishers, Dordrecht, Netherlands

Orski, C. K. (1986) Transportation Management Associations: Battling Suburban Traffic Congestion. *Urban Land 45*, 2-5

Pisarski, A. E. (1987) *Commuting in America.* Eno Foundation for Transportation, Inc., Westport

Richardson, H. W. (1988) Monocentric vs. Policentric Models: The Future of Urban Economics in Regional Science. *The Annals of Regional Science 22*, 1-12

Taylor, L. J. (1969) Development Patterns: A Simulation Study. *Quarterly Journal of Economics 83*, 220-241

Taylor, L. J. and Black, S. (1974) Practical General Equilibrium Estimation of Resource Pulls Under Trade Liberalization. *Journal of International Economics 4*, 37-58

The United Nations (1963) A Study of Industrial Growth, ST/ECA/74. Department of Economic and Social Affairs, New York

Urban Land Institute (1987) Market Profiles 1987, Urban Land Institute, Washington, DC

______________. (1986) Development Trends 1986, Urban Land Institute, Washington, DC

Walras, L. (1954) *Elements of Pure Economics.* Homewood, Illinois

Part D. Summing Up

SUMMING UP

John R. Roy and Yoshitsugu Hayashi*

The final panel discussion summarised the main results of the seminar and examined future research directions to include 'the environment' more completely within the modelling and evaluation of land use and transport development. The findings and progress achieved in this seminar are summarised as follows:

1. Trends in Urban Land Use, Transport and the Environment in Developed and Developing Countries

Energy consumption in the transport sector is increasing and the environment is becoming degraded due to low-density suburbanisation, growing car use and congestion [Phiu-nual (Chap 2)]. This is caused by the processes of economic development, urbanisation and motorisation [Hayashi (Chap 1)]. While energy consumption in developed countries is more sensitive to transport conditions than to the more durable land use patterns [Wegener (Chap 6)], rapidly sprawling suburbanisation due to poor land use control is the determining factor in developing countries.

Reduction in private car use is now clearly defined as a national policy objective in several developed countries, such as within the Better Cities Program of the Australian Government [Telford (Chap 3)]. Cities such as in Australia are trying to move towards higher density housing through urban consolidation in order to achieve better transport and energy efficiency [Black (Chap 4)].

---

*With contributions as panellists from: David Boyce, John Brotchie, Hisayoshi Morisugi, and Michael Wegener.

Y. Hayashi and J. Roy (eds.), Transport, Land-Use and the Environment, 415–420.

Considering that the dominant thrust of urbanisation is anticipated to be occurring in developing countries in the future, it is particularly important to observe and analyse future trends in land use changes, transport demand and energy consumption, and to evaluate appropriate countermeasures.

2. Requirements and Strategies for Better Cities and Regions

To make cities better, the following requirements may be considered [Brotchie (Chap 5) and Wegener (Chap 6)]: 1) Economy: competitiveness through interegional/international superiority and agglomeration, and attractiveness via new jobs and labour skills, is required. 2) Equity: social and spatial disparities should be reduced. 3) Ecology: the city activities should be kept within a sustainable level of energy consumption and emissions.

It has been observed in developed countries that policy priorities are gradually changing, through Economy to Equity and to Environment.

Possible strategies to meet these requirements include close connections of work places and residences, appropriate modal balances between road and rail, promotion of teleworking and teleshopping, improvement of light rail transit and demand-responsive bus systems, etc. [Brotchie (Chap 5)]. There should be also a concern for harmony of the natural and built environment.

3. Analysis Tools

3.1 Modelling Techniques

In this seminar, a variety of alternative modelling approaches have been shown, including a) equilibrium in location and transport networks [Martínez (Chap 12), Ueda, Nakamura and Shimizu (Chap 16), Taylor (Chap 17), Lundqvist (Chap 18), Seo, Moore II and Gordon (Chap 19)], b) optimisation in activity location [de la Barra (Chap 14)], c) an expert system [Mackett (Chap 15)] and d) preliminary design [Roy, Marquez, Taylor and Ueda (Chap 7)]. Martínez (Chap 12) gives a most sophisticated formulation of characterisation of urban land use, transport and the environment in the context of random utility and bidding theory. Miyamoto and Rungsun (Chap 13) present a practical model system consisting of similar models as given by Martínez (Chap 12). Ueda, Nakamura and Shimizu (Chap 16) use a formulation of equilibrium of land and constructed floorspace combined with a GIS system. Taylor (Chap 17) presents a most comprehensive model framework of land use, transport and the environment, which can examine various transport policies. Lundqvist (Chap 18) reports on the development of modern combined network equilibrium models and their applications to environmental assessment, supported by a user friendly software. Seo, Moore II and Gordon (Chap 19) describes a simulation model accounting for interactions between a priori profitability, transport costs in a congested network, externalities, relocation costs and technological changes, under an estimated growth of regional economy by an I/O model.

De la Barra (Chap 14) describes a comprehensive optimisation model of land use-transport which can treat a variety of variables. Mackett (Chap 15) develops a pioneering artificial intelligence approach to identify an appropriate new public transport system, considering its interactions with land use and the resulting environment. A cohort survival model is used for simulating ages of vehicles which influence on the emission rate of pollutants from engines in Morisugi and Ohno (Chap 9). Emission models were also applied by Wegener (Chap 6), Roy, Marquez, Taylor and Ueda (Chap 7), Young and Gu (Chap 8) and Morisugi and Ohno (Chap 9).

3.2 Computer Aided Tools

Computer aided tools were also widely introduced, such as GIS [Ueda, Nakamura and Shimizu (Chap 16)], GUI (Graphical User Interface) [Young and Gu (Chap 9)], window systems on the micro computer [Miyamoto and Rungsun (Chap 13)] and computer graphics [Taylor (Chap 17)]. Such techniques are being applied during the past two or three years as powerful and cheap tools for land use-transport models.

3.3 Modularisation and Model System Packages

Several attempts have been conducted to structure sets of submodels into integrated packages on the micro computer [Roy, Marquez, Taylor and Ueda (Chap 7), Miyamoto and Rungsun (Chap 13)]. This is a substantial improvement, made possible by progress in micro computer techniques. Modularisation of models is also applied to develop an educational tool [Young and Gu (Chap 9)] which is a new direction in land use transport interaction models where development is ongoing.

4. Modelling of Implications of Alternative Policies

Roy, Marquez, Taylor and Ueda (Chap 7) and Young and Gu (Chap 8) can illustrate the effects of different strategies on total vehicle trip length, public transport trip length, petrol consumption and CO_2 emissions in simplified cities. Kim, Eash and Hanley (Chap 10) can also examine short-term impacts of pricing strategies on VMT reduction.

The Dortmund Model [Wegener (Chap 6)] has one of the most comprehensive structures, and can examine dynamically the effects on vehicle CO_2 emissions of a variety of policy options for reorganisation of urban activities.

Morisugi and Ohno (Chap 9) can forecast the effects on NO_x emissions of fuel tax policies. This could be built in as a submodel in integrated land use-transport models.

Doi and Okamoto (Chap 11) formulates the process of deforestation in a suburb of the Tokyo metropolitan area. This is a pioneering work in that it formulates the changing balance of urban and agricultural land uses.

5. Requirements for a New "LTE" Framework

As the inclusion of the environment in this field is not seen merely as an appendage to the existing models and their application, but as an influence which permeates these models and adds new options, this summary is presented in two parts. The first part describes recent improvements, as well as limitations of current land use/transport (*LT*) models, both in their modelling of the interaction of land use and transport and in their corresponding portrayal of the environment. In the second part, some required changes in the land use/transport evaluation framework are discussed in order to include the environment as an intrinsic influence, both active and passive, thereby creating a new *LTE* (Land Use, Transport and Environment) framework.

5.1 Improvements and Limitations of the LT Framework

The improvements and limitations of LT models will be treated in succession, for the modelling of transport, land use and land use/transport interaction.

Recent improvements in LT modelling

Transport

- Nesting within the 4-step model
- Consideration of road pricing
- Interfacing with GIS
- Attachment of improved fuel consumption sub-models and emissions sub-models
- Some attempts at network design

Land Use

- Improved models of land price and housing density
- Interfacing with GIS
- Evaluation of deforestation
- Increased disaggregation of industry sectors
- Integrated analysis of land and building markets

Land Use/Transport Interaction

- Consideration of housing and job mobility
- Market segmentation of households and jobs
- Inclusion of timelags in land use adjustment
- Some consideration of market disequilibrium
- Examination of integrated policies
- Stronger techniques of economic evaluation
- Object-oriented code for sub-models

Some limitations in LT modelling

Transport

- Static assignment models without departure times
- Inadequate treatment of trip frequency and car ownership
- More work required on shared ride modes
- Noise and air pollution dispersion models need improvement
- Assessments needed of telecommuting potential
- Need household activity analysis and time budgets

Land Use

- Require better models for location of knowledge-based firms
- More optimisation needed in land use zoning
- Fuller integration with GIS required
- Better integration with waste disposal and water supply networks
- Better evaluation of trade-offs of urban vs. agricultural land
- More study on urban consolidation vs. job dispersion

Land Use/Transport Interaction

- Need improved consideration of market disequilibrium
- Distinguish urban growth from moves within existing stock
- More exchangeable models required
- More comparative analysis between different cities
- Greater consideration of equity of access
- More study of intercity competition and infrastructure supply
- Validation of models in assessing impacts of large changes
- Flexible treatment of integrated policy scenarios

5.2 Requirements for a New LTE Framework

In addition to the limitations described above in the LT framework, many of which influence on environmental impact assessment, further efforts are required to create a LTE framework, as summarised below:

- Better specification of 'sustainability' for the land use/transport system
- Assessment of decentralised concentration strategies
- Effects of job dispersion on job choice behaviour
- Putting a value on the environment
- Inclusion of natural drainage methods
- Evaluation of environmental carrying capacity of regions
- Introduction of more aesthetics into the built environment
- Assessment of abatement costs of CO_2 reduction
- Comparison of environmental performance between cities of different sizes
- Assessment of costs of failure to control CO_2 emissions
- Environmental feedback on consumer/firm location preferences
- Output of water pollution and waste disposal consequences of scenarios
- Need to evaluate embodied energy of transport infrastructure in materials and construction
- Impacts of urban sprawl on deforestation
- Evaluate effects of tax policies on local CO and NO_x emissions
- Specification of new social welfare functions
- Optimisation with respect to policy variables

Many of the above items require significant conceptual work and corresponding model developments. Others require an expanded use of existing models. The above represent a subset of the measures required for our models to play a significant role in advising governments on the integrated policy scenarios necessary for the transport sector to contribute equitably to meeting the 2005 Greenhouse targets. Even if the onset of the Greenhouse effect is slower than expected, or is mitigated by countervailing influences, the transport sector is a major consumer of many non-renewable energy resources. Our efforts to reduce CO_2 emissions will in many cases be pursuing the parallel goal of reducing our consumption of non-renewable resources, as well as yielding improved quality-of-life in cities and regions.

INDEX

access measures 255
activities 110
activities-transport model 281, 285
activity location model 318, 321
accessibility improvement, impact of 223
Aidelaide, Southern Areas Area Strategy 49
air pollution 15, 37
  in Bangkok 15, 39
  in Nagoya 15
  phenomena 233
  problems, severity of 38
air quality 40
  standards 37
Akcelik's function 343
alternative policies, modelling of implications 417
Alonso model 268
Alonso's school 237
ample fuel use, *see* fuel use
American cities, contrast between European and 11
analysis tools 260, 262, 263, 416
analysis zones
  average distances 197
  Chicago 194, 195
annual vehicle use 57, 58
artificial intelligence 297
  approach 297, 307
  use of 307
automobile choice 200
  estimating probability 200
Australian urban development 56, 57
Australian model code for urban housing 56, 72
Australia
  better cities program 47
  October 1990 Special Premiers' Conference 55
  transport in the cities 55
average speed model, *see* emission/consumption models

Bangkok (Thailand) area 3
  air pollution 37
    data 39
  air quality in 40
  car ownerships, level of 15
  CO level 15
  comparison with London 5
    with Nagoya 5
    with Tokyo 5
  demolition and land-take associated with new infrastructure projects 37
  development in 32
    ribbon 32
  from water-based to road-based city 27
  Greater Area (GBA) 29
    traffic congestion in 30
    traffic management in 31
    travel time in 30
  location of activities 34
  major environmental issues 35
  noise 37, 41
  pedestrian/vehicle conflict 37
  present transport system in 27
  railways in 29
  road accidents in 32
  severance of communities and agricultural land 37
  transport in 32, 34
  trip distances in 31
  visual impact 37
BART (Bay Area Rapid Transit system), in San Francisco 302
behavioural modelling 233
best-bid rule 237, 238
better cities
  program 47, 52
    goals of 48
    in Australia 47

requirements of 416
strategies for 416
bid-choice model 237, 238, 239, 243, 252
empirical 252
theoretical 252
Bosch method 37
Boso peninsula region, Tokyo 214
recreational development in 214
Brotchie Triangle 111
BTSC 29

Canberra
student projects, comparison of 164
Canberra Network 154
development 154
Canberra strategies 154, 158
comparison of 158
car ownership 6
growth 297
in France 303
in Germany 303
in Bangkok 15
in Nagoya 15
in Tokyo 15
levels 57, 58
car engines, advanced innovative technology in 23
car trip length 13
long 14
total 14
car use 12
centralized strategy 157, 166
Chicago area 191
compliance issues 192
transportation study 193, 194
trip tables for 1990, by 194
choice model 179
choice tree 273
cities
compactness of 11, 12
in competition 81
CITY 136
definition of 137
hierarchy of objects in 138
Clean Air Act 1970 191
Amendments of 1990 (CAAA 1990) 191, 192
CMAWIN 368
CO, pollution levels of 15
CO_2 emissions 126, 134
reduction of 103
transport 103
transport-related, in Dortmund 107
CO_2 production 140
Cobb-Douglas function 225
Cobb-Douglas technology of developer's production 224
CODMA 369
Cohort model
flow chart 176
of predicting number of diesel automobiles 174
community perspectives 66
computer aided tools 417
computer-based information systems 68
computer-based method, on impact assessment of transport policies 338
computable general equilibrium (CGE) models 391
conceptual model 61, 67
congestion 6, 27, 30
causes of 15
in Bangkok 275
models 343
traffic 15
control mechanism 308
cost-benefit analysis 233
Council Directives 42
81/334/EEC 42
84/424/EEC 42
cross-sectioned analogous analysis 23

data converter 266
data synthesis 402
decentralized strategy 157, 166
defenitions
of developer's profit 324
of landowner's profit 325
of surplus function 323

Dennis package 360, 362, 375, 376, 379
developer's production
  Cobb-Douglas technology of 224
developer's profit, definition of 324
developer's strategy 221
development strategy 157, 166
diesel automobiles 171
  cohort model of predicting number of 174
domain 308
do-nothing strategy 157, 166
Dortmund, Germany 107, 108
  study area 107
dual networks 288
dynamic data exchange function 266

ecologically sustainable develepment (ESD) 55, 61
  recommendations 62

econometric modelling 383
economic cause-effect mechanism 235
economic development 6
  and its influence on the environment 3
  and the environment nexus 23
  cycle 21
  relationship with environmental changes 6
    mechanism 7
economic growth patterns 402
eco-tourism 50
efficiency 120
elasticity model, derivation of 212
elastic travel demand, consideration of 355
elemental model, *see* emission/consumption models
empirical location model 238
emission
  reduction strategies 210
  transportation control measures 198
emission/consumption models for traffic streams
  average speed model 345
  elemental model 345
  instantaneous model 345
  running speed model 345, 347
emissions
  carbon monoxide (CO) 341, 377
  carbon dioxide (CO_2) 341, 377
    hydro carbons 377
  from mobile sources 191
  lead 341
  nitrogen oxides (NOx) 171, 341, 377
  of road traffic 341
  Volatile Organic Compounds (VOCs) 191, 192, 341
  water vapour 341

EMME/2 364, 367, 371, 374, 376
employer trip reduction (ETR) requirements 192
employment 64, 65, 108
  distribution 160
  spatial distribution of 129
energy 68
  consumption 3, 4, 13, 57, 297
    increased 6
  crisis, 1973 125
  resources 126
    consumption 126, 134
  saving 11
  use 131
environment 147
  changes in 23
  in a metropolis 259
  in developed countries 415
  in developing countries 415
  nexus 23
environmental assessments
  in the 1990 regional plan outline 374
  in the 1991 regional plan proposal 374
  of the Dennis package agreement 376
environmental changes 6

environmental degradation problems
6
environmental issues 35
in Bangkok *see* Bangkok
environmental impacts
of road traffic 338, 339
of transport policies 65
prediction 342
traffic related 372

environmental implications 313
environmental planning decisions
in transport planning 337
a modelling framework 337
environmental policies 244
analysis of 247
equilibrium
applications to environmental
assessments 359
computable (CGE) 391
conditions 327
general 392
models
combined network 359
in Stockholm 364
programmability of an 330
ESTRAUS, Santiago's transport
model 256
European cities, contrast between
American and 11
evaluation procedure 289
expert system 308
Expressway and Rapid Authority
(ETA) 29

fast transport networks 81
FORTRAN 368
freight rates and prices
increasing ratio of 185
forecast of 186
fuel consumption, in road traffic 341
fuel/emissions model, for transport
network analysis 348
fuel prices 10, 12, 172, 173
gasoline 10
in developed countries 172
influence on car use 10
influence on suburbanisation 10
light oil 10
fuel use, ample 15
fully-automatic metro, Lille 302

gasoline, price of, *see* fuel prices 10
GIS (Geographical Information
System) 6, 122, 264, 332
design of 319
for urban transport 317, 318
major components of 319
general interactions mechanism 244
geometrical shape 128
government policy 67
graphical user interface 264
Greater Bangkok Area (GBA), *see*
Bangkok
green area, attractiveness 222
Green Belt, act of 1938 20
Greenpeace urban strategy for Sydney
67
growth
managed, policy of 34
modern economic 386
theory 383
GRP (Gross Regional Product) 10

Hartridge method 37
higher density housing, in Australian
cities 55
highway network improvement 214
its impact on green conservation
214
housing 64, 68
distribution 129
higher density, in Australian cities
55
market 108
preference survey 67
hypothetical metropolitan region,
spatial representation 402

ICD (Intelligent Cities Database) 311
impact/assessments, methodology 372
IMREL (Integrated Model of

Residential and Employment Location) 379
industrial revolution 82
information economy 82
transition to 82
infrastructure 3
investment 18
level 18
input-output techniques 383
instantaneous model, *see* emission/consumption models
integrated land-use transport economic model 234
microeconomic framework of 234
integrated model, structure of 278
integrated transport strategy 65
for Greater Sydney 65
ISQUTI study 112, 121, 125

job choice 131

knowledge base 308

labour market 108
LAND (Location of Activities and Network Development) package 147, 148, 154, 163
application to Canberra 154
environmental module 152
land use module 150
model 148
structure 149, 150
transport module 151
land and construction market 109
land demand, derived from random utility analysis 268
landowner's profit, definition of 325
land planning system 20
land prices 244
land rents 239
Santiago case study 239
land supply, derived from random rent-bidding analysis 268
land use 6, 110, 135, 147, 165, 259, 313
change in 213, 216, 244
causality analysis of 217
development 213
in a metropolis 259
in the Santiago case 248
natural, index of 217
rearrangement 244
land use categories
commercial land 217
forest land 217
grassland and farm land 27
industrial land 217
residential land 217
suburban recreational land 217
urban recreational land 217
land use configuration, change of 216
land use control
in the UK 20
strict 11
land use economics 236
land use/location strategies 210
land use models 260, 263, 271, 321
Lowry type 321
RURBAN 271, 274, 278
land use simulation 256
land use parameter 85
land use planning system 18
land use system 110, 120
land use taxes 248
land use/transport coordination 59
land use transport feedback cycle 110
land use transport models 107, 359, 383
4-stages 243
5-stages 243
large scale networks 359
in Stockholm region 359
light oil, price of, *see* fuel prices
light rail system, Grenoble 302
LILT (Leeds Integrated Land use Transport) model 309
LISREL, causality analysis, based on 218
location choice behaviour 325
location equilibrium 328
location, residential 134
location utility 273

logit model 321
long car trips, *see* car trip length
longitudinal (time-serial) analysis 23
Lowry-type land use model, *see* land use model
LTE framework 418, 420
LT framework 418

Malar region 360, 363
managed growth 34
  policy of 34
market equilibria 359
measure of the benefit 255
metro
  in Lille 302
  in Lyon 302
  in Marseille 302
  in Paris 302
metropolitan dispersal 83
microeconomic framework 234, 235
mobility sub-model 240
mode choice 131, 241
  strategies 210
mode split sub-model 241
model estimation 226
model operation, principle of 140
modelling techniques 416
model system packages 417
modern economic growth, *see* growth
motor car fuel consumption 57, 58
motorisation 20
MRTA 29
multidimensional networks 288
multinomial logit form 239, 254
mutual enhancement 12

NAAQS (National Ambient Air Quality Standards) 191
Nagoya
  car ownership in 15
  CO level in 15
natural land use, index of, *see* land use
network specifications 130
networks, small scale in Stockholm region 359
NMNL (Nested Multinomial Logit) models 285, 287
noise 37, 41
  levels 43
  problems 43
  standards 41

non-residential buildings 108
  market for 108
NOx emissions, *see* emissions
NOx reduction 171
  effects 171
  simulation analysis on 180
    policies 180, 181

object-oriented design (OOD) 136
oil shock of 1973/74 60
optimising procedures 292
optimising techniques
  in activities-transport models 281
  need for 281
original data base 265
output presentation 265
overlapping control 287

pedestrian/vehicle conflict 44
person trips, estimation of changes in 204
Pivot point analysis 199
  model 199, 208
PLUTO model of Bonsall 127
policy implications 408
policy measures
  analysis system 259
  integrated sets of 262, 265
pollutants
  from road traffic 342
  NOx 171
pollution 297
  in Bangkok 27
  levels of CO, *see* CO
population 108
  density 57, 58
  distribution 159
prices 110
  average, increasing ratio of 187

change 171
predicting model of 183
pricing strategies 211
short-term impact analysis of 191
private/public funds 23
private motor vehicles, use of 52
public transport 69, 134, 302, 306
comparison 162, 165
in British cities 297
the Leeds model 306
in French cities 297
in German cities 297
in United States' cities 297
modelling decisions 306
modern 297
need for new systems 298
network 162
supply, subsidies to 247
technology 297
use of 52, 69
Puget Sound Transportation Panel Data 1989 71

railways 9, 29
in Bangkok 29
BTSC 29
expressway and rapid transit authority (ETA) 29
MRTA 29
skytrain mass transit system 29
RAMTUFS 29
random rent-bidding analysis 268
random utility analysis 268
random utility theory 237
recreational resources
indoor 215
outdoor 215
recreational travel demand, modelling 221, 222
recreational visitors, trends of 215
regional economic analysis 385
regional economic growth/urban spatial structure link 383
regional spatial structure 385
re-location 134
residential 134
rent-bidding function 274
residential buildings 108
residential density 68, 69
Restriction of Ribbon Development Act of 1935 20
ribbon development 20, 32
roads 27
freeways 133
local 133
urban arterials 133
road improvement
in the suburban recreational area 213
spatial impact of 213
road network 133, 162, 165
comparisons 161
in Bangkok 28
road pricing 247, 294, 297
road traffic
emissions of 341
environmental impact of, estimation 338
fuel consumption in 341
pollutants from 342
road transport 29
fuels 341
diesel 341
liquid petroleum gas (LPG) 341
super-graded 341
unleaded petrol (ULP) 341
route assignment 133
route choice model 364
routes 241
running speed model, *see* emission/consumption models
RURBAN
(Random Utility/Rent-Bidding Analysis) 260, 263, 271, 274, 278
brief explanation of 266
structural equations of 268
structure of 270

salt mining industry, of south Australia 50
Santiago

air pollution conditions in 234
environmental pollution in 234
land rents, case study of 239
land use in 248
transport model ESTRAUS 256
S-bahn, in Cologne 302
scenario analysis method 21
its phenomenal interpretation 21
of environmental influences 21
scenarios 112
SENE model 329
programming formulation of 330
practical use of 331
sensitivity analysis, transportation costs 406
sequential urban land use model 387
severance 44
short-term impact analysis of pricing strategies 191
simulation bath file 266
simulation data file 266
simulation models 282
simultaneous equilibrium
of market and location 328
with externalities (SEE) 329
with no externalities (SENE) 329
SIP (State Implementation Plan) 192
skytrain mass transit system 29
small scale networks 359
in Stockholm region 359
smog 233
South Australia, salt mining industry 50
Southern areas area strategy in Adelaide 49
spatial structure 383
regional, trends in 384
strict land use control, *see* land use control
structural transformation model 391, 397
algorithmic elements of 397
Stockholm region 359, 360
combined network equilibrium models 364
CMA 364
EMME/2 364, 367, 371, 374
suburbanisation 6, 10, 11, 12, 13
influence on car trip length 13
suburban recreational areas 213
transport improvement in 213
SUN SPARC 1 workstation 334
surplus functions, definition of 323
survival model 177
SUSTAIN (Sustainable Urban Structure and Interaction Networks) model 96, 125, 127, 129, 131, 133, 134, 135, 136, 140
structure of 136
sustainability 81, 120
SYNERGETIC 369, 370, 371
system building
background of 263
basic concepts of 263
system development
basic concepts for 264
stages of 334
system parts, functions of 264
system structure 264

TALA (Transport and Activity Location Analysis) subsystem 320
input-output relation of 320
tax 172
in developed countries 172
TCMs (Transportation Control Measures) 194, 208
impacts of 194
TDM (Travel Demand Management) 337
technological change, impacts of 81
telecommunications 81, 96
test run simulation, results of 334
Thailand 3
Thai environmental agencies/institutions 36
Thai transport agencies/institutions 36
TOPAZ model 93, 96, 136
Tokyo, car ownership in 15
total car trip length, *see* car trip length
total transport energy consumption,

see transport energy consumption
Town and Country Planning Act of 1919 20
traffic 68
congestion, *see* congestion
equilibria 359
management 31
speeds 297
transport 64, 68, 72, 147, 313
availability 27
demand model 242
demand modelling 337
development, Ultimo-Pyrmont area strategy 52
emissions
private 234
public 234
energy 140
energy consumption 11, 14
total 14
in a metropolis 259
infrastructure 9
adequate 23
interaction 6
market 109
model 260, 271, 287
analysis 263
network
analysis 348
planning, applications in 353
networks 130
parameter 85
planning 282, 337
policy instruments 135
public 69, 134, 302, 306
use 52, 69
-related CO_2 emission reduction 107
systems 110, 120, 121, 240
interaction 243
technology 84
changes in 84
trends in 415
in developed countries 415
in developing countries 415
Transport Working Group, final report 1991 61
recommendations 63
transportation
costs 406
market for 388
TRANUS 121, 284, 285, 289
travel time 30
trend model 178
trip
destination 240
distances 31, 140
times 140
trips
distribution sub-model 241
number of 240
Tri-State regional Planning Commission Study of the New York region 1963 71

Ultimo-Pyrmont area strategy 52
undefined values 284
undefined variables 289
United Nations Stockholm Conference 1972 61
urban activities 103
reorganisation of 103
system 236
urban affairs, Australian federal government role 61
urban consolidation 61
urban density 68
urban environmental policies, analysis of 233
urban form 84, 128
changes in 84
urban housing 68, 128
model code for 68
urban land market 238, 388
urban land use
trends in 415
in developed countries 415
in developing countries 415
urban productivity 81
urban public transport improvement 297
urban radius 13

urban rail transit systems 12
urban simulation model, elements of 128
urban sprawl control 210
urban structures 120
urban system
  changes to 84
  model of 85
urban transport 318, 321
urbanisation 3, 20
  development, stages of 6
    deurbanisation 6
    reurbanisation 6
    suburbanisation 6
    urbanisation 6
URPOL model 153
USA, 1990 National Personal Transportation Survey 69, 71
user interface 308
utility function 274
UTOPIA (Urban Transport Planning and Operations using Intelligence Analysis) 298, 309, 311, 313

vehicles 130
  emission 41, 42, 341
    noise limits 42
    standards 37
  flows, segmentation of 344
  types 131
visual impact 37, 44
VMT (Vehicle Miles Travelled)
  reductions 191, 192, 194, 198, 204, 208
  strategies 210
VOC emissions, *see* emissions

Walrasian (general) multimarkets equilibrium 321, 327
Wardrop's equilibrium laws 241, 245
Wardrop's principles 353
water and sewerage treatment programs 49
wetlands, role of 51
Wigan's methodology 339

Transportation Research, Economics and Policy

1. I. Salomon, P. Bovy and J-P. Orfeuil (eds.): *A Billion Trips a Day.* Tradition and Transition in European Travel Patterns. 1993
   ISBN 0-7923-2297-5
2. P. Nijkamp and E. Blaas: *Impact Assessment and Evaluation in Transportation Planning.* 1994 ISBN 0-7923-2648-2
3. B. Johansson and L.-G. Mattson (eds.): *Road Pricing: Theory, Empirical Assessment, and Policy.* 1995 ISBN 0-7923-3134-6
4. Y. Hayashi and J. Roy: *Transport, Land-Use and the Environment.* 1996 ISBN 0-7923-3728-X

Kluwer Academic Publishers – Dordrecht / Boston / London